Cell Interactions and Development

CELL INTERACTIONS AND DEVELOPMENT

Molecular Mechanisms

Edited by

KENNETH M. YAMADA

National Institutes of Health
Bethesda, Maryland

A Wiley-Interscience Publication

JOHN WILEY & SONS

New York Chichester Brisbane Toronto Singapore

Library of Congress Cataloging in Publication Data:

Main entry under title:
Cell interactions and development.

"A Wiley-Interscience publication."
Includes index.
Contents: Fertilization/Paul M. Wassarman—Sexual differentiation and mating strategies in the yeast *Saccharomyces* and in the green alga *Chlamydomonas*/Ursula W. Goodenough and Jeremy Thorner—The biochemistry of cell adhesion in vertebrates/Stephen Roth—[etc.]
1. Cell interaction. 2. Developmental cytology.
I. Yamada, Kenneth M. [DNLM: 1. Cell communication.
2. Cells—Physiology. 3. Molecular biology.

QH604.2.C444 1982 574.3 82-11050
ISBN 0-471-07987-1

Printed in the United States of America

10 9 8 7 6 5 4 3 2 1

Contributors

Samuel H. Barondes, Department of Psychiatry, School of Medicine, University of California—San Diego, and Veterans Administration Medical Center, San Diego, California

Frank B. Dazzo, Department of Microbiology and Public Health and Department of Crop and Soil Sciences, Michigan State University, East Lansing, Michigan

Ursula W. Goodenough, Department of Biology, Washington University, St. Louis, Missouri

Albert K. Harris, Department of Zoology, University of North Carolina, Chapel Hill, North Carolina

Klaus Kratochwil, Institut für Molekularbiologie der Österreichischen Akademie der Wissenschaften, Salzburg, Austria

Stephen Roth, Department of Biology, University of Pennsylvania, Philadelphia, Pennsylvania

Bryan P. Toole, Department of Anatomy and Cellular Biology, Tufts University School of Medicine, Boston, Massachusetts

Jeremy Thorner, Department of Microbiology and Immunology, University of California, Berkeley, California

Charles B. Underhill, Department of Anatomy and Cellular Biology, Tufts University School of Medicine, Boston, Massachusetts

Paul M. Wassarman, Department of Biological Chemistry, Harvard Medical School, Boston, Massachusetts

James A. Weston, Department of Biology, University of Oregon, Eugene, Oregon

Kenneth M. Yamada, Membrane Biochemistry Section, Laboratory of Molecular Biology, National Cancer Institute, National Institutes of Health, Bethesda, Maryland

Preface

Embryonic development is one of the last major areas of modern biology still poorly understood in terms of mechanisms. An exciting recent advance, however, has been the analysis of cell interactions in development at the molecular level.

The goal of this book is to summarize and evaluate current knowledge in a number of highly active areas of research concerning the mechanisms of cell interactions in development. This book also provides a framework for possible future investigations by providing examples of successful approaches in other areas that can now be applied to specific experimental systems. It is written for advanced undergraduates, graduate students, and postdoctoral fellows, and for scientists and teachers interested in areas other than their own specialized area of expertise.

Each chapter attempts to provide a general introduction to an area of research into developmental interactions, and then discusses both general and detailed examples of experimental analyses at the molecular level. As will be apparent, progress in some areas has been primarily in biochemical analyses of specific molecules while in others it has been in genetic or more physiological approaches. In still others, little is known at the molecular level, yet the great importance of the events makes it imperative that we consider what is known at a descriptive level and define key questions so we can determine the most fruitful experimental approaches for the future. For those wishing to learn more about each research area, a substantial reference list is provided in each chapter to provide access to both the original literature and to recent reviews.

This book begins with a review by Wassarman of fertilization, a cell–cell interaction of universal importance that has become understood in more molecular detail in only the past few years. Goodenough and Thorner then consider another type of interaction between single cells, the surprisingly sophisticated sexual differentiation and mating strategies of certain unicellular organisms. The book then turns to cell adhesion, an essential process in tissue formation and organization; Roth provides a review of adhesion molecules, especially those of vertebrates.

Once cells and tissues adhere to one another, they must reorganize and differentiate in precise patterns. Kratochwil reviews embryonic induction and how this classical problem can be approached by molecular methods. The rearrangements of cells by directed embryonic migration is then examined by Harris, and the analysis of general questions involving both induction and migration in a single embryonic system is considered in the review of the neural crest by Weston.

Specific cytoplasmic and extracellular molecules that are often precisely developmentally regulated and may play important regulatory or adhesive roles are the subjects of Chapters 7–9. Barondes describes developmentally regulated animal lectins, which are often precisely regulated in development yet whose functions are not entirely clear. Extracellular molecules, such as hyaluronic acid, which appear to be important in developmental events, are reviewed by Toole and Underhill. The adhesive glycoprotein fibronectin is reviewed by Yamada as an example of a molecule that may have multiple biological activities mediated by specific molecular mechanisms. The book then concludes with a review by Dazzo of current knowledge about a series of plant developmental systems that involve highly specific molecular mechanisms, including pollination and the establishment of symbiosis for nitrogen fixation.

Each chapter shows the importance of obtaining a molecular understanding of cell interactions in order to begin to explain complex developmental interactions, which have been the source of much investigation and speculation since the time of Aristotle. More detailed investigations of each system and applications of the approaches shown to be successful in other systems should result in an area of intense and fruitful investigation for many years to come.

KENNETH M. YAMADA

Bethesda, Maryland
October 1982

Contents

Cell Interactions and Development

1

Fertilization

PAUL M. WASSARMAN
Department of Biological Chemistry
Harvard Medical School
Boston, Massachusetts 02115

Almost all multicellular organisms are capable of sexual reproduction, although this process may be supplemented by various forms of asexual reproduction, such as fission, budding, or spore formation. Sexual reproduction involves two kinds of gametes or germ cells, sperm and eggs, which unite with each other to form a zygote—the true beginning of a new individual. Fertilization, the union of sperm and egg, may occur outside the parents or within the female parent and constitutes the first instance of cell–cell interaction in the development of a new individual.

As a consequence of its fundamental role in the life history of an organism, the process of fertilization has been of great interest to scientists for more than a century. However, only in recent years have molecular aspects of fertilization received adequate attention in the laboratory. Our understanding of the molecular bases of fertilization-related phenomena is expanding rapidly, and this knowledge will undoubtedly have great impact on both medical and social issues in ensuing years.

In this chapter, some of the basic features of fertilization in mammalian and nonmammalian animal species will be described. For the most part, the sea urchin and mouse will serve as examples of experimental systems for investigating various aspects of fertilization. This chapter is not intended to be a comprehensive review of this area of research; rather, it should serve only as an introduction to some of the recent advances in fertilization research. In the interest of brevity, many important contributions have not been cited and it has been necessary to adopt specific points of view on issues that may be the subject of some controversy among workers in the field. It is hoped that, because of these and other shortcomings, the reader will be stimulated to refer to some of the more detailed accounts of fertilization research (1–10).

1. EGG AND SPERM COMPONENTS INVOLVED IN FERTILIZATION

The following cursory description of egg and sperm components is simply intended to provide a vocabulary with which to discuss various aspects of fertilization. A much more detailed description can be found elsewhere (1–11).

1.1. Eggs

Eggs of both mammalian and nonmammalian animal species are very large. For example, mouse and sea urchin eggs are about 270,000 μm^3 in volume ($\simeq$80 μm in diameter), as much as 1400 times the volume of most somatic cells, but they are very small in the realm of animal eggs. The relatively enormous size of eggs is related to their role as depots for storage of various metabolites, macromolecules, and organelles that accumulate during growth of eggs. In general, the genetic, nutritional, synthetic, energetic, and regulatory requirements of early embryos are met to varying degrees by eggs of both mammalian and nonmammalian animal species (12).

Common to nearly all eggs are one or more acellular layers that completely surround the plasma membrane. In sea urchins, these are the vitelline envelope and jelly coat, whereas in mice (and humans) there is one layer, the *zona pellucida* (Fig. 1). Tens of thousands of microvilli protrude from the surface of the egg (plasma membrane), penetrate these layers, and, in some cases, establish lines of communication with other cell types via junctional complexes. The jelly coat ($\simeq$30 μm), vitelline envelope ($\simeq$20 nm), and zona pellucida ($\simeq$8 μm) can all be considered as glycocalyces composed of glycoproteins, mucopolysaccharides, and/or sulfated polysaccharides and arranged as a three-dimensional network of interconnected fibrils and granules. For example, the mouse egg's zona pellucida is composed of three different glycoproteins, designated ZP1, ZP2, and ZP3, which have apparent molecu-

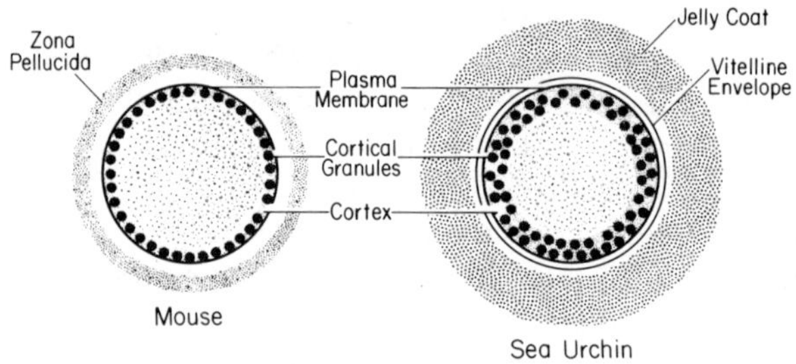

FIGURE 1. Diagrammatic representation of the morphology of sea urchin and mouse eggs.

lar weights of 200,000, 120,000, and 83,000 daltons, respectively (13). All three glycoproteins are synthesized, glycosylated, and secreted during the growth phase of mouse eggs (14). Whereas the vitelline envelope and zona pellucida appear during growth of eggs, the jelly coat either appears during oogenesis or is deposited after fully grown eggs are ovulated. Since sperm must bind to and penetrate these extracellular layers in a species-specific manner, they are intimately involved in regulating sperm-egg interactions both before and after fertilization.

1.2. The Cortical Reaction

Cortical granules are also involved in regulating sperm-egg interactions following fertilization. These small ($\simeq$500 nm in diameter), membrane-bounded organelles populate the region of the egg's cytoplasmic cortex, a fibrous matrix about 2 μm in diameter just beneath the plasma membrane, and are often found attached to the cytoplasmic face of the plasma membrane (Fig. 1). Cortical granules first appear during growth of eggs as a product of the Golgi complex; they can number in the tens of thousands in sea urchin eggs and are responsible in large part for preventing polyspermy. As a consequence of fertilization, cortical granules fuse with the plasma membrane (*cortical reaction*), empty their contents (which include glycoproteins, mucopolysaccharides, proteases, peroxidases, and oligosaccharases) into the perivitelline space, and thereby modify the extracellular layer. In sea urchins, the vitelline envelope is converted into the fertilization envelope as a result of the cortical reaction, and in mice, the zona pellucida undergoes the so-called zona reaction. Modification of the egg's extracellular layers enables them to act as barriers to further sperm binding or penetration (*block to polyspermy*), an event of great importance, since polyspermy inevitably leads to abnormal development or death of the embryo.

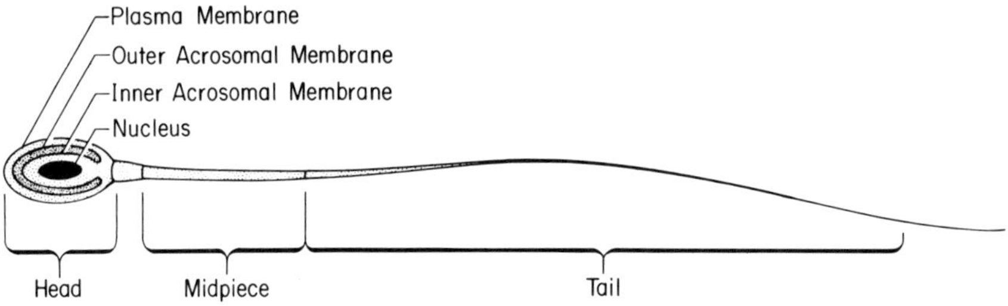

FIGURE 2. Diagrammatic representation of the morphology of sperm.

1.3. Sperm

Sperm size and shape exhibit considerable variation among different animal species. For example, the length of mammalian sperm ranges from less than 40 μm to more than 250 μm, whereas echinoderm sperm vary in length from about 1 μm to 10 μm. Despite highly variable overall morphology, certain common structural features ensure that sperm will be able to reach, recognize, bind to, penetrate, and fertilize eggs. Among these features are a plasma membrane completely enclosing the sperm, an acrosome located at the anterior region of the head of the sperm, and a tail or flagellum at the rear (Fig. 2). The latter is responsible for sperm motility, which is nearly always a characteristic of sperm and is a requirement for fertilization.

The sperm's acrosome is strikingly similar to the egg's cortical granules with respect to its development, structure, and function; in fact, both organelles are thought to be closely related to lysosomes. Like cortical granules, the acrosome is a product of the Golgi complex, is membrane-bounded, and contains an arsenal of lytic enzymes. As a consequence of sperm-egg interaction, the outer acrosomal and plasma membranes fuse at many sites

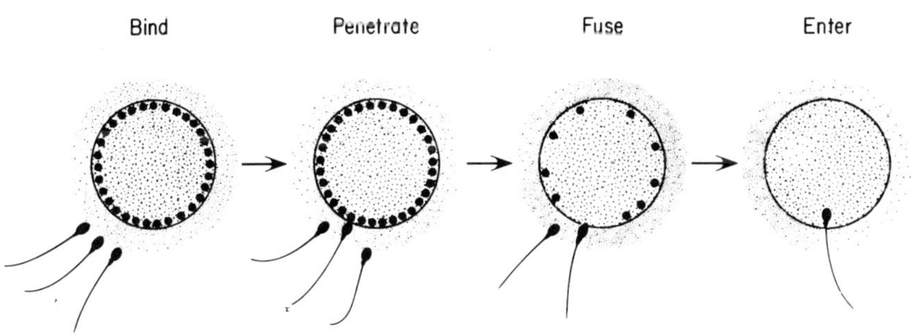

FIGURE 3. Diagrammatic representation of sperm-egg interactions during fertilization.

(*acrosome reaction*) and small vesicles are released, thereby exposing the inner acrosomal membrane.

Only minutes elapse from the initial interaction of sperm and egg to the fusion of the two gametes and establishment of the block to polyspermy (Fig. 3). Of the hundreds of sperm that interact with the outermost layer of each egg, only one sperm normally fuses with the plasma membrane of the egg and enters the cytoplasm. Fusion of a single sperm with less than 0.001% of the egg's plasma membrane is sufficient to set in motion a series of events that ultimately leads to the development of a new individual. This truly remarkable example of cell–cell interaction, embodying features common to many other biological phenomena, represents a crowning achievement of nature.

2. RECEPTOR-MEDIATED SPERM–EGG INTERACTION

For more than 50 years, it has been recognized that the species-specificity of sperm–egg interactions is probably attributable to sperm receptors present in the extracellular layers of eggs. Subtle variations among different animal species in the macromolecular structures of such receptors could conceivably account for the observed specificity of sperm-egg interactions. Naturally, the sperm receptor on eggs must be recognized by a complementary macromolecule on sperm, and its structure would change in concert with that of the sperm receptor. In fact, various kinds of experimental evidence are now available that strongly support the idea of receptor-mediated sperm–egg interactions in both mammalian and nonmammalian animal species.

Among mammals, penetration of the egg's zona pellucida by sperm is nearly always species-specific (Table 1). The failure of heterologous sperm to penetrate the zona pellucida is undoubtedly attributable to the incompatibility of receptors and complementary macromolecules on eggs and sperm of different species. However, this incompatibility can be overcome in many cases by removing the zona pellucida and exposing zona-free eggs to acrosome-reacted sperm; under these conditions, heterologous sperm are often able to penetrate and fertilize eggs (Table 1). These and other observations strongly suggest that the zona pellucida is the site of species-specificity in mammalian fertilization.

2.1. Molecular Mechanisms in Mammals

Recently, we investigated whether or not any of the three glycoproteins (i.e., ZP1, ZP2, and ZP3) identified in mouse egg zonae pellucidae (13, 14) possessed sperm-receptor activity (20, 21). These glycoproteins constitute 90%

TABLE 1. Penetration or Fertilization of Mammalian Eggs by Heterologous Sperm[a]

	Eggs With or (Without) Zonae Pellucidae[b]				
Sperm	Hamster	Guinea pig	Mouse	Rat	Rabbit
Hamster	+	–	–(+)	–(+)	nd[c]
Guinea Pig	–(+)	+	–	–	nd
Mouse	–(+)	nd	+	–(+)	–(+)
Rat	–(+)	nd	nd	+	–(+)
Human	–(+)	–(+)	–	–	–

[a] Compiled from data in references 15–19.
[b] Penetration of eggs with zonae pellucidae or fertilization of eggs without zonae pellucidae.
[c] nd, not determined.

or more of the dry weight of zonae pellucidae, are all acidic (pI, <6.0), and possess inter- or intramolecular disulfide bonds (Fig. 4). We were encouraged by the results of initial experiments, demonstrating that sperm exposed to solubilized zonae pellucidae exhibited greatly diminished binding to eggs as compared with untreated sperm (Fig. 5). These results suggest that zonae pellucidae from mouse eggs contain a component that behaves in the manner expected for a sperm receptor. These experiments involved the use of an in vitro binding assay (*competition assay*) in which mouse sperm were first cultured in the presence of solubilized zonae pellucidae that were isolated individually from mouse eggs. Ovulated eggs were then added to the cultures, and the number of sperm that bound to the eggs was quantitated by counting sperm tightly associated with eggs after gentle pipetting to remove loosely associated sperm. Sperm receptors in the solubilized zonae pellucidae

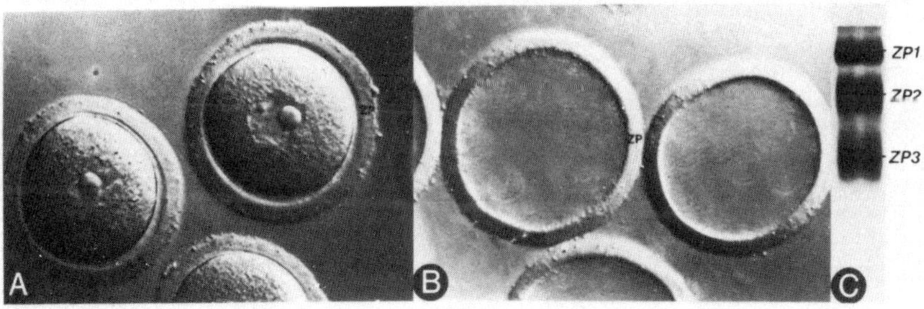

FIGURE 4. Photomicrographs of mouse eggs (*A*), isolated zonae pellucidae (*B*), and an autoradiogram of [^{125}I]iodosulfanilic acid-labeled zona pellucida glycoproteins following SDS-polyacrylamide gel electrophoresis (*C*). zp, zona pellucida. (See reference 13 for details.)

would be expected to bind to complementary sites on sperm and reduce binding of these sperm to eggs.

To evaluate which, if any, of the mouse egg zona pellucida glycoproteins served as a sperm receptor, ZP1, ZP2, and ZP3 were purified by sodium dodecylsulfate-polyacrylamide gel electrophoresis (SDS-PAGE) from individually isolated zonae pellucidae. A small portion of the starting material was radiolabeled with [^{125}I]iodosulfanilic acid in order to be able to monitor the effectiveness of the purification procedure and quantitate recovery of the three glycoproteins. Since each zona pellucida contains only about 4 ng of protein, standard methods for the detection and quantitation of protein were too insensitive for use in these experiments. The positions of ZP1, ZP2, and ZP3 on the gels were identified by autoradiography, the appropriate sections of the gels were excised, and the glycoproteins were each electroeluted from the gel slices. After extensive dialysis against urea (to remove SDS) and distilled water (to remove urea), and lyophilization, the purity of each glycoprotein was assessed by SDS-PAGE. ZP1, ZP2, and ZP3 were then tested individually for their ability to prevent binding of sperm to eggs by using the competition assay described above. Whereas ZP1 and ZP2 did not affect the extent of sperm binding, ZP3 was nearly as effective as solubilized zonae pellucidae in reducing the number of sperm bound to eggs (Fig. 5). Two

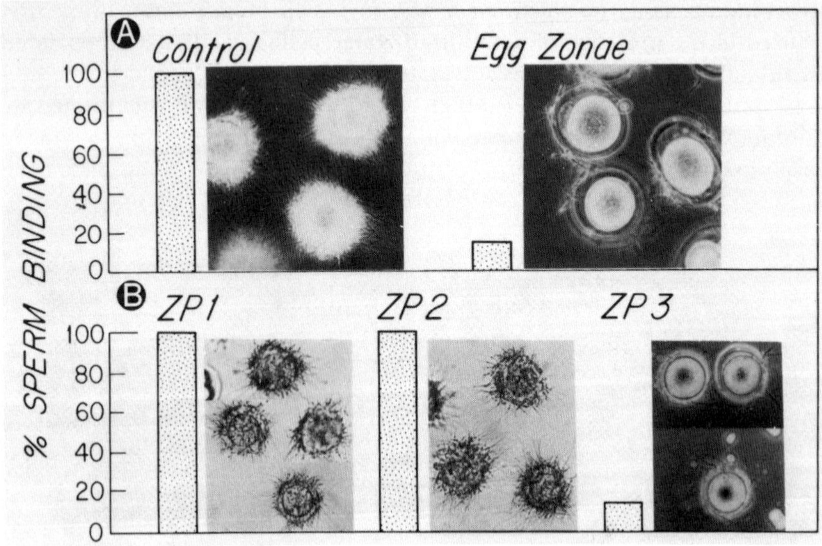

FIGURE 5. Binding of sperm to mouse eggs before (*A*) or after exposure of sperm to solubilized zonae pellucidae (*A*) or individual zona pellucida glycoproteins (*B*) from unfertilized mouse eggs. (See reference 20 for details.)

other glycoproteins, mucin and transferrin, were subjected to the same preparative procedures as the zona pellucida glycoproteins, tested for their ability to reduce sperm binding, and found to have no effect on binding of sperm to eggs. These and other results suggest that ZP3 is the component of the unfertilized mouse egg's zona pellucida that serves as a sperm receptor. It can be calculated from our measurements that there are more than 10^9 molecules of ZP3 per zona pellucida, and results of binding and synthesis studies indicate that ZP3 is not only located on the external surface, but also is distributed throughout the zona pellucida ($\simeq$40,000 molecules ZP3/μm^3).

2.2. Molecular Mechanisms in Invertebrates

Studies of sperm–egg interactions in sea urchins have also implicated the involvement of a glycoprotein, located in the vitelline envelope, as a species-specific sperm receptor (22–28). Sperm bind to the outer surface of vitelline envelopes in a species-specific manner, and fragments of vitelline envelopes compete with the binding of sperm to unfertilized eggs, also in a species-specific manner. The sea urchin sperm receptor, like ZP3 in mice, is an acidic glycoprotein (pI, $\simeq$4.0; $\simeq$30% carbohydrate) that is sensitive to various proteases (e.g., trypsin), but has a much higher molecular weight (>5,000,000 daltons) than ZP3 ($\simeq$83,000 daltons). Various kinds of plant lectins bind to mouse and sea urchin eggs and prevent fertilization (26, 29–32)—a result consistent with the glycoprotein nature of sperm receptors, but not necessarily attributable to lectin : sperm receptor complex formation.

The sea urchin sperm receptor is recognized by *bindin*, a major protein component of the sperm's acrosome (28, 33–36). Bindin is a highly insoluble protein (no carbohydrate), of about 31,000 molecular weight, that can act as a species-specific agglutinin of unfertilized sea urchin eggs and can be considered as a lectin specific for lactose and galactose. The aminoterminal region of bindin is extremely rich in hydrophobic amino acid residues, especially glycine, suggesting that this region of the protein is associated with the acrosomal membrane. Since bindin is localized within the acrosomal vesicle, it can only interact with sperm receptors in the vitelline envelope following completion of the acrosome reaction; this is unlike the situation in mice, since ZP3 is recognized only by sperm with intact acrosomes (discussed below).

In both sea urchins and mice, binding of sperm to eggs exhibits saturation kinetics, with a maximum of about 1500 sperm bound per egg at high sperm concentrations (21, 22). Such kinetics provide additional evidence for the presence of specific sperm receptors in vitelline envelopes and zonae pellucidae of sea urchin and mouse eggs, respectively. It can be calculated from the results of various binding studies, that there is a single sperm bound per

25 μm^2, or so, of vitelline envelope (sea urchin) or zona pellucida (mouse) surface area. It is not apparent why more sperm are not able to bind to each egg.

2.3. Summary

In summary, the experimental evidence currently available suggests that species-specific binding of sperm to eggs is mediated by a lectinlike protein (sperm) : glycoprotein receptor (egg) interaction. It should be noted that lectinlike interactions are thought to be responsible for intercellular recognition in a wide variety of other biological systems (37, 38; Chapter 7). In the mouse, the recognition-event between sperm and egg involves the zona pellucida, whereas in the sea urchin it involves the vitelline envelope.

3. RESPONSE OF SPERM TO SPERM–EGG INTERACTIONS

In order for sperm to penetrate the extracellular layers of eggs and reach the plasma membrane, they must undergo the acrosome reaction (Fig. 6). In both mammalian and nonmammalian animal species, this reaction involves fusion and extensive vesiculation of the sperm's plasma and outer acrosomal membranes and results in exposure of the inner acrosomal membrane. Mouse sperm that have undergone the acrosome reaction prior to exposure to unfertilized eggs fail to bind to the zona pellucida (21, 39, 40). This suggests that, in mammals, it is the sperm's plasma membrane, not the inner acrosomal membrane, that interacts with the zona pellucida in a species-specific manner; immunological evidence is available to support such a proposal (41). This is quite unlike the situation in sea urchins, since species-specific interaction of bindin, which is associated with the inner acrosomal membrane, with sperm receptors in the vitelline envelope can only occur after completion of the acrosome reaction. Consequently, there are distinctive differences in the pathways of fertilization in sea urchins and mice.

3.1. The Acrosome Reaction

The molecular basis of the acrosome reaction has been extensively investigated in sea urchins and many of the features elucidated appear to apply to mammalian sperm as well. The acrosome reaction is characterized by calcium influx and proton efflux through the plasma membrane surrounding the head of the sperm. Presumably, a change in permeability of the plasma membrane allows movement of ions to take place when sperm interact with either the jelly coat or zona pellucida. There is an absolute requirement for

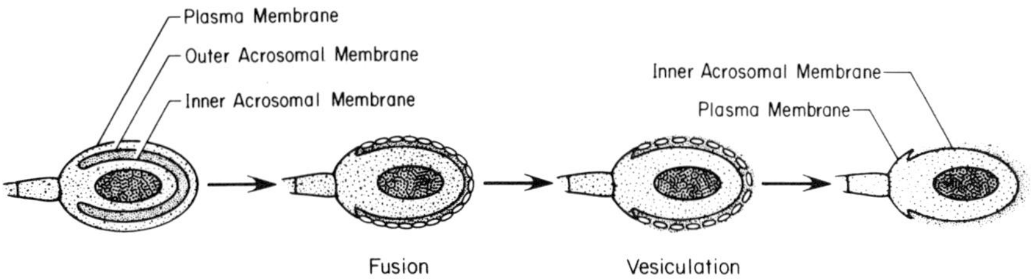

FIGURE 6. Diagrammatic representation of sperm in the process of undergoing the acrosome reaction.

extracellular calcium in order for the acrosome reaction to occur, and the antibiotic calcium ionophore A23187, which abolishes selective permeability of a cell's plasma membrane to calcium and increases cytoplasmic free calcium (42, 43), acts as an inducer of the acrosome reaction in both mammalian and nonmammalian sperm (44–49). Recently, a role has been proposed for calmodulin, a calcium-binding protein found in nearly all eukaryotic cells (50), in regulating the acrosome reaction (51, 52). Calmodulin is found in mammalian sperm in the anterior region of the head (postacrosomal region) and may serve as the link between calcium influx and calcium-dependent fusion of the plasma and outer acrosomal membranes.

Acid is released concomitant with fusion of the sperm's plasma and outer acrosomal membranes, and it is likely that an increase in intracellular pH is required for the calcium-dependent acrosome reaction to take place (45, 47). In addition, in sea urchins and other echinoderms, proton efflux is probably involved in actin polymerization during extension of the acrosomal process, a thin, membrane-bounded process as much as 90 μm in length that penetrates the jelly coat and attaches to the vitelline envelope. It has been proposed that a rise in intracellular pH causes dissociation of actin from specific binding proteins, the form in which it is stored in the sperm head (*profilactin*) prior to the acrosome reaction; actin can then be incorporated into the filaments of the acrosomal process (53–55).

In mammals, the acrosome reaction does not include extension of an acrosomal process, but results in exposure of lytic enzymes associated with the inner acrosomal membrane (56, 57). One of these enzymes, *acrosin*, has trypsinlike activity and is able to digest the zona pellucida after conversion from a zymogen form (*proacrosin*). It takes less than 10 min for sperm to traverse the entire zona pellucida, traveling obliquely, and leaving behind a clearly defined region that describes its path. The diameter of the path is no greater than that of the sperm head, suggesting that lysins responsible for sperm penetration do not diffuse away from the inner acrosomal membrane; such diffusion would presumably cause widespread digestion of the zona pellucida. It has been found that a variety of naturally occurring or synthetic trypsin inhibitors can prevent fertilization, consistent with the proposed involvement of sperm proteases in zona pellucida penetration. However, this remains a controversial aspect of fertilization research, and more definitive experiments will be required to demonstrate a direct link between lysins associated with acrosomes and penetration of the zona pellucida. It should be noted that passage of sea urchin sperm through the vitelline envelope may also be facilitated by a membrane-bound, trypsinlike protease found in acrosome-reacted sperm (58); however, once again, a direct link between proteolysis and penetration has not been demonstrated.

3.2. Egg Components Triggering the Acrosome Reaction

Under normal conditions, it is the jelly coat, the outermost layer of sea urchin eggs, that induces sperm to undergo the acrosome reaction (23, 24, 59–61). The jelly coat consists of sialoprotein ($\simeq$20% of total mass) and fucose sulfate polysaccharide ($\simeq$80% of total mass), and, in several different species of sea urchins, it appears that the polysaccharide is sufficient to induce the acrosome reaction (59, 60). The question of whether or not there is some species specificity associated with the induction of the acrosome reaction in sea urchins, has not been satisfactorily resolved. It is possible that the species specificity reported in certain cases does not represent a true "recognition event" between sperm and jelly coat, but is related to some other parameter, such as the calcium dependence of the acrosome reaction.

Since only sperm with intact acrosomes bind to zonae pellucidae of mouse eggs, the possibility arises that a component of the zona pellucida is responsible for triggering the acrosome reaction. Using an electron microscopic assay for the acrosome reaction, mouse sperm were examined before and after exposure to either solubilized zonae pellucidae or to individual zona pellucida glycoproteins (21). The average background level of acrosome-reacted sperm prior to any treatment was about 20% of the population, but exposure of sperm to the antibiotic calcium ionophore A23187 increased the level of acrosome-reacted sperm to about 70% of the population. Exposure of sperm to solubilized zonae pellucidae, isolated individually from unfertilized mouse eggs, induced the acrosome reaction to an extent comparable to that observed with ionophore A23187 (Fig. 7). These results suggest that zonae pellucidae from mouse eggs, like jelly coats from sea urchin eggs, contain a component that triggers the acrosome reaction. To identify the zona pellucida component responsible, each of the zona pellucida glycoproteins—ZP1, ZP2, and ZP3—was purified by preparative electrophoresis (as described above) and added to sperm. Electron microscopic examination of sperm treated in this manner revealed that only ZP3 induced the acrosome reaction to greater-than-background levels. In fact, ZP3 was as effective as either solubilized zonae pellucidae or ionophore A23187 in triggering the

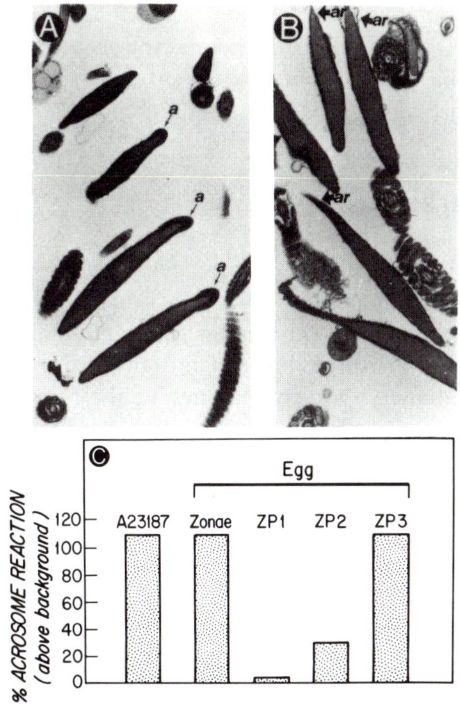

FIGURE 7. Induction of the acrosome reaction in mouse sperm by calcium ionophore A23187, solubilized zonae pellucidae from unfertilized mouse eggs or individual zona pellucida glycoproteins from unfertilized mouse eggs (*C*). Electron micrographs of unreacted and reacted mouse sperm are shown in (*A*) and (*B*), respectively, a, acrosome; ar, acrosome reacted. (See reference 21 for details.)

reaction in mouse sperm (Fig. 7). Since ZP1 and ZP2 have no effect on either the binding of sperm to eggs (Fig. 5) or on the extent of the acrosome reaction (Fig. 7), it would appear that ZP3 serves as both a sperm receptor and an inducer of the acrosome reaction. It is tempting to conclude from the results of these experiments that ZP3 has evolved so as to discharge the functions of macromolecules located in the jelly coat (induce the acrosome reaction) and vitelline envelope (species-specific sperm receptor) of nonmammalian eggs.

3.3. Summary

In summary, the acrosome reaction involves a calcium-dependent fusion of the plasma and outer acrosomal membranes of sperm. The reaction is triggered in response to sperm interaction with the jelly coat (sea urchin) or zona pellucida (mouse) and results in exposure of bindin (sea urchin) or acrosin (mouse) associated with the inner acrosomal membrane. In sea urchins, a fucose sulfate polysaccharide induces the acrosome reaction, whereas in mice, an acidic glycoprotein, that also serves as a sperm receptor, induces the reaction.

4. FUSION OF SPERM AND EGG

Once through the vitelline envelope or zona pellucida, the sperm head makes contact with the plasma membrane of the egg and adheres to it. Incorporation of the sperm's genome into the egg, restoring it to a diploid state, occurs as a result of fusion of sperm and egg membranes. In most species, fusion occurs almost immediately upon sperm reaching the surface of the egg. For example, in sea urchins, fusion is completed within about 15 sec and the sperm enters the egg in only a few more seconds, moving at a linear rate of about 8 μm/min (62). In mammals, fusion of sperm and egg occurs within about 3 min of insemination of zona pellucida-free eggs, and the entire sperm head is incorporated into the egg's cytoplasm within about 15 min (63). The initiation of membrane fusion in both sea urchins and mice is characterized by distinct changes in the behavior of sperm. Particularly noticeable is the cessation of sperm tail movement, probably due to changes in the properties of the plasma membrane of sperm.

In sea urchins, the plasma membrane of the egg fuses with the inner acrosomal membrane surrounding the apical region of the acrosomal process, whereas in mice, the plasma membrane of the egg fuses with the plasma membrane posterior to the inner acrosomal membrane (Fig. 8). This represents a major difference in fertilization pathways in mammalian and non-

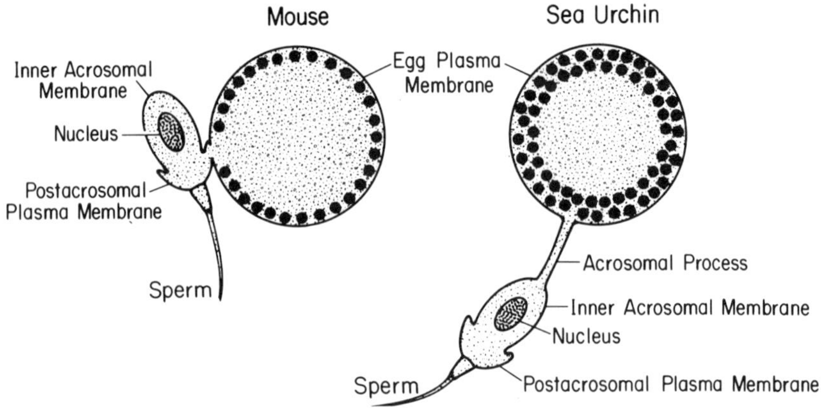

FIGURE 8. Diagrammatic representation of sperm-egg fusion in sea urchins and mice.

mammalian animal species since, in one case, the fusion bridge involves the inner acrosomal membrane (sea urchin) and, in the other, the plasma membrane (mouse) of acrosome-reacted sperm. Perhaps, the inner acrosomal membranes of acrosome-reacted mouse and sea urchin sperm differ in their degree of "fluidity," and it is this factor that determines the site of fusion. In this connection, it is noteworthy that only acrosome-reacted mammalian sperm fuse with eggs freed of their zonae pellucidae (64). This suggests that the properties of the postacrosomal membrane are altered as a result of the acrosome reaction; that is, the membrane is rendered fusible. Therefore, for different reasons, fusion of sperm and eggs in both mammalian and nonmammalian animal species requires that sperm first undergo the acrosome reaction—in one case, to modify the postacrosomal plasma membrane (mammals), and in the other, to expose the inner acrosomal membrane (nonmammals).

The fusion of sperm and eggs is greatly facilitated by the presence of numerous microvilli on the surface of eggs. The acrosomal membrane and postacrosomal plasma membrane of sperm are relatively smooth. Fusion nearly always involves microvillar membranes, probably because microvilli have a relatively low radius of curvature (<0.1 μm) that permits maximum apposition (<1.5 nm apart) of sperm and egg membranes. In sea urchins, the involvement of microvilli in sperm-egg fusion and sperm penetration has been carefully documented by using scanning electron microscopy to follow these events (65). Immediately following membrane fusion, a group of about 20 elongated microvilli cluster tightly around and interdigitate over the fused sperm head; as the sperm penetrates into the egg, these microvilli are resorbed (Fig. 9). These observations suggest that it is the egg that draws the sperm inside, so that sperm motility, which ceases at the time of fusion, is not required for penetration. It is likely that the driving force for engulfment of the fused sperm comes from the fibrous cortex of eggs, which contains the contractile proteins actin and myosin and undergoes a contraction at the time of fertilization (66–68). Since microvillar filaments are embedded in the cortex, such a contraction could "draw" the microvilli, and the associated sperm, into the cytoplasm of the egg.

Apparently, there is very little species specificity associated with sperm–egg fusion. Sperm from different species will fuse with heterologous eggs freed of their zonae pellucidae or vitelline envelopes, but, in most cases, will not fertilize these eggs prior to removal of the extracellular layer (Table 1). Nor does there appear to be a specific site on the egg's plasma membrane where fusion must occur, although in many animal species there are large regions where fusion tends to take place (e.g., vegetal pole). Various evidence suggests that eggs and acrosome-reacted sperm are simply highly fusible cells, and the lack of site- or species-specific sperm–egg fusion is consistent

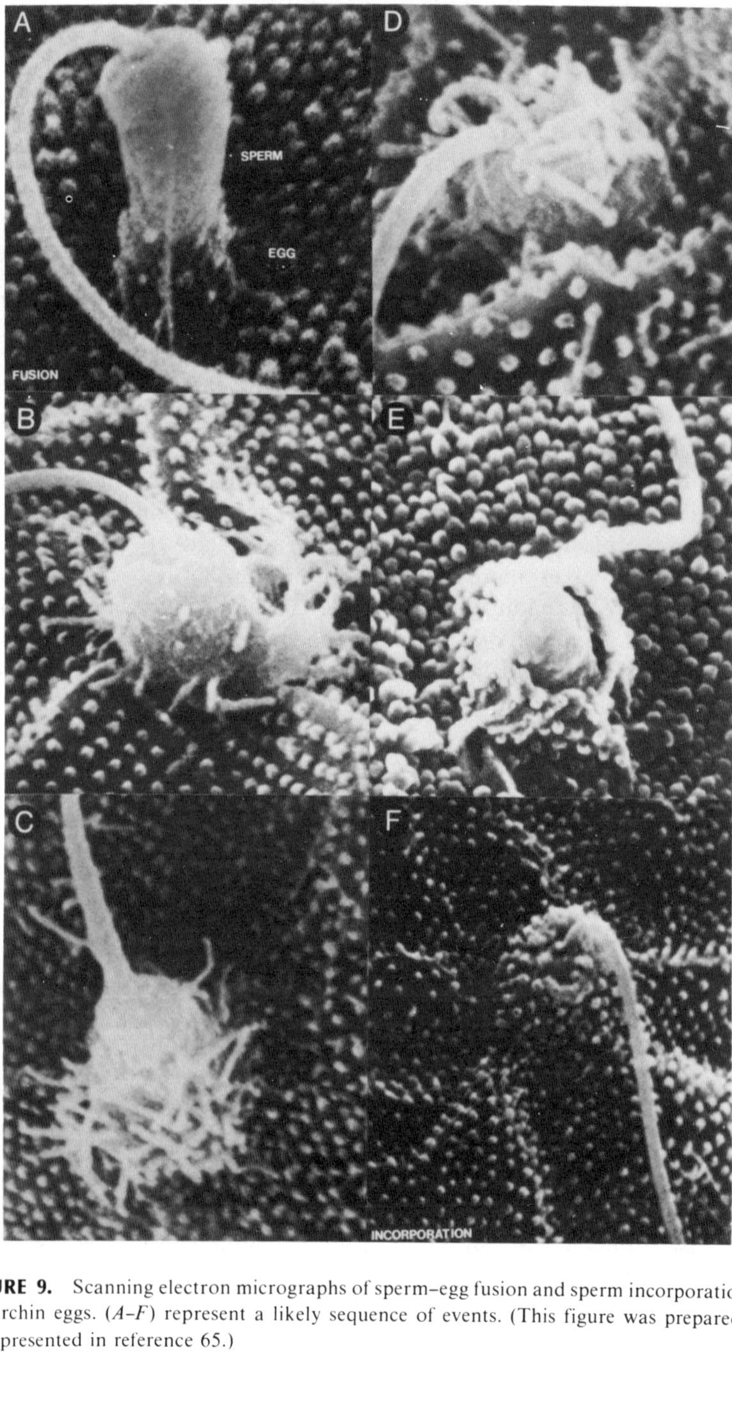

FIGURE 9. Scanning electron micrographs of sperm–egg fusion and sperm incorporation into sea urchin eggs. (*A–F*) represent a likely sequence of events. (This figure was prepared with data presented in reference 65.)

with results of studies of somatic cell fusions. Such studies have demonstrated that interspecific hybrid cells can be created by fusion of cells from diverse species and taxonomic orders (69). Since the plasma membrane of eggs resembles that of somatic cells, that is, it consists of a phospholipid bilayer with intercalated proteins, it is likely that the same factors affect the fusibility of these cells. Considerable evidence suggests that the degree of membrane "fluidity" (i.e., mobility of macromolecules within the membrane), especially the mobility of the lipids, is the major factor that affects somatic cell fusion (70–72). In fact, the results of experiments on mammalian and nonmammalian eggs indicate that the lipid moieties of the plasma membrane are, indeed, of primary importance in sperm–egg fusion (73, 74).

In summary, sperm–egg fusion occurs very rapidly between the microvillar plasma membrane of eggs and either the inner acrosomal membrane (sea urchin) or postacrosomal plasma membrane (mouse) of sperm. There is very little site- or species-specificity associated with fusion of sperm and egg. Following the fusion event, the sperm is surrounded by a group of elongated microvilli and then is drawn into the cytoplasm of the egg, probably by contraction of the cytoplasmic cortex.

5. RESPONSE OF EGGS TO FERTILIZATION

Fertilization of an egg by a single sperm triggers a series of events that are responsible for activation of a developmental program laid down during oogenesis and for establishment of a block to polyspermy. The former includes increases in cellular respiration, transport, and macromolecular synthesis (e.g., DNA, RNA, and protein) and the onset of cell division, while the latter includes changes in the properties of the zygote's plasma membrane and extracellular layers. Many of the factors involved in activation of the egg and establishment of the block to polyspermy are known, since this has proved to be one of the more fruitful areas of fertilization research in recent years.

5.1. Changes in Ionic Composition

Three changes involving the ionic composition of eggs takes place within the first minute after fertilization, and these changes serve as the signals for both activation of eggs and prevention of polyspermy:

1. Within about 3 sec of fertilization, sea urchin eggs undergo a large transient change in membrane potential, or depolarization (75–78). The increase in egg membrane potential, which results from changes

in permeability to certain ions (e.g., Na^+, K^+, and Cl^-), can be as much as +80 mV and lasts as long as 1 min or so, before declining to the original negative potential. Rapid depolarization of the egg's plasma membrane following fertilization appears to serve as the first line of defense against polyspermy (*fast block*). This was demonstrated most clearly (76) by applying a voltage clamp to unfertilized eggs, such that the membrane potential was kept above zero (i.e., positive). Under these conditions, added sperm attached to, but did not fuse with the egg's plasma membrane; when the membrane potential was dropped below zero, sperm–egg fusion took place. On the other hand, polyspermy was induced by maintaining an artificial negative holding potential on eggs in the presence of sperm.

2. Within about 10 sec of fertilization, sea urchin eggs undergo a large increase in cytoplasmic free calcium (79, 80). The calcium is derived from intracellular stores and the concentration of free calcium in the cytoplasmic cortex may exceed 10 μM during this period. The sudden rise in free calcium is apparently responsible for inducing the cortical reaction, which establishes the second line of defense against polyspermy (*slow block*).
3. Within about 30 sec of fertilization, sea urchin eggs undergo a major influx of sodium ions and efflux of protons, which eventually ($\simeq$1–5 min after fertilization) leads to a large increase in intracellular pH ($\simeq$0.4 pH units) (81–84). The transient rise in pH results from induction of the sodium-proton exchange system by increased intracellular calcium levels. Induction of this system appears to be necessary for activation of eggs (e.g., increased macromolecular synthesis and transport), including the onset of cell division.

As described, fertilization by a single sperm leads to two independent responses by the egg: (1) depolarization of the plasma membrane and (2) release of calcium from intracellular stores. While depolarization of the egg's plasma membrane serves as a fast, but transient block to polyspermy, the release of calcium induces the cortical reaction that establishes a slow, but permanent block to polyspermy.

5.2. The Cortical Reaction

In both mammalian and nonmammalian animal species, the cortical reaction (85) involves fusion of cortical granules and plasma membranes and is propagated as a wave emanating from the point of sperm–egg fusion, bidirectionally, over the entire surface of the egg (Fig. 10). For example, in sea

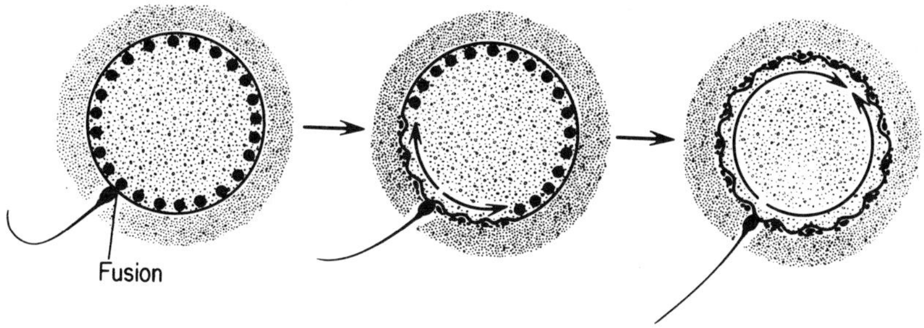

FIGURE 10. Diagrammatic representation of the cortical reaction following sperm–egg fusion.

urchin eggs, fusion of cortical granules is initiated within about 30 sec of fertilization and is completed within about another 30 sec (22, 86); in mammalian eggs, several minutes are required for completion of the cortical reaction (87, 88). In both mammalian and nonmammalian eggs, there is a clear relationship between the onset of the cortical reaction and inhibition of the binding of sperm to eggs (Fig. 11). It would appear that, following the initial cortical reaction in the vicinity of the fertilizing sperm, the wave of cortical reaction over the entire surface of the egg is self-propogated (*auto-*

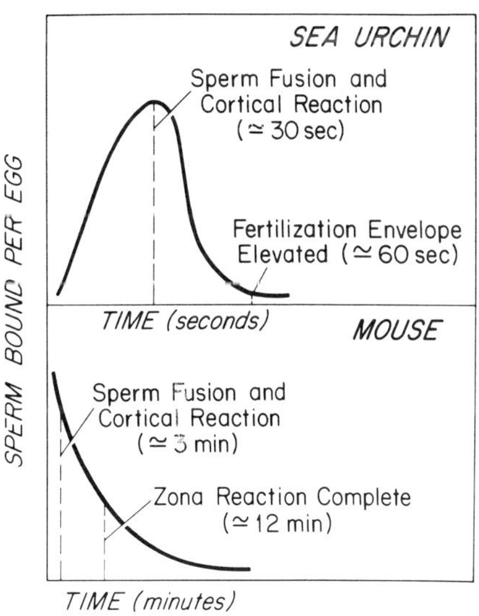

FIGURE 11. Diagrammatic respresentation of the temporal relationship between sperm binding and the cortical reaction, fertilization envelope elevation, and the zona reaction in sea urchin or mouse eggs.

catalytic) by release of calcium from the granules themselves (90). Fusion of cortical granules and the plasma membrane results in exocytosis of cortical granule contents and in the formation of a "mosaic" membrane around the fertilized egg (Fig. 10). In sea urchins, the contribution of cortical granule membrane to plasma membrane increases the surface area of the egg from about 40,000 μm^2 (unfertilized egg) to about 90,000 μm^2 (fertilized egg) within 1 min of fertilization; this increase is due in part to considerable elongation of the more than 125,000 microvilli on the egg's surface (91). However, during the next 15 min or so after fertilization, much of this membrane is apparently resorbed, such that there is less than a 10% increase in overall surface area as compared with the unfertilized egg. Both the lipid and protein composition of the plasma membrane change following the cortical reaction (92, 93), and it is possible that a change in membrane organization ("fluidity") also takes place upon fertilization (94).

The cortical granule contents that are released into the perivitelline space as a result of membrane fusion are responsible for major changes in the extracellular layers of mammalian and nonmammalian eggs. In the sea urchin, the vitelline envelope is converted into a fertilization envelope and, in the mouse, the zona pellucida undergoes the so-called zona reaction. In both cases, the extracellular layers become considerably less soluble (i.e., become "harder") in the presence of denaturants, reducing agents, or proteases and, in addition, will no longer bind sperm.

5.3. Conversion of the Vitelline Envelope

In sea urchins, the vitelline envelope is modified and begins to detach and elevate away from the plasma membrane within about 30 sec of fertilization; the fertilization envelope is completely elevated within about another 30 sec (62, 95, 96). At least two enzymes, present in the cortical granule exudate, are involved in modification of the vitelline envelope of fertilized eggs: (1) trypsinlike protease that is responsible for specific inactivation of sperm receptors in the vitelline envelope (97–101), and (2) an ovoperoxidase that catalyzes the crosslinking of tyrosyl residues (one dityrosine crosslink per 55,000 daltons of protein) and, as a result, is responsible for the hardening of the fertilization envelope (102–104). It is likely that other enzymes are involved in detachment of the vitelline envelope from the plasma membrane and that other cortical granule components contribute to the structural reorganization of the vitelline envelope following fertilization.

5.4. The Zona Reaction

In mammals, the contents of the cortical granules are responsible for altering the properties of the zona pellucida following fertilization. In particular, sperm do not bind to the zonae pellucidae of fertilized eggs or cleaving embryos, suggesting that sperm receptors are inactivated as a result of the zona reaction (88, 89, 105–108). In order to understand the molecular basis of this aspect of the zona reaction, we carried out experiments, identical in approach to those described above (Sections 2 and 3), but using zonae pellucidae and individual zona pellucida glycoproteins from two-cell mouse embryos, as well as from unfertilized eggs (21). In one series of experiments, using the competition assay described above (Section 2), we compared the effects of solubilized zonae pellucidae from eggs and embryos on the binding of sperm to ovulated eggs. As expected, whereas egg zonae pellucidae were very effective in preventing sperm binding, embryo zonae pellucidae were totally ineffective (Fig. 12). Furthermore, comparable experiments carried out with purified ZP1, ZP2, and ZP3 from eggs and embryos, demonstrated that only ZP3 from egg zonae pellucidae was able to depress the extent of sperm binding; ZP3 from embryo zonae pellucidae had no effect on sperm binding (Fig. 12). Consistent with these observations, we also found, using the electron microscopic assay described above (Section 3), that ZP3 purified from embryo zonae pellucidae, unlike egg ZP3, was unable to induce sperm to undergo the acrosome reaction (Fig. 12). These results strongly suggest that ZP3 is modified following fertilization ($ZP3_f$), such that it is no longer able to function as either a sperm receptor or an inducer of the acrosome reaction. The conversion of ZP3 to $ZP3_f$ does not noticeably alter its mobility on SDS-PAGE (20, 21), and it is possible that the change involves the

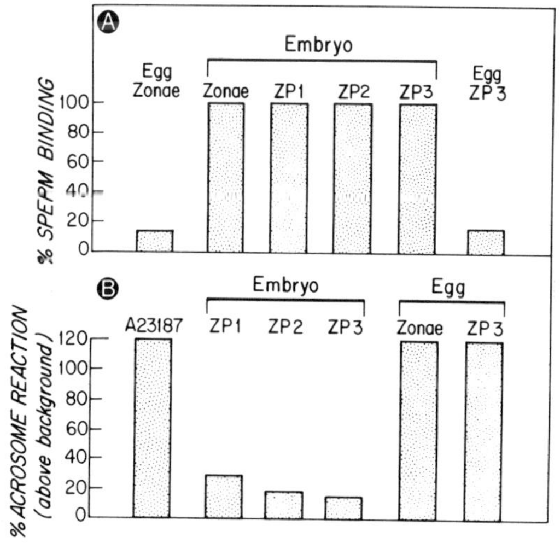

FIGURE 12. Effect of solubilized zonae pellucidae and individual zona pellucida glycoproteins from unfertilized eggs and two-cell embryos on sperm binding (*A*) and the acrosome reaction (*B*). A23187, calcium ionophore. (For details see references 20 and 21.)

carbohydrate moiety of ZP3 rather than the polypeptide portion itself. Such a situation would be compatible with the idea that binding of sperm to eggs is mediated via a lectin (sperm): glycoprotein receptor (egg) complex (discussed in Section 2). At fertilization, elimination or modification of the carbohydrate moiety involved in the recognition event would prevent further sperm binding to the zona pellucida which, in turn, would preclude the acrosome reaction.

In addition to inactivatio.ı of sperm receptors, the zona reaction results in dramatic changes in the biochemical and physical properties of the zona pellucida. For example, zonae pellucidae from fertilized eggs or embryos are considerably less soluble ("harder") than zonae pellucidae from unfertilized eggs in the presence of acidic buffers or buffers containing certain proteases or reducing agents (109–111). Although an ovoperoxidase may be involved in bringing about the change in zona pellucida solubility (112, 113), we have

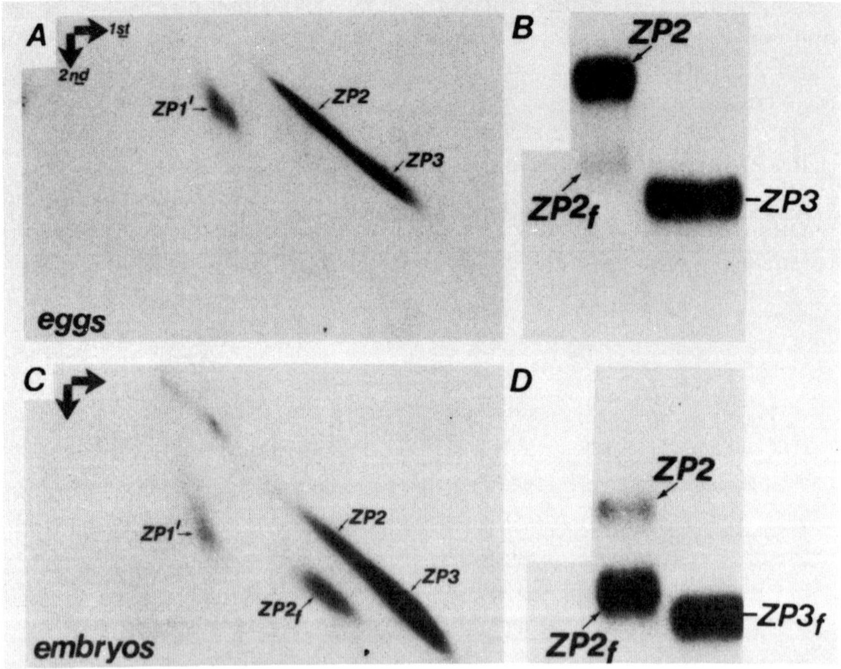

FIGURE 13. Electrophoretic analyses of the conversion of zona pellucida glycoprotein ZP2 to $ZP2_f$ following fertilization of mouse eggs. (*A*) and (*C*) are autoradiograms of [^{125}I]iodosulfanilic acid-labeled zonae pellucidae from unfertilized eggs and two-cell embryos, respectively. The radiolabeled zonae pellucidae were subjected first to SDS-PAGE under nonreducing conditions ("1st") and then to SDS-PAGE under reducing conditions ("2nd"). (*B*) and (*D*) are autoradiograms of [^{125}I]iodosulfanilic acid-labeled zona pellucida glycoproteins, purified following SDS-PAGE under nonreducing conditions, and then subjected to SDS-PAGE under reducing conditions. (For details see references 21 and 114.)

not obtained any evidence indicative of crosslinking of the zona pellucida glycoproteins following fertilization. On the other hand, we have found that ZP2, the major glycoprotein of mouse zonae pellucidae, is subjected to proteolysis following fertilization (21, 114). Under reducing conditions, ZP2 from embryos, but not from unfertilized eggs, migrates on SDS-PAGE with an apparent molecular weight of 90,000 daltons instead of 120,000 daltons, and has been designated $ZP2_f$ (Fig. 13). Conversion of ZP2 to $ZP2_f$ following fertilization involves proteolysis of the glycoprotein by an enzyme in the cortical granule exudate; however, in the absence of a reducing agent, intramolecular disulfide bonds prevent the release of peptide fragments. These and other results suggest that the zona pellucida, like the vitelline envelope of sea urchin eggs, undergoes a structural reorganization following fertilization such that it becomes quite callous.

5.5. Summary

In summary, fertilization triggers membrane depolarization, release of free calcium, and stimulation of the sodium-proton exchange system in the egg. These changes are responsible for activation of eggs and establishment of a block to polyspermy. The latter includes the cortical reaction, conversion of the vitelline envelope into a fertilization envelope (sea urchin), and the zona reaction (mouse). The slow block to polyspermy involves inactivation of sperm receptors and a general hardening of the egg's extracellular layers by components of the cortical granule exudate.

6. RETROSPECTIVE

The preceding is intended to provide an overview of fertilization in mammalian and nonmammalian animal species. While certainly not a comprehensive survey of the field, it is an attempt to inform students and practicing scientists about this exciting example of cell–cell interaction during animal development. In the century following the landmark publications of Van Beneden (115) and Hertwig (116), fertilization research has made the transition from morphology to molecular biology, resulting in a deeper understanding of a biological process that is perhaps more fundamental to the maintenance of life than any other. For, as has been pointed out, "Sex, which is merely fun for the individual, is indispensable for the species" (117).

ACKNOWLEDGMENTS

I am grateful to the National Institute of Child Health and Human Development, the National Science Foundation, and the Rockefeller Foundation for supporting the research carried out in my laboratory that is described here. It is also a great pleasure to acknowledge the many contributions of my student, Dr. Jeffrey Bleil, to this research.

REFERENCES

1. R. Yanagimachi, in A. A. Moscona and A. Monroy, eds., *Current Topics in Developmental Biology*, Vol. 12, *Fertilization*, Academic Press, New York, 1978, p. 83.
2. C. B. Metz, in A. A. Moscona and A. Monroy, eds., *Current Topics in Developmental Biology*, Vol. 12, *Fertilization*, Academic Press, New York, 1978, p. 107.
3. D. Epel, in A. A. Moscona and A. Monroy, eds., *Current Topics in Developmental Biology*, Vol. 12, *Fertilization*, Academic Press, New York, 1978, p. 186.
4. R. B. L. Gwatkin, *Fertilization Mechanisms in Man and Mammals*, Plenum Press, New York, 1977.
5. R. Yanagimachi, in M. Edidin and M. H. Johnson, eds., *Immunobiology of Gametes*, Cambridge University Press, Cambridge, 1977, p. 255.
6. D. Epel and V. D. Vacquier, in G. Poste and G. L. Nicolson, eds., *Cell Surface Reviews*, Vol. 5, *Membrane Fusion*, Elsevier/North Holland, Amsterdam, 1978, p. 1.
7. C. R. Austin, in C. R. Austin and R. V. Short, eds., *Reproduction in Mammals*, Vol. 1, *Germ Cells and Fertilization*, Cambridge University Press, Cambridge, 1972, p. 103.
8. B. M. Shapiro and E. M. Eddy, *Int. Rev. Cytol.* **66**, 257 (1980).
9. C. R. Austin and A. Walton, in A. S. Parkes, ed., *Marshall's Physiology of Reproduction*, 3rd ed., Vol. 1, Pt. 2, Longmans, Green and Co., Ltd., London, 1960, p. 310.
10. C. B. Metz and A. Monroy, eds., *Fertilization: Comparative Morphology, Biochemistry and Immunlogy*, Academic Press, New York, 1969.
11. W. Bloom and D. W. Fawcett, *A Textbook of Histology*, 10th ed., W. B. Saunders Co., Philadelphia, 1975.
12. E. H. Davidson, *Gene Activity in Early Development*, 2nd ed., Academic Press, New York, 1976.
13. J. D. Bleil and P. M. Wassarman, *Dev. Biol.* **76**, 185 (1980).
14. J. D. Bleil and P. M. Wassarman, *Proc. Natl. Acad. Sci. USA* **77**, 1029 (1980).
15. A. Hanada and M. C. Chang, *Biol. Reprod.* **6**, 300 (1972).
16. R. Yanagimachi, *J. Reprod. Fertil.* **28**, 477 (1972).
17. A. Hanada and M. C. Chang, *J. Reprod. Fertil.* **46**, 239 (1976).
18. R. Yanagimachi, H. Yanagimachi, and B. J. Rogers, *Biol. Reprod.* **46**, 239 (1976).
19. J. M. Bedford, *Anat. Rec.* **188**, 477 (1977).
20. J. D. Bleil and P. M. Wassarman, *Cell* **20**, 873 (1980).
21. P. M. Wassarman and J. D. Bleil, in L. Glaser, W. Frazier, and D. Gottlieb, eds., *Cellular Recognition*, ICN-UCLA Symposia on Molecular and Cellular Biology, Alan R. Liss, New York, 1982.
22. V. D. Vacquier and J. E. Payne, *Exp. Cell Res.* **82**, 227 (1973).
23. R. G. Summers and B. L. Hylander, *Exp. Cell Res.* **96**, 63 (1975).
24. R. G. Summers and B. L. Hylander, *Exp. Cell Res.* **100**, 190 (1976).

25. C. G. Glabe and V. D. Vacquier, *J. Cell Biol.* **75**, 410 (1977).

26. E. Schmell, B. J. Earles, C. Breaux, and W. J. Lennarz, *J. Cell Biol.* **72**, 35 (1977).

27. C. G. Glabe and V. D. Vacquier, *Proc. Natl. Acad. Sci. USA* **75**, 881 (1978).

28. V. D. Vacquier and G. W. Moy, in E. R. Dirksen, D. M. Prescott, and C. F. Fox, eds., *Cell Reproduction: In Honor of Daniel Mazia*, ICN-UCLA Symposium on Molecular and Cellular Biology, Vol. 12, Academic Press, New York, 1978, p. 379.

29. T. Oikawa, R. Yanagimachi, and G. L. Nicolson, *Nature* **241**, 256 (1973).

30. G. L. Nicolson, R. Yanagimachi, and H. Yanagimachi, *J. Cell Biol.* **66**, 263 (1975).

31. K. Aketa, *Exp. Cell Res.* **90**, 56 (1975).

32. M. Veron, and B. M. Shapiro, *J. Biol. Chem.* **252**, 1286 (1977).

33. V. D. Vacquier and G. W. Moy, *Proc. Natl. Acad. Sci. USA* **74**, 2456 (1977).

34. N. F. Bellet, J. P. Vacquier, and V. D. Vacquier, *Biochem. Biophys. Res. Commun.* **79**, 159 (1977).

35. C. G. Glabe and V. D. Vacquier, *Nature* **267**, 836 (1977).

36. C. G. Glabe and W. J. Lennarz, *J. Cell Biol.* **83**, 595 (1979).

37. W. Frazier and L. Glaser, *Annu. Rev. Biochem.* **48**, 491 (1979).

38. F. Grinnell, *Int. Rev. Cytol.* **53**, 65 (1978).

39. P. M. Saling and B. T. Storey, *J. Exp. Zool.* **209**, 229 (1979).

40. P. M. Saling and B. T. Storey, *J. Cell Biol.* **83**, 544 (1979).

41. R. N. Peterson, L. Russell, D. Bundmann, and M. Freund, *Science* **207**, 73 (1980).

42. P. W. Reed and H. A. Lardy, *J. Biol. Chem.* **247**, 6970 (1972).

43. D. R. Pfeiffer and H. A. Lardy, *Biochem.* **15**, 935 (1976).

44. R. Yanagimachi and N. Usui, *Exp. Cell Res.* **89**, 161 (1974).

45. G. L. Decker, D. B. Joseph, and W. J. Lennarz, *Dev. Biol.* **53**, 115 (1976).

46. P. Talbot, R. G. Summers, B. L. Hylander, E. M. Keough, and L. E. Franklin, *J. Exp. Zool.* **198**, 383 (1976).

47. F. Collins and D. Epel, *Exp. Cell Res.* **106**, 211 (1977).

48. D. P. L. Green, *J. Cell Sci.* **32**, 137 (1978).

49. R. W. Schackmann, E. M. Eddy, and B. M. Shapiro, *Dev. Biol.* **65**, 483 (1978).

50. C. B. Klee, T. H. Crouch, and P. G. Richman, *Annu. Rev. Biochem.* **49**, 489 (1980).

51. H. P. Jones, M. M. Bradford, R. A. McRorie, and M. J. Cormier, *Biochem. Biophys. Res. Commun.* **82**, 1264 (1978).

52. H. P. Jones, R. W. Lenz, B. A. Palevitz, and M. J. Cormier, *Proc. Natl. Acad. Sci. USA* **77**, 2772 (1980).

53. L. G. Tilney, *J. Cell Biol.* **69**, 73 (1976).

54. L. G. Tilney, D. P. Kiehart, C. Sardet, and M. Tilney, *J. Cell Biol.* **77**, 536 (1978).

55. L. G. Tilney, J. G. Clain, and M. S. Tilney, *J. Cell Biol.* **81**, 229 (1979).

56. R. A. McRorie and W. L. Williams, *Annu. Rev. Biochem.* **43**, 777 (1974).

57. E. F. Hartree, *Biochem. Soc. Trans.* **5**, 375 (1977).

58. A. E. Levine and K. A. Walsh, *Dev. Biol.* **72**, 126 (1979).

59. G. SeGall and W. J. Lennarz, *Dev. Biol.* **71**, 33 (1979).

60. W. H. Kinsey, J. A. Rubin, and W. J. Lennarz, *Dev. Biol.* **74**, 245 (1980).

61. W. H. Kinsey, G. K. SeGall, and W. J. Lennarz, *Dev. Biol.* **71**, 49 (1979).

62. M. J. Tegner and D. Epel, *J. Exp. Zool.* **197**, 31 (1976).
63. R. Yanagimachi and Y. D. Noda, *Am. J. Anat.* **128**, 429 (1970).
64. R. Yanagimachi and Y. D. Noda, *J. Ultrastruct. Res.* **31**, 486 (1970).
65. G. Schatten and D. Mazia, *J. Supramolec. Struct.* **5**, 343 (1976).
66. T. Miki-Noumoura and H. Kondo, *Exp. Cell Res.* **61**, 31 (1970).
67. I. Mabuchi, *J. Biochem.* **76**, 47 (1974).
68. Y. Hiramoto, *Exp. Cell Res.* **89**, 320 (1974).
69. N. R. Ringertz and R. E. Savage, *Cell Hybrids*, Academic Press, New York, 1976.
70. G. Poste and A. C. Allison, *Biochim. Biophys. Acta* **300**, 421 (1973).
71. Q. F. Ahkong, D. Fisher, W. Tampion, and J. A. Lucy, *Nature* **253**, 194 (1975).
72. R. B. Spertell, *J. Theor. Biol.* **60**, 197 (1976).
73. Y. Hirao and R. Yanagimachi, *Gamete Res.* **1**, 3 (1978).
74. A. F. Conway and C. B. Metz, *J. Exp. Zool.* **198**, 39 (1976).
75. R. A. Steinhardt, R. A. Ludin, and D. Mazia, *Proc. Natl. Acad. Sci. USA* **68**, 2426 (1971).
76. L. A. Jaffee, *Nature* **261**, 68 (1976).
77. J. D. Johnson, D. Epel, and M. Paul, *Nature* **262**, 661 (1976).
78. L. A. Jaffee and K. Robinson, *Dev. Biol.* **58**, 483 (1977).
79. R. Steinhardt and D. Epel, *Proc. Natl. Acad. Sci. USA* **71**, 1915 (1974).
80. R. Steinhardt, R. Zucker, and G. Schatten, *Dev. Biol.* **58**, 185 (1977).
81. M. Paul and D. Epel, *Exp. Cell Res.* **94**, 1 (1975).
82. J. D. Johnson, D. Epel, and M. Paul, *Nature* **262**, 661 (1976).
83. A. Lopo and V. D. Vacquier, *Nature* **269**, 590 (1977).
84. S. Shen and R. A. Steinhardt, *Nature* **272**, 253 (1978).
85. B. Gulyas, *Int. Rev. Cytol.* **63**, 357 (1980).
86. M. Paul and D. Epel, *Exp. Cell Res.* **65**, 281 (1971).
87. C. Barros and R. Yanagimachi, *J. Exp. Zool.* **180**, 251 (1971).
88. Y. Fukuda and M. C. Chang, *Biol. Reprod.* **19**, 261 (1978).
89. K. Sato, *J. Exp. Zool.* **210**, 353 (1979).
90. V. D. Vacquier, *Dev. Biol.* **43**, 62 (1975).
91. T. E. Schroeder, *Dev. Biol.* **70**, 306 (1979).
92. M. S. Barber and J. F. Mead, *Wilhelm Roux's Arch.* **177**, 19 (1975).
93. B. M. Shapiro, *Dev. Biol.* **46**, 88 (1975).
94. D. E. Wolf, W. Kinsey, W. Lennarz, and M. Edidin, *Dev. Biol.* **81**, 133 (1981).
95. M. J. Tegner and D. Epel, *Science* **179**, 685 (1973).
96. M. Veron, C. Foerder, E. M. Eddy, and B. M. Shapiro, *Cell* **10**, 321 (1977).
97. V. D. Vacquier, M. J. Tegner, and D. Epel, *Nature* **240**, 352 (1972).
98. V. D. Vacquier, M. J. Tegner, and D. Epel, *Exp. Cell Res.* **80**, 111 (1973).
99. H. Schuel, W. L. Wilson, K. Chen, and L. Lorand, *Dev. Biol.* **34**, 175 (1973).
100. E. J. Carroll and D. Epel, *Dev. Biol.* **44**, 22 (1975).
101. E. J. Carroll, E. W. Byrd, and D. Epel, *Exp. Cell Res.* **108**, 365 (1977).
102. S. Katsura and A. Tominaga, *Dev. Biol.* **40**, 292 (1974).
103. C. A. Foerder and B. M. Shapiro, *Proc. Natl. Acad. Sci. USA* **74**, 4214 (1977).

104. H. G. Hall, *Cell* **15**, 343 (1978).

105. C. Barros and R. Yanagimachi, *Nature* **233**, 268 (1971).

106. R. B. L. Gwatkin, D. T. Williams, J. F. Hartmann, and M. Kniazwuk, *J. Reprod. Fertil.* **32**, 259 (1973).

107. J. F. Hartmann, R. B. L. Gwatkin, and C. F. Hutchinson, *Proc. Natl. Acad. Sci. USA* **69**, 2767 (1972).

108. D. P. Wolf and A. Hamada, *Biol. Reprod.* **17**, 350 (1977).

109. R. B. L. Gwatkin, *J. Reprod. Fertil.* **7**, 99 (1964).

110. M. Inoue and D. P. Wolf, *Biol. Reprod.* **10**, 512 (1974).

111. M. Inoue and D. P. Wolf, *Biol. Reprod.* **11**, 558 (1974).

112. B. J. Gulyas and E. D. Schmell, *Gamete Res.* **3**, 267 (1980).

113. E. D. Schmell and B. J. Gulyas, *Gamete Res.* **3**, 279 (1980).

114. J. D. Bleil, C. F. Beall, and P. M. Wassarman, *Dev. Biol.* **86**, 189 (1980).

115. E. Van Beneden, *Bull. Acad. Belg. Cl. Sci.* **40**, 686 (1875).

116. O. Hertwig, *Morphol. Johrb.* **1**, 347 (1876).

117. E. Borek, *The Sculpture of Life*, Columbia University Press, New York, 1973.

2

Sexual Differentiation and Mating Strategies in the Yeast *Saccharomyces* and in the Green Alga *Chlamydomonas*

URSULA W. GOODENOUGH
Department of Biology
Washington University
St. Louis, Missouri 63130

JEREMY THORNER
Department of Microbiology and Immunology
University of California
Berkeley, California 94720

The biological realm contains two major groups of organisms, the prokaryotes and the eukaryotes. Bacteria, such as *Escherichia coli*, are the most familiar prokaryotes and humans are the most conspicuous eukaryotes. An important "middle" group are simple eukaryotes, like the yeast *Saccharomyces cerevisiae* (Fig. 1) and the green alga *Chlamydomonas reinhardi* (Fig. 2). These organisms resemble bacteria in that they are single-celled microbes, but they are more like human cells because they possess membrane-enveloped nuclei, histone-containing chromosomes that undergo mitosis and meiosis, and a full complement of such typical eukaryotic organelles as mitochondria, endoplasmic reticulum, and Golgi complexes. Since yeast and *Chlamydomonas* are vigorous occupants of present-day habitats, it is quite meaningless to speak of them as more or less "primitive" than *Escherichia coli* or humans. Nonetheless, since it is likely that the first eukaryotes were walled, unicellular organisms, the study of similarly designed modern counterparts will hopefully give us clues about our eukaryotic origins.

Primordial single-celled eukaryotes must have attained the capacity to undergo cell–cell fusion at an early stage in their evolution because the practice is ubiquitous among all groups of modern eukaryotes, but is an

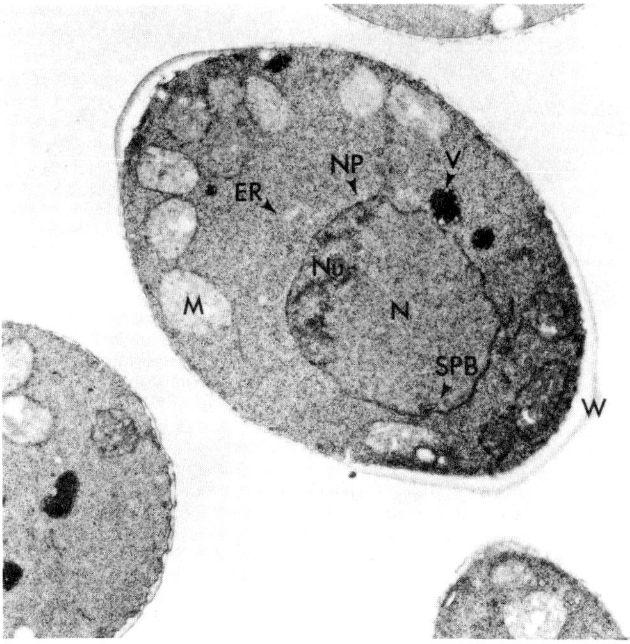

FIGURE 1. Ultrastructure of *Saccharomyces cerevisiae*. The subcellular components indicated by the letters are: ER, rough endoplasmic reticulum; M, mitochondrion; N, nucleus; NP, nuclear pore; Nu, nucleolus; SPB, spindle pole body; V, vacuole; and W, cell wall (which has been partially removed in this preparation by lytic enzyme digestion which facilitates fixation and staining of the rest of the cell). [Reprinted, with permission, from Thorner (4).]

unknown mode of sexual exchange among the prokaryotes. As with all sexual processes, the first step entails a species-specific recognition to ensure that fusion occurs between the correct cell types. For yeast, recognition is mediated by the cell walls themselves, while *Chlamydomonas* gametes recognize one another by flagellar membranes that poke out through holes in the cell walls. For both organisms, the second step entails a modification or shedding of the cell walls so that pairs of cells can make membrane-to-membrane contact; the final step is membrane fusion itself.

An important difference between yeast and *Chlamydomonas* mating lies in the factors that trigger mating responses. Yeast cells can be induced to differentiate into gametic cells, even during normal mitotic growth, if presented with a pheromone secreted by cells of the opposite sex. In the wild, of course, the presence of this signal indicates to a cell that cells of the opposite type are in the vicinity and that sexual fusion is an imminent possibility; but in the laboratory, cells can be "fooled" into displaying sexual responses by

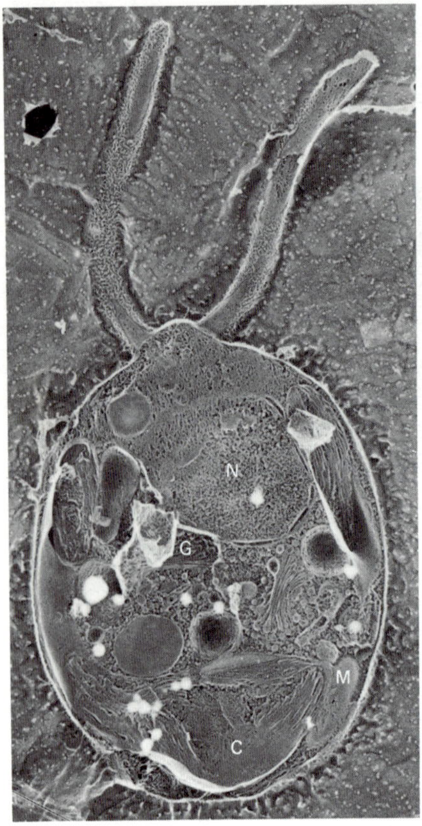

FIGURE 2. *Chlamydomonas reinhardi* gamete, mating-type *minus*. Conspicuous eukaryotic organelles include the nucleus (N), chloroplast (C), mitochondrion (M), and Golgi (G). The two anterior flagella display surface coats with numerous knobbly protuberances, which are thought to represent glycosylated moieties. The cell was rapidly frozen without prior fixation, fractured and deep-etched according to the method of Heuser and Salpeter (157). Transmission electron micrograph by Dr. John E. Heuser. Magnification, 24,000×.

presenting them with a solution of purified pheromone in the absence of cells of the opposite type. The trigger for sexual differentiation in *Chlamydomonas*, on the other hand, is nitrogen starvation. Cells of one sex, finding themselves without a nitrogen source, acquire gametic traits even if cells of the opposite sex are not present. The differentiated cells then "wait" until receptive partners arrive, and proceed to mate within seconds after recognition takes place.

These two sexual strategies can, in fact, be considered paradigms for the two predominant forms of cell–cell interaction that are thought to occur during development in multicellular organisms. In some cases, cells are thought to induce differentiative events in other cells by the release of signal molecules; in other cases, cells apparently differentiate in response to independent or internal cues and then subsequently join together in specific intercellular associations (see Chapters 3 and 4).

In the accounts of yeast and *Chlamydomonas* mating that follow, we first describe what is known (and speculated) about sexual differentiation (which is perhaps better understood in yeast); we then consider what is known (and speculated) about the mating reaction itself (which is perhaps better understood in *Chlamydomonas*).

1. SACCHAROMYCES CEREVISIAE

Unravelling the complex interactions that occur among individual cells in multicellular organisms during development remains one of the major problems of modern biology. How can such complicated events be attacked experimentally? One kind of system for studying the biochemical basis of intercellular signalling and developmental control would be one in which the interactions between just two kinds of cells could be examined in detail. This situation occurs in the life style of the simple unicellular eukaryote *Saccharomyces cerevisiae*, common baker's yeast. This organism has two haploid cell types, designated **a** and α, which have the capacity to fuse with one another to form a diploid cell. It has been appreciated only relatively recently that this conjugation, or "mating," represents for each haploid a simple developmental alternative to continued vegetative multiplication (1–4). The initial events in this process are triggered by the reciprocal action of diffusible oligopeptide pheromones secreted into the medium by each haploid cell type (5–7). Production of and response to these pheromones are strictly cell-type-specific; the pheromone secreted by α cells specifically affects **a** cells, and vice versa. Diploid cells neither produce nor respond to pheromones. In spite of the apparent structural differences between the **a** and α pheromones, discussed below, both molecules evoke seemingly identical responses in their respective target cells. These responses may be regarded as initiating a transient differentiation of the vegetative cells into cellular forms possessing the characteristic features of gametes. Specifically, this "gametogenesis" includes induction of cell-surface agglutinins, which allow specific recognition between mating partners, synchronization of the conjugating cells by arrest of their cell-division cycles in the G_1 phase, and localized changes in the structure of the cell wall that apparently are a prerequisite for cell fusion.

Because yeast is a eukaryote that is particularly amenable to biochemical analysis (8,9) and to both traditional genetic studies (10) and genetic manipulations involving recombinant DNA methodology (11,12), it has many experimental advantages that permit an examination of these cell–cell interactions at the molecular level. In particular, the pheromones released by **a** and α haploids have been well characterized and provide uniquely tractable probes for investigating the expression and function of developmentally related genes in these yeast cells. Thus, the objectives of much of the current research on yeast cells are:

1. To discover how production of these intercellular signal molecules is regulated.
2. To understand by what specific biochemical mechanisms these extracellular signal molecules control the entrance of haploid cells into their mating "program."
3. To determine what particular biochemical processes are responsible for the ensuing morphogenetic responses of the target cells, like the induction of cell-surface agglutinins and the imposition of specific cell-cycle arrest.

Reviewed below is our present state of knowledge about these processes. It will be clear to the reader that developmental events in this "simple" organism display many of the intricacies typical of "higher" eukaryotic cells in metazoans.

1.1. Sexual Differentiation and Its Genetic Control

1.1.1. Life Cycle of *Saccharomyces cerevisiae*. *Saccharomyces cerevisiae* can exist either as a haploid or as a diploid (Fig. 3). In either state, vegetative growth proceeds by an asymmetric cell division process, termed *budding*. This mode of duplication provides a handy morphological indicator of the stage of the cell cycle an individual cell is in at any given time because the size of the bud enlarges progressively as the cell cycle proceeds (13, 14). Upon exhaustion of the nutrients in a medium, meiosis and sporulation of diploid cells occurs. Each diploid yields an ascus (the old cell wall of the original diploid cell) containing a tetrad of two each of two different types of haploid spores (**a** and α), as expected for the segregation of a typical Mendelian trait. Each spore, if isolated by micromanipulation after enzymatic digestion of the ascus coat, can be germinated on solid medium and will produce a clone of progeny, all of the same haploid cell type as the initial spore from which they were derived. Each haploid cell type alone is unable to sporulate. Reestablishment of the diploid condition occurs only when cultures of the two different haploid cell types are mixed together (**a** with α), but not when cultures of the same haploid cell type are mixed (**a** with **a**, or α with α). Because only haploids of unlike cell type can conjugate, the **a** and α cells are referred to as "opposite mating types." Most strains give rise to **a** and α cells that can each be stably maintained as the same haploid cell type and are termed *heterothallic* yeasts.

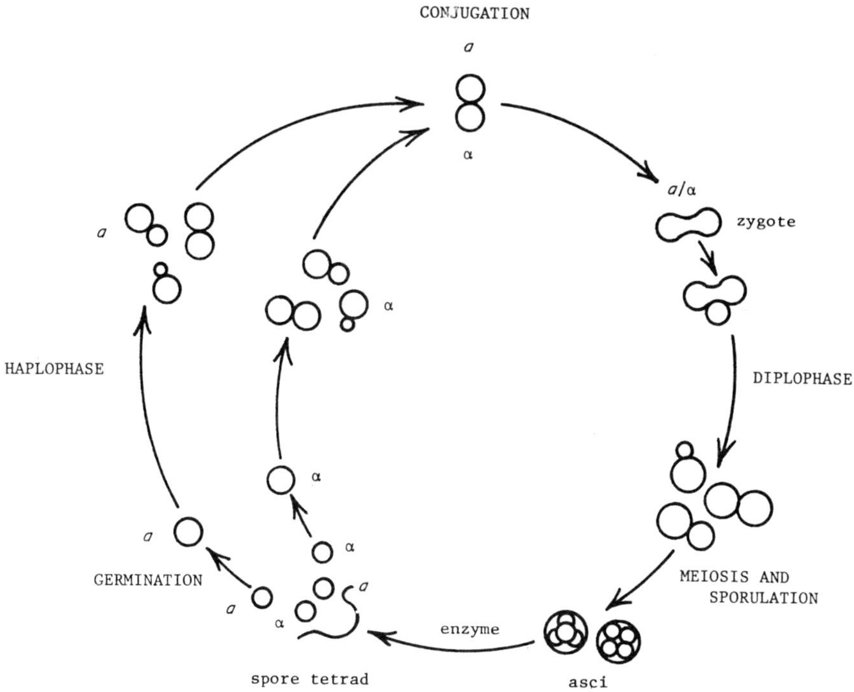

FIGURE 3. Life cycle of *Saccharomyces cerevisiae*. Shown here in schematic form are the developmental options open to haploid cells (vegetative growth or conjugation) and diploid cells (vegetative growth or sporulation) of a heterothallic strain of baker's yeast. [Reprinted, with permission, from Thorner (4).]

In certain genetic backgrounds, a single haploid spore of either mating type can give rise to haploid progeny of the opposite mating type during mitotic growth. Since cells of both types appear in a culture under such conditions, mating occurs between them, eventually yielding a population consisting mainly of diploid cells. This kind of strain is called *homothallic*.

1.1.2. Genetic Control of Cell Type in Yeast. Genetic studies of the mechanism by which a haploid cell of a homothallic strain can change (*switch*) to its opposite mating type have provided essential information about the behavior of the genes that control cell type in yeast (15). The formal genetic analyses have been confirmed by direct physical studies of these chromosomal regions through the use of recombinant DNA methodology (16, 17, 18). The yeast genome has three loci (*MAT*, *HML*, and *HMR*), each at a different site on chromosome 3, which can harbor either **a** or *α* information (Fig. 4). When either information is resident at *MAT* (the *ma*ting *t*ype locus) near the centromere, it is expressed. This region determines the

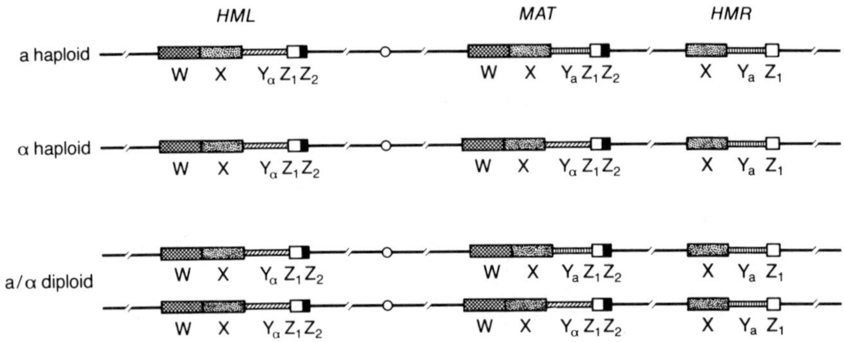

FIGURE 4. Structure of chromosome 3 in the three yeast cell types. Shown here in diagrammatic fashion are the expressed (*MAT*) and "silent" copies (*HML* and *HMR*) of mating-type information resident on chromosome 3. The mating-type locus *MAT* is near the centromere, indicated by the open circle; the silent loci are near the telomeres of the chromosome arms. The cross-hatched (W) and solid (Z_2) areas represent DNA sequences that are completely homologous in *MAT* and *HML*. The stipled (X) and open (Z_1) areas represent DNA sequences that are identical in all three regions. The DNA sequences specific to α cells (Y_α) and to **a** cells (Y_a) are indicated by the diagonal and vertical bars, respectively. An α haploid is *MAT*α, i.e., it has Y_α genetic information present at *MAT*. An **a** haploid is *MAT*a, i.e., it has Y_a genetic information present at *MAT*. An **a**/α diploid cell has two copies of chromosome 3, one with *MAT*a and one with *MAT*α. A model for how these regions control genes that determine the characteristics of each cell type is given in Figure 5. [Adapted from Nasmyth et al. (24).]

cell's characteristics because the functions coded for by *MAT*α and *MATa* are master regulatory proteins. These regulatory proteins control the expression of cell-type-specific functions required for mating and sporulation that are encoded elsewhere in the genome (Fig. 5). Examples of the kinds of functions whose expression is regulated by *MAT* are certain cell-type-specific *STE* (*ste*rile) genes. *STE* genes of this type were initially recognized by mutations unlinked to *MAT* (19, 20), which caused mating-deficiency only in one haploid and not in the other (Table 1).

When **a** or α information is resident at either *HML* or *HMR* (*silent*, *cryptic*, *storage*, or *cassette* loci), it is not expressed due to the action of at least four different unlinked loci, the *SIR* or *MAR* genes (21, 22). The *SIR* (*s*ilent *i*nformation *r*egulator) or *MAR* (*ma*ting type *r*egulator) genes code for functions that somehow negatively control transcription of *HML* and *HMR* by acting at *cis*-dominant regulatory sites that are present in the DNA that flanks *HML* and *HMR*, but are not present in the DNA surrounding *MAT* (23, 24).

There are three ways by which the normally silent information at *HML* and *HMR* can become expressed. One way to allow expression of the cryptic loci is to eliminate by mutation the function of the *SIR/MAR* regulators (25)

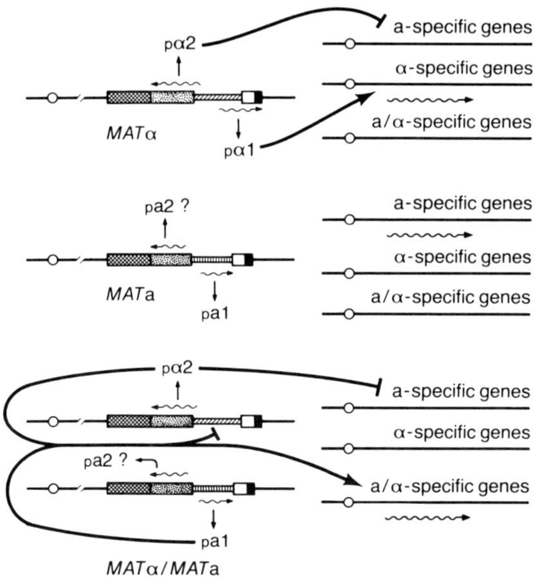

FIGURE 5. Control of cell-type-specific gene expression in *Saccharomyces cerevisiae*. Negative control ("turn-off") is indicated by a T bar; positive control ("turn-on") is indicated by an arrowhead. Wavy lines represent poly(A)$^+$ mRNAs. A large body of genetic and biochemical studies support the following model for regulation of cell type-specific gene expression by *MAT*. The **a** cell-specific functions are expressed constitutively. In an *α* cell, therefore, expression of **a** cell-specific genes must be prevented and, in addition, *α* cell-specific functions must become turned on. Thus, the protein product (p*α*2) coded for by the *α*2 transcript from *MATα* is a "repressor" of **a** functions and the product (p*α*1) of the *α*1 transcript is an "inducer" of *α* functions. In a diploid, the presence of the product (pa1) of the a1 transcript from *MAT*a changes the nature of p*α*2 (perhaps by forming a protein–protein complex) such that it represses expression of all haploid-specific functions, including *α*1 transcription, and permits expression of functions required for meiosis, sporulation, and other diploid-specific processes. The regulatory role of the product (pa2) of the a2 transcript from *MAT*a is not yet known. [Adapted from Klar et al. (23).]

(Table 1). The *SIR/MAR* gene products appear to be involved in the control of both *HML* and *HMR*, since their elimination by mutation "tricks" otherwise normal haploid cells into behaving like nonmating diploid cells because they are expressing both **a** (from *HMRa*) and *α* (from *HMLα*) information simultaneously.

Another means by which expression of *HML* and *HMR* can be achieved is as the result of rare intrachromosomal recombination events, like large deletions (26) and the formation of ring chromosomes (27). These chromosomal rearrangements connect either the *HML* site or the *HMR* site to the *MAT* locus and, in the process, alter or eliminate the regulatory regions that are normally recognized by the *SIR/MAR* gene products. Because genetic in-

TABLE 1. Genes Involved in the Control of Cell Type in *Saccharomyces cerevisiae*

Locus	Map Position[a]	Phenotype of the Mutation[b]	Possible Function
		Mating Type Controls	
*MAT*α1	3-C	Mating$^-$, α factor$^-$, agglutination$^-$, barrier^{-c}, sporulation$^+$ in **a**/α	Positive regulator of α-specific genes
*MAT*α2	3-C	Mating$^-$ (but mates weakly with α cells), α factor$^-$, agglutination$^-$, barrier$^+$, sporulation$^-$ in **a**/α	Negative regulator of **a**-specific genes; positive regulator of **a**/α-specific genes
*MAT*a1	3-C	Mating$^+$, **a** factor$^+$, agglutination$^+$, sporulation$^-$ in **a**/α	Positive regulator of **a**/α-specific genes
*MAT*a2[c]	3-C	High-level constitutive agglutination, sporulation$^+$ in **a**/α	Unknown
		Functions Affecting Expression of HML and HMR	
*SIR*1	ND	Mating$^-$ in either **a** or α	Negative regulator of *HML* and *HMR*
*SIR*2[d]	4-C	Mating$^-$ in either **a** or α	Negative regulator of *HML* and *HMR*
*SIR*3[e]	12-R	Mating$^-$ in either **a** or α	Negative regulator of *HML* and *HMR*
*SIR*4[f]	ND	Mating$^-$ in either **a** or α	Negative regulator of *HML* and *HMR*
		α Cell-specific Functions	
*STE*3	ND	Mating$^-$ in α, α factor$^+$, agglutination$^+$, **a** factor^R	**a**-Factor "receptor"
*STE*13	ND	Mating$^-$ in α, α factor$^-$, agglutination$^?$, **a** factor^S	Enzyme involved in processing α-factor precursor
		α Cell-related Functions	
*KEX*2	14-L	Killer toxin$^-$ in **a** or α, mating$^-$ in α, α-factor$^-$, **a**-factor^S	Secretion function
*TUP*1[g]	3-R	Mating$^?$ in α, α factor$^?$	Unknown

		a *Cell-specific Functions*	
STE2	ND	Mating$^-$ in **a**, **a** factor$^+$, agglutination$^+$, **a** factor^R	α-Factor "receptor"
*STE*6	ND	Mating$^-$ in **a**, **a** factor$^-$, agglutination$^?$, α factor^S	Unknown
*STE*14[h]	ND	Mating$^-$ in **a**, **a** factor$^-$, agglutination$^?$, α factor^S	Unknown
*SST*1[i]	9-C	Mating$^+$, barrier$^-$, α-factor$^{super\text{-}S}$	Membrane associated α-factor-degrading peptidase
		a *Cell-related Functions*	
None yet known			
		Functions Involved in Mating in Both Haploid Cell Types	
*STE*4	15-R	Mating$^-$ in **a** or α, pheromone^R, agglutination$^-$	Unknown
*STE*5	4-R	Mating$^-$ in **a** or α, pheromone^R, agglutination$^-$	Regulation of membrane-bound adenylate cyclase
*STE*7	4-L	Mating$^-$ in **a** or α, pheromone^R, agglutination$^-$	Unknown
*STE*11	ND	Mating$^-$ in **a** or α, pheromone^R, agglutination$^-$	Unknown
*STE*12	ND	Mating$^-$ in **a** or α, pheromone^R, agglutination$^-$	Unknown
*STE*15[j]	8-C	Mating$^-$ in **a** or α, pheromone^R, agglutination$^+$	Unknown
*STE*20[k]	ND	Mating$^-$ in **a** or α, pheromone$^?$, agglutination$^-$	Unknown
*SST*2	12-R	Mating$^+$, pheromone$^{super\text{-}S}$	Pheromone receptor turnover

TABLE 1. *(continued)*

Most of these loci, or representatives of a particular class of mutations, are discussed in the text where appropriate.

[a] Number refers to the chromosome and the letter refers to location (C = centromere-linked; L = on left arm; R = on right arm) on the most recent linkage map of yeast genes (158).

[b] − = unable to carry out the indicated process; + = able to carry out the indicated process; R = insensitive to pheromone action; S = sensitive to pheromone action; ? = property not yet tested or variable; barrier = ability of **a** cells, but not α cells, to destroy α-factor activity; killer toxin = a low molecular weight glycoprotein, secreted by yeast cells carrying endogenous viruslike double-stranded RNA-containing particles ("killer" particles) in their cytoplasm, which lyses susceptible yeast strains that lack the killer particles.

[c] Also called *SAG1* (88).

[d] Also called *MAR1* (22).

[e] Also called *STE8* (50), *CMT1* (159), or *MAR2* (A. Klar, personal communication).

[f] Also called *STE9* (50).

[g] It has been reported by several laboratories that *tup1* (160) and other mutations purported to be allelic, including *cyc9* (161), *umr7* (162), and *flk1* (163), are pleiotropic lesions that affect permeability to dTMP, iso-2-cytochrome *c* production, radiation-induced mutability, etc., in both haploid cell types and, in addition, cause α cells, but not **a** cells, to be mating deficient and unable to produce their pheromone. However, recent studies (164) have shown that the original phenotype by which *tup1* was initially selected (d*T*MP *up*take) segregates like a single nuclear gene, but does not cosegregate with α-cell sterility or with "clumpiness" (self-agglutination), another purported phenotype of these mutations. Thus, the actual defect causing α-cell mating deficiency in these strains is unknown.

[h] L. C. Blair, unpublished results.

[i] Also called *BAR1* (20).

[j] A. Burkeholder and L. Hartwell, unpublished results.

[k] N. Yanagishima, unpublished results.

formation essential for cell viability often is lost by such massive chromosomal changes, in most cases the affected chromosome can be maintained only in diploids that possess a second intact copy of chromosome 3. Indeed, such chromosomal aberrations were first recognized as lesions of chromosome 3 that always caused diploids to segregate meiotically two inviable spores and two viable spores that were both of the same mating type.

The third mechanism whereby the mating type information at *HML* or *HMR* can be activated is by directed transposition to *MAT* (15). In other words, under certain conditions, *HML* and *HMR* can act like mobile genetic elements. The transposition event does not affect any other part of chromosome 3 and is a nonreciprocal transfer or "transmission" of the information specifically from *HML* or *HMR* to the mating-type locus, since the sequences formerly at *MAT* seem to be discarded, while those at *HML* and *HMR* remain intact. The frequency with which such a transposition event will occur depends on the presence or absence of a functional copy of the *HO* gene, which resides on chromosome 4 (28). The transposition process occurs at quite a low frequency ($<10^{-6}$) in typical heterothallic strains that lack an operative *HO* gene (*MATa HMLα HMRa ho* or *MATα HMLα HMRa ho*). Hence, cell type is quite stable in such strains. The transposition of information to *MAT* occurs at a very high frequency (as often as once per generation) in typical homothallic yeasts that do possess a functional *HO* gene (*MATa HMLα HMRa HO* or *MATα HMLα HMRa HO*), explaining the interconversion (*switching*) of mating type in these cells and the subsequent diploidization of the culture due to conjugation between the opposite mating types so generated. The *HO* gene product may be some sort of "transposase," an enzyme involved in catalyzing the process of DNA transposition (29). Other cellular functions must also be involved in the transposition process. For example, *rad52* mutations that map near the centromere of chromosome 13 and were first recognized as deficient in the repair of DNA damage, are unable to undergo mating type switching even if *HO* is present (30).

The movements of the genetic regions controlling mating type in yeast cells from one chromosomal location to another appear to have counterparts in higher cell development, for example, the chromosomal rearrangements of the genetic information for immunoglobulin synthesis that occur during lymphocyte differentiation (31).

MATa and *MATα* are codominant in diploids; diploid cells behave neither like **a** haploids nor like *α* haploids. Furthermore, genetic studies of *mat* mutations suggested originally that the *MAT* loci are not simply alternative alleles, but are distinct functionally nonequivalent blocks of DNA (15, 19, 32). Indeed, heteroduplex analysis of cloned *MATa* and *MATα* DNA demonstrates that there are homologous segments shared between them, but this complementarity resides in regions (0.33 and 1.5 kb) that flank a stretch of completely nonhomologous DNA (0.65 kb in *MATa* and 0.75 kb in *MAT*α) (Fig. 4) (16–18). Examination of cellular RNA species by probing with cloned *MATa*-specific or *MAT*α-specific DNA indicates that each *MAT* locus produces two discrete RNA species (23, 24) that appear to code for separate and distinct sets of regulatory proteins (15) essential for controlling expression of the genes that distinguish one cell type from another (Fig. 5). The entire nucleotide sequences for the *MAT*, *HML*, and *HMR* DNAs have been determined (33; M. Smith, personal communication) and confirm the existence of long open reading frames in the regions covered by the *MAT* transcripts. Furthermore, direct in vitro mutagenesis of the *MAT* DNAs cloned on plasmids (K. Tatchell, personal communication), followed by their reintroduction into *mat* mutants using spheroplast transformation, has further corroborated the regulatory functions ascribed to these proteins from previous genetic studies in vivo (15, 34).

Of particular importance for understanding the intercellular interactions of **a** and α haploids, the particular genes resident at *MAT* direct whether or not a cell will produce a mating pheromone and, if it does so, which pheromone it will secrete. To date, however, none of the **a** cell-specific *STE* genes has been identified as the structural gene for **a** factor; nor have any of the α cell-specific *STE* genes been demonstrated to be the structural gene for α factor (Table 1).

1.2. Intercellular Interactions and the Mating Process

1.2.1. Mating Pheromones and Intercellular Signalling. As already mentioned above, one characteristic of a haploid cell is its secretion of a specific oligopeptide pheromone. The molecule secreted by α haploids is called *α factor* (35, 36). Exposure of **a** haploids to this α pheromone initiates biochemical and morphological changes in the **a** cells that appear to be preparatory to conjugation, even if α cell partners are absent. The **a** cell responses include, in order of their appearance:

1. Cell surface alterations that enhance the strength and selectivity of adhesion to α cells (37).
2. Transient arrest of cell division, specifically as an unbudded cell in the G_1 phase of the cell cycle, prior to duplication of the nuclear spindle pole body (a microtubule-organizing center embedded in the yeast nuclear membrane [Fig. 1]) (38) and prior to the initiation of chromosomal DNA synthesis (39).
3. Anisotropic cell wall and membrane synthesis resulting eventually in the formation of markedly elongated cells (35, 36, 40) (Fig. 6).

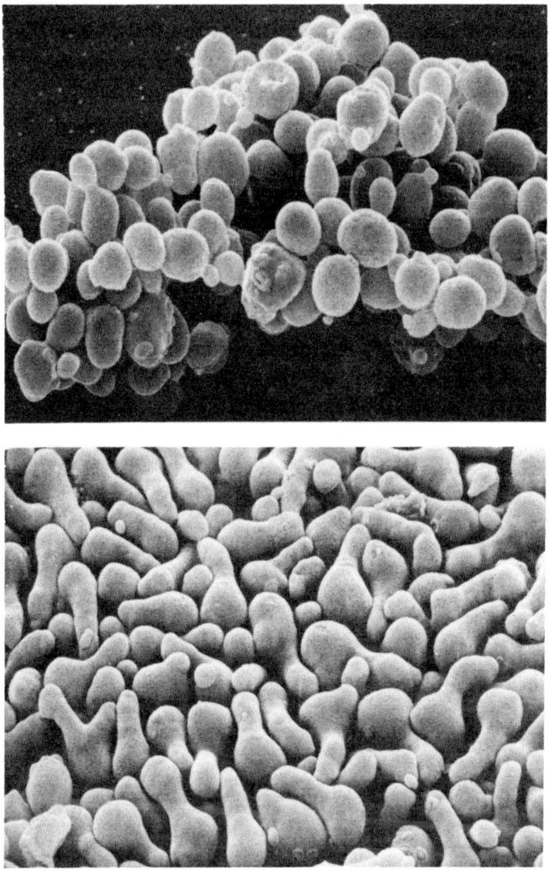

FIGURE 6. Marked morphological changes induced in **a** cells by α-factor treatment. Cells in an exponentially growing culture of **a** mating-type haploids were exposed to 5 units/ml of a preparation of purified α-factor pheromone for 4 hr at 30° C in nutrient medium, then harvested and processed for scanning electron microscopy. Cells from an untreated control culture are shown in the upper panel. The pheromone-treated cells are shown in the lower panel. Note the markedly elongated appearance of the cells from the α-factor treated culture, compared to the normally spherical or ovoid shape of vegetatively growing yeast cells. [Reprinted with permission, from Thorner (4).]

In a similar fashion, **a** cells release a pheromone, called **a** *factor*, which has precisely analogous effects on α haploids. Both the α and **a** pheromones have been purified and completely characterized.

The α factor is a linear tridecapeptide with the following primary structure (41):

H_2N–Trp–His–Trp–Leu–Gln–Leu–Lys–Pro–Gly–Gln–Pro–Met–Tyr–COOH

A dodecapeptide lacking the N-terminal Trp is also found in preparations of α factor, presumably resulting from limited aminopeptidase action, and has full biological activity. Both forms of α factor have been chemically synthesized (42). These synthetic molecules elicit the entire array of biological effects attributed to the natural α pheromone and thus prove unambiguously that α factor is the sole primary signal for inducing the mating program in **a** cells.

The **a** factor is a mixture of two highly related linear undecapeptides with the following amino acid sequences (43; W. Duntze, personal communication):

H_2N–Tyr–Ile–Ile–Lys–Gly–Val –Phe–Trp–Ala–Asx–Pro–COOH
H_2N–Tyr–Ile–Ile–Lys–Gly–Leu–Phe–Trp–Ala–Asx–Pro–COOH.

The reason for there being two very similar molecules with **a** factor activity is beginning to become clear from studies of the pheromone genes using recombinant DNA techniques, as discussed below.

Earlier work using both inhibitors and temperature-sensitive mutations that block macromolecular synthesis suggested that α factor biosynthesis involves ribosomal translation of an mRNA (44). This finding implied that the information for the primary structure of α factor is coded directly in the yeast genome, in contrast with the peptide antibiotics produced by certain bacteria whose structures are not coded for directly in DNA but are dictated by the substrate specificities of the multienzyme complexes that assemble them (45). The fact that cells auxotrophic for amino acids not found in the mature α factor molecule rapidly ceased producing the pheromone when starved for these required amino acids further suggested that α factor probably is produced from a larger precursor polypeptide. Essentially all the peptide hormones made by higher eukaryotic cells are generated through proteolytic processing of a larger precursor molecule, for example, insulin from preproinsulin (46). Indeed, the recent isolation of the α factor structural gene by recombinant DNA methodology supports and greatly extends these earlier observations. Analysis of the nucleotide sequence of this DNA segment indicates that it can be translated into one long precursor polypeptide in which four exact copies of the mature α factor sequence are linked in tandem array by spacer amino acids (Fig. 7). Thus, to produce the final α pheromone, the α factor sequences must be properly excised from the precursor by precise proteolytic processing events. In fact, α cells bearing the α-specific mutation *ste13* secrete apparently improperly processed forms of α factor of greatly reduced biological activity with the following structures (D. Julius, L. Blair, T. Brake, and J. Thorner, in preparation):

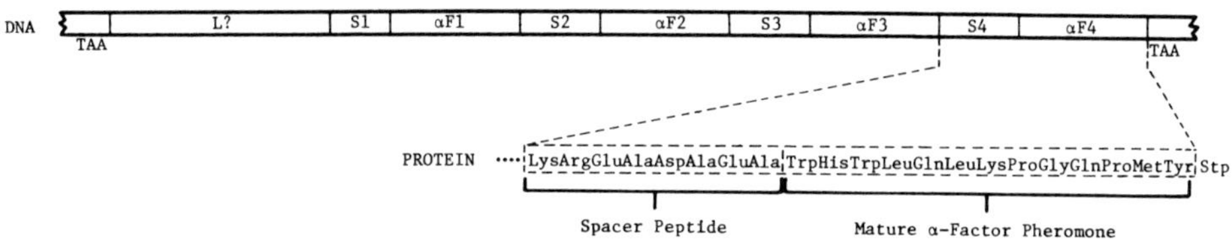

FIGURE 7. Structure of the α-factor gene in *Saccharomyces cerevisiae*. In the chromosomal DNA of the yeast cell, there is a region in which the nucleotide sequence codes for four tandem repeats (αF1-αF4) of the amino acid sequence of the mature α-factor molecule. These repeats are embedded in nucleotide sequence that specifies a larger precursor protein, such that the repeats of pheromone-coding information are flanked by sequences coding for shorter spacer peptides (S1-S4). Spacers S3 and S4 have identical amino acid sequences; S2 is very similar to S3 and S4, but has a Glu residue in place of the central Asp residue found in S3 and S4; and S1 is also highly related to S2-S4, but is shorter by two amino acids, having the sequence Lys–Arg–Glu–Ala–Glu–Ala. αF4 is apparently at the C-terminus of the precursor protein, and hence is the only repeat that is not flanked on both sides by the spacer peptide. The N-terminus of the precursor has not yet been identified and, therefore, the length and structure of any additional leader peptide (L?) in the pheromone precursor is not yet known. [Based on the findings of Kurjan and Herskowitz (47).]

H_2N–Glu–Ala–Glu–Ala–Trp–His–Trp–Leu–Gln–Leu–Lys–Pro–Gly–Gln–Pro–Met–Tyr–COOH
H_2N–Asp–Ala–Glu–Ala–Trp–His–Trp–Leu–Gln–Leu–Lys–Pro–Gly–Gln–Pro–Met–Tyr–COOH

The additional amino acids not removed from α factor in *ste13* mutants corroborate the existence of the larger pheromone precursor predicted from the DNA sequence and furthermore suggest that *ste13* mutants may be defective in one of the proteolytic processing functions (48). Expression of the α factor gene is an α-cell-specific function; poly(A)$^+$ mRNA complementary to the α-factor structural gene is only found in α haploids, not in **a** haploids or **a**/α diploids (J. Thorner, unpublished results). If the *ste13* mutation indeed represents a lesion in a special processing protease, this enzyme, which is involved in choosing the appropriate pheromone segments to cleave from the precursor, may also be a cell-type-specific function under *MAT* control.

If the **a**-factor structural gene is similar to that for α factor, then it may contain at least two tandem repeats of the mature **a**-factor sequence that differ by only a single base change in the DNA, explaining the production of both the Val- and Leu-containing **a**-factor molecules by a single **a** haploid strain.

In support of the important role of the pheromones in the physiology of conjugation are the findings that many mating-defective mutants do not produce their particular pheromone and/or do not respond to that produced by the opposite mating type (19) and that **a** cell mutants selected as resistant to α factor are almost invariably mating deficient (49, 50). Despite the clear correlation of pheromone production and response with cell type and the mating process, little is currently known about the actual biochemical mechanism whereby the pheromones act.

In the case of α factor, the fact that the dose-dependence of **a** cell response is first-order with respect to α pheromone concentration and shows saturation kinetics is compatible with the existence of a specific receptor on **a** cells for α factor (51, 52). However, because there is a high degree of nonspecific binding of α factor to the outermost cell wall layer (4), called the phosphomannan layer, and rapid proteolysis of the α-factor molecule by an **a** cell-bound endopeptidase (53, 54, 55), direct measurements of the binding of α factor usually fail to reveal any marked differential binding to **a** haploids.

On the basis of analogy to the features of peptide hormone action in higher eukaryotes, the effects of α factor on the activity of the membrane-bound adenylate cyclase of yeast cells were examined recently. It was found (56) that α factor markedly inhibits the enzyme in a dose-dependent manner. The physiological relevance of this effect was supported by two additional observations. First, one class (*ste5*) of **a**-cell mutants that grow normally at either 23 or 34°C, yet neither responds to α factor nor mates at the higher temperature (50), possessed an adenylate cyclase activity that was not inhibitable by α factor at 34°C but was fully sensitive to pheromone inhibition at 23°C. Second, addition of cyclic AMP (cAMP) or cAMP phosphodiesterase inhibitors (57) to **a** cell culture medium shortened the duration of, or almost completely prevented, pheromone-induced G_1 arrest of the **a** cells. These results indicate that the biochemical mode of action of α factor may be to lower the intracellular cAMP concentration in **a** cells.

Many of the actions of cAMP in animal cells appear to be mediated through cAMP-dependent protein kinases (58, 59). In this same regard, it is an extremely significant finding that the *src* gene products of RNA tumor viruses (*v-src*) and the T antigen proteins produced by DNA tumor viruses, possess cAMP-independent protein kinase activity (60, 61). These viral enzymes are self-phosphorylating and also phosphorylate certain cellular target proteins on tyrosine residues, a recently discovered posttranslational modification of proteins (62, 63). The result of viral infection of appropriate hosts is malignant growth, that is, continued uncontrolled cell division. In normal animal cells, a protein kinase activity (called *sarc* or *c-src*) with the ability to phosphorylate tyrosine residues is also present, but only at a very low level (64). Thus, the entrance into the cell of the virally coded enzymes presumably overrides the controlled phosphorylation normally conducted by *c-src*, and perhaps by other Tyr-phosphorylating activities (65), as well as by cAMP-dependent protein kinases (which phosphorylate Ser and Thr residues), thereby subverting proper regulation of animal cell proliferation.

The action of a yeast pheromone represents the converse situation to that of a tumor virus: the role of the pheromone is to impose a controlled arrest of the yeast cell division cycle. It is not yet known whether yeast cells contain Tyr-phosphorylating activities similar to those seen in higher organisms. However, it is possible that the lowering of cAMP level in **a** cells in response to α factor specifically inhibits the activity of cAMP-dependent protein kinase(s). The inhibition of cAMP-dependent protein kinase would reduce the degree of phosphorylation of enzymic and/or structural components that may normally be needed, in their phosphorylated form, for a yeast cell to initiate its division cycle. Hence, in the presence of a pheromone, cell division would be arrested. This hypothesis predicts, conversely, that other yeast proteins (like those responsible for diverting cellular metabolism and gene expression into the mating program) are inactive in their phosphorylated form, and hence "turned on" when cAMP-dependent protein kinase is inhibited.

In this regard, it is clear that yeast cells do possess a cAMP-based regulatory system that appears very analogous to that seen in higher eukaryotic cells. These functions include: adenylate cyclase (66); two (high and low K_m) cyclic nucleotide phosphodiesterases (67, 68); protein kinase activities (69–71); and a cAMP binding protein that is a protein kinase regulatory subunit (72–74). Moreover, a number of yeast proteins are already known to be targets for reversible phosphorylation and dephosphorylation, for example, ribosomal proteins (75, 76), RNA polymerase subunits (77), and NAD-glutamate dehydrogenase (78). In addition, calcium-dependent regulatory components, which interlock with cyclic nucleotide-based control systems (79), have been demonstrated in other lower eukaryotes. For example, the protozoan *Tetrahymena pyriformis* (80) and the cellular slime mold *Dictyostelium discoideum* (81) have calmodulin-like proteins with properties very similar to those of their counterpart in bovine brain. A similar calmodulin-like protein is present in *Saccharomyces cerevisiae* cells (G. Stetler and J. Thorner, unpublished results).

Therefore, the events that occur in controlling the cell cycle and initiating development in relatively simple eukaryotic cells like yeast should have important features in common with those operating in higher organisms. In fact, a haploid yeast cell is very much like an undifferentiated or embryonic cell, growing and dividing until a signal (in this case, a mating pheromone) diverts it into a specific developmental pathway. In the same way, α-factor-resistant **a** cell mutants could be considered to have escaped from their normal developmental regulation, and therefore resemble neoplastic cells. Hence, elucidation of the pathway of α-factor action should help us to understand how entrance into alternative developmental fates is controlled and, possibly, how the cell cycle is regulated in eukaryotes.

1.2.2. Cell-Surface Glycoproteins and Cell–Cell Recognition. Although G_1 arrest facilitated by the pheromones appears to be a necessary condition for conjugation, the opposite mating types must also be brought into close proximity to permit their fusion. Since yeast cells are nonmotile, the efficiency of mating depends on the opportunity for cell–cell contact provided by random collision. Because successful mating requires close cell–cell contact, it is very significant that the earliest detectable response of **a** cells to exposure to α factor is some cell-surface alteration that markedly increases their agglutinability toward α cells (37, 82). This increase in intercellular adhesion presumably increases the probability that opposite mating types will remain in close contact after they have met by chance collisions.

When they are mixed under nongrowing conditions, **a** and α cells that have not previously been exposed to each other (so-called "naive" cells) bind to each other only weakly, although they do interact with each other more strongly than they cohere to cells of like mating type (37, 83), suggesting that the cells normally express only a basal level of mating-specific cell-surface agglutinins. In contrast, within 30–60 min after incubation of mass mating mixtures in growth medium or immediately after mixing haploids that have been preincubated individually for 30–60 min in growth medium containing purified pheromone from the opposite mating type, the **a** and α haploids (so-called "conditioned" cells) adhere to each other quite strongly (37, 84, 85), suggesting that expression of mating agglutinins has been highly induced. In fact, two classes were observed initially among various available **a**-cell stocks with regard to this particular response (86): (1) those **a** strains that agglomerate with α cells immediately upon admixture (*constitutive strains*), and (2) those that require a period of mutual incubation with α cells before a strong agglutination reaction is obtained (*inducible strains*). The competence of most α haploids to agglutinate is enhanced much less dramatically, or not at all, by prior exposure to **a** cells or to **a** factor (37, 85, 87). It has been found that the high degree of intrinsic agglutinability of many constitutive **a** strains is only observed when the cells are grown at 23–28°C. If cultured at 34–36°C, the formerly constitutive **a** strains behave like inducible cells, i.e., they are only weakly agglutinative when naive, but after being conditioned with α factor they display markedly increased agglutinability toward α cells (88–90).

One possible interpretation of these results is that intrinsic agglutinability and inducible mating-specific adhesion are separable functions carried out by different cell-surface macromolecules that differ in their thermostability. This seems unlikely because partially purified glycoprotein fractions that appear to mediate mating-specific intercellular recognition, as discussed below, have very similar properties whether they are prepared from constitutive **a** strains or from conditioned inducible **a** strains (90). Furthermore, the agglutinin made by constitutive strains cultured at low temperature is lost only slowly, by growth dilution, after shift to 36°C (88).

An alternative possibility to explain the two types of **a** stocks is that in inducible **a** cells synthesis of mating-specific agglutinin is normally repressed at all temperatures by a negative regulatory element. However, the action of this repressor can be prevented somehow by exposure of the cells to α factor, resulting in induction of the synthesis of mating agglutinin. By this model, in the so-called constitutive **a**-strain expression of mating-specific agglutinin is allowed at 23–28°C because the repressor bears a defect that makes its function (or synthesis) cold-sensitive. Hence, at lower temperatures where the repressor is inactive (or not present), mating agglutinin is made whether or not cells are exposed to α factor (constitutive phenotype); but at higher temperatures, where the repressor is operative, mating agglutinin can be made only if α factor is present (inducible phenotype). This model predicts that total elimination of the repressor (a null mutation) would make an **a** cell constitutive for expression of mating agglutinin at all temperatures. Indeed, a mutation (*sag1*) conferring such a completely constitutive phenotype has been found among **a**-cell stocks (88). If the *sag1* mutation represents the loss of a negative regulatory element that operates in *trans*, then its constitutivity should be recessive to inducibility. This was, in fact, found to be the case by constructing **a/a** diploids (which behave like **a** haploids) through protoplast fusion of an inducible **a** strain and the *sag1* constitutive **a** mutant. Moreover, the *sag1* mutation segregated like a single nuclear gene in crosses to an inducible α strain and was inseparable from the *MATa* locus in all 59 tetrads examined. These results were confirmed in a similar study made by another group (91). It has been suggested (J. Shuster, personal communication) that the *sag1* mutation may define a negative regulatory role for the protein encoded by the *a2* transcript of *MATa* (Fig. 5), a gene previously without an identifiable function (Table 1).

At the chemical level, the cell wall synthesis that occurs in **a** cells in the presence of α factor seems altered in several ways. The newly made wall material contains more glucan and less mannan than in unexposed **a** cells (92). Furthermore, the mannan coat of the walls of α-factor-treated **a** cells contains an increased proportion of shorter oligosaccharide side chains and unsubstituted $\alpha(1\rightarrow6)$ backbone, which probably explains the greater binding of fluorescently labeled concanavalin A (93) and antibody directed against $\alpha(1\rightarrow6)$ mannosyl haptens (94) to pheromone-treated **a** cells. Also, cells treated with α factor have higher chitin synthetase activity than untreated cells and appear to deposit chitin diffusely throughout the elongating portion of the cells (40). Chitin is normally a minor cell wall component usually found only in a narrow ring at the septum formed between the bud and mother cell at the time of cell division. Because of all these cell wall changes, α-factor-treated **a** cells are more susceptible than normal cells to digestion with glucanases (lytic enzymes that attack the inner cell wall polysaccharide layer called the glucan layer). Electron micrographs reveal that the protuberance that forms on elongating **a** cells that have been exposed to α factor in fact has a thinner wall layer, with a "fuzzy" exocellular coat, and contains subtending small vesicles (92), which presumably carry the enzyme activities and precursors for cell wall and membrane growth (95).

Conceivably, cell-surface alterations like those discussed above may reflect or be part of the synthesis, activation, or exposure of the cell-surface-associated agglutinins responsible for mating-specific intercellular adhesion. The induction of increased agglutinability caused by exposure to pheromone requires active growth and energy metabolism and therefore will not occur at 0°C or in medium lacking either a carbon or nitrogen source (87). This clearly distinguishes the yeast mating response from that of *Chlamydomonas*: in the alga, conjugation is triggered by nutritional limitation. It has been demonstrated by the use of inhibitors of macromolecular synthesis that new RNA and protein synthesis are required for the pheromone-induced acquisition of increased agglutinability of yeast haploids (37, 82, 85). Whether the new protein(s) made are themselves the agglutinins or are involved in the assembly, modification, or unmasking of preexisting agglutinin molecules is not yet known. However, inhibitors of polysaccharide and glycoprotein biosynthesis (including tunicamycin, 2-deoxyglucose, and amphomycin) fail to prevent α-factor-induced enhancement of agglutination, even at concentrations of these inhibitors that would be inhibitory to subsequent growth. Hence, the protein synthesis that is involved in agglutinin induction does not appear to require de novo glycoprotein assembly (C. Atcheson, G. Fehrenbacher, K. Perry, and J. Thorner, in preparation).

Attempts have been made to isolate cell-surface molecules from haploid cells that may correspond to mating agglutinins, most notably by Yanagishima and his coworkers (87). For example, by degrading **a** cell walls or α cell walls with Glusulase (a mixture of hydrolytic enzymes in the juice from the digestive tract of land snails) (96) or by autoclaving haploid cells (97, 98), glycoprotein fractions are released that appear to act as univalent blockers of the aggregation of **a** and α cells. These molecules are only incompletely characterized as yet (Table 2). It should be mentioned here that aggregation of the haploid cells of another yeast species, *Hansenula wingei*, also appears to be mediated by cell wall-bound glycoproteins (2, 3) (Table 2). These components have been more extensively characterized (99–101) than the putative *Saccharomyces cerevisiae* agglutinins; however, in order to release the *Han-*

senula wingei agglutinins from the cell walls of the haploids, treatments have been used that also destroy portions of the molecules to some degree. Such cell-surface glycoproteins also seem to be involved in mating-type-specific agglutininations in at least two other species of ascomycetous yeasts (101). The cell-surface agglutinins in all these yeast species are quite heat stable in situ; thus, agglutination between cells of opposite mating type still occurs even if the cells have been killed by brief boiling. As discussed below, successful mating-specific adhesion of *Chlamydomonas* cells requires that at least one of the two cell types be viable.

It is clear that only unbudded G_1-arrested yeast cells actually mate (13, 39). In contrast, it appears that the induction of agglutinability by a pheromone can occur when the target haploid is at any stage of its cell cycle. For example, after conditioning with pheromones, haploids at all stages of the cell division cycle are found in mating aggregates (37, 84) and conditioned haploids at various stages of the cell cycle obtained by density gradient fractionation of pheromone-treated cultures are all agglutination competent (90). Furthermore, budded cells blocked in S phase by hydroxyurea treatment (88) or cells blocked at many stages of the cell cycle by raising various temperature-sensitive cell division cycle (*cdc*) mutants to a nonpermissive temperature (L. Hartwell, personal communication) can be fully induced for agglutination by mating pheromones. These findings agree with the fact that, kinetically, the induction of increased agglutinability is an early response to pheromone treatment, detectable after 15–20 min and maximal by 45–60 min. In fact, if the unbudded fraction of a vegetative population of **a** haploids is isolated, the cells will not agglutinate tightly with unbudded α haploids and subsequently yield zygotes unless the unbudded **a** cells have first been incubated with α factor (or with the α cells themselves) for 20–60 min (102). This indicates that merely being in G_1 is *not* sufficient to confer agglutination competence, at least in **a** cells, and suggests therefore that α factor is essential for efficient mating.

These results, taken together, indicate the following time course for initial mating events. First, the pheromones induce cell-surface changes that increase and stabilize cell–cell contacts. While being held in close apposition in mating aggregates, the population accumulates in G_1 because the presence of the pheromones also prevents the cells from progressing through the cell division cycle. Concomitantly, subsequent functions required for cell fusion (or to circumvent budding) are called into play, again presumably because of the continued presence of the pheromones. Whether the primary signal generated by a pheromone (for example, lowering the intracellular level of cAMP) has many separate effects, or sets off a cascade of dependent steps, is unknown.

TABLE 2. Properties of the Agglutination Process and of the Isolated Glycoprotein Agglutinins from Two Yeast Species[a]

	Organism			
	S. cerevisiae		*H. wingei*	
	Cellular Agglutination (in situ)			
	a *cells*	*α cells*	*5 cells*	*21 cells*
Pheromone induction[b]	+	+/−	−	−
Pretreatment:				
boiling	+	+	+	−/+
reducing agent (−SH)	−	+	−	+
chelating agent (EDTA)	+	+	−/+	−/+
Pronase	−	−	−	−
subtilisin	−	−	−	−
trypsin	+	−	?	?
chymotrypsin	−	−	?	?
carboxypeptidase Y	−	+	?	?
pH 3.0	−	−	−	−
pH 9.5	−	−	?	?
Victoria Blue dye	−	+	?	?
concanavalin A — tetravalent	+	+	?	?
concanavalin A — univalent	−	−	?	?

Isolated Agglutinins (in vitro)

	a *agglutinin*	α *agglutinin*	*5 factor*	*21 factor*
"Native" molecular weight[c]	$\sim10^5$–10^{6}[d]	$>10^{6}$[d]	$\sim0.2 - 1 \times 10^{6}$[e]–$10^{8}$[d]	23[e]–27,000[f]
"Subunit" molecular weight	~10[g]–23,000[h]	130,000[h]	12,500[g]	23–27,000[g]
Carbohydrate content	~61%	~47%	85%	5%
Ser- or Thr-linked	?	?	+++	+
Asn-linked	?	?	+	+++
pI	4.5	4.3	?	3.8
Binding	Univalent	Univalent	Multivalent	Univalent
Activity	Masks agglutinability of α cells; binds to α agglutinin	Masks agglutinability of **a** cells; binds to **a** agglutinin	Agglutinates 21 cells; binds to 21 factor	Blocks ability of 5 factor to agglutinate 21 cells; binds to 5 factor
Pretreatment				
boiling	+	−/+	+	−
reducing agent (−SH)	−	+	−	+
exo-α-mannosidase	?	?	−	+
pH 9.5	−	−	?	?
trypsin	+	−	+	−/+
Pronase	?	?	−	−
chymotrypsin	?	?	+	−/+
subtilisin	−	−	+	−/+

[a] Compiled, in the main, from the data of Yanagishima and coworkers (84, 87, 97, 98) and Ballou and coworkers (99–101). + = still functional, i.e., shows agglutination response or retains binding ability; − = not functional, i.e., no agglutination response or unable to bind.

[b] *H. wingei* cells display a strong, constitutive sexual agglutination. No pheromones secreted by *H. wingei* have yet been detected.

[c] Estimated by gel filtration chromatography.

[d] Released by Glusulase digestion.

[e] Released by subtilisin or trypsin digestion.

[f] Released by Zymolyase digestion.

[g] Released by mercaptan treatment.

[h] Released by French pressure cell disruption, followed by brief boiling and/or autoclaving.

1.2.3. Conjugation: Cell and Nuclear Fusion. Even less is understood at the biochemical level about the actual steps of zygote formation. Once in firm contact, the cell walls and plasma membranes separating the two G_1-arrested cells must be removed to permit cell fusion or "plasmogamy." The initial breakdown of the wall and membrane is presumably accomplished by autolytic enzymes. The endogenous mannanases and glucanases of yeast, their physiological role during normal growth, and their regulation by pheromones during mating, have been studied to only a limited extent (103). Both endo-$\beta(1\rightarrow3)$- and exo-$\beta(1\rightarrow3)$-glucanase activities have been demonstrated in *Saccharomyces cerevisiae*. Both enzymes have been reported to be glycoproteins that are released into the medium upon spheroplasting of yeast cells, which suggests that they are cell wall-associated activities. However, at least a significant fraction (~10%) of the total activities of these enzymes have been found associated with endoplasmic reticulum-derived small vesicles, presumably those involved in normal cell wall and membrane growth at the bud site.

The frequency with which a haploid will fuse with more than one other cell of opposite mating type is very low (less than one multiple mating per 10^4 matings that result in normal diploids), even under conditions where it can be shown that a haploid cell is in intimate contact with as many as six cells of the opposite mating type. Presumably, some other process distinct from simple cell–cell contact must restrict cell fusion to the appropriate cell pair. Perhaps fusion is restricted because of the requirement for some site or organelle to be in proper orientation in both haploids.

In electron micrographs of conjugating pairs, the glucan layers of opposing parental cell walls do seem to become progressively thinner, similar to what is observed for **a** cells treated with α factor. Fragmentation of the plasma membrane by vesiculation occurs concomitantly, until finally both the wall material and the membrane become sufficiently open to allow cytoplasmic continuity of the apposed haploid partners. Electron microscopic observations (38) also show that during the time of cell fusion, microtubules that radiate from the outer surface of the spindle-pole body of each haploid nucleus often appear to reach the region of contact between the two cells, and frequently pass through the aperture formed once the two membranes have merged with each other. The extranuclear microtubules that project off of the cytoplasmic face of each of the spindle-pole bodies lengthen and eventually interconnect the spindle-pole bodies of the two nuclei. Therefore, the site for cell fusion, and hence the restriction on partner formation, may be directed by the microtubules that radiate from the spindle-pole body toward a specific place on the cell membrane. In this regard, it should be noted that during vegetative growth haploids always tend to bud from the apex of the cell near their birth scar, a chitin-containing structure left in the rigid bud cell wall when a newborn cell divides from its mother cell. Indeed, scanning electron microscopy reveals that the abnormal cell wall extension that occurs in **a** cells treated with α factor also seems to extend from this same cell pole (4). Consistent with this finding, it has been reported that the fewer the number of buds a haploid cell has already produced, the greater its mating potential (104). Perhaps, too many bud scars (chitin-containing structures left in the mother cell wall when it divides off daughter buds) occlude the construction of a site for cell fusion. Equally striking is the observation (J. Rine, personal communication) that on a solid surface the cell wall outgrowth of α-factor-treated **a** cells tends to form most readily when the birth-scar pole is facing the source of the pheromone. In the brown algae *Fucus* and *Pelvetia*, it has been reported that a gradient of a calcium ionophore, a compound that makes cell membranes permeable to this divalent cation, will cause eggs to send out their rhizoidal processes from the side of the cell exposed to the highest calcium concentration. Similarly, in the water mold *Blastocladiella*, outgrowth of the rhizoidal tips can be directed toward the highest concentrations of membrane-permeable agents that can carry protons. Perhaps α-factor treatment results in the generation of such ion currents in **a** cells that serve to set the direction of assembly or orientation of the morphogenic determinants (like the spindle-pole body-associated microtubules perhaps) that dictate the restriction on partner formation. The pairing problem is not so acute in *Chlamydomonas* because specialized extracellular projections, the flagella, direct the cell–cell recognition process.

In addition to cell wall and membrane breakdown, continued resynthesis of all cell envelope components must be occurring during mating in order to prevent lysis of the mating pairs. While the biosynthesis of the complex cell envelope macromolecules involved, including glucan, mannan, chitin, and phospholipids, are being investigated by many groups in some detail in vegetatively growing cells, very little is known about the restructuring of these components in conjugating cells. The bizarre elongated shape eventually formed by **a** cells in the presence of α pheromone, but in the absence of α cells, probably reflects a chronic condition of being unable to complete the normal agglutination and fusion process.

Mutants that still respond to pheromones by arresting in G_1, but that are nevertheless mating deficient (21; L. C. Blair, personal communication) are perhaps blocked in other aspects of the mating process or at a late stage of conjugation. For example, such mutants may be defective in agglutinin synthesis, or in functions necessary for the cell wall dissolution or the cell wall resynthesis required for conjugation. It has been shown, however, that a mutant (*exb1*-1) apparently lacking an exo-$\beta(1\rightarrow3)$-glucanase activity shows no growth impairment and is not detectably mating deficient (105).

The last step of conjugation, nuclear fusion or *karyogamy*, is initiated by the drawing together of the two haploid nuclei, a mechanism that somehow involves the depolymerization of the microtubules that have become interconnected between the nuclei during cell fusion (38). The fusion of the nuclei begins specifically at the edges of the spindle-pole bodies of each nucleus, resulting in the formation of a single large spindle-pole body. Since the new diploid nucleus has a complete spindle-pole body, it is therefore ready to resume a normal vegetative cell cycle, or to commence meiosis. Just how the two nuclear envelopes sort themselves out during nuclear fusion is not known, although in higher eukaryotes nuclear membrane dissolution and reformation has been correlated with the phosphorylation and dephosphorylation of the nuclear membrane-associated lamina proteins (106). Indeed, several "nuclear-limited" mutations of *Saccharomyces cerevisiae* have been identified that prevent nuclear fusion (107, 108). Thus, the consequence of mating a strain carrying such a mutation is the formation of a zygote that remains binucleate, that is, a *heterokaryon*. These mutations have been called *kar* for *kar*yogamy-defective. Since such a zygote is able to bud off either a heterokaryon or cells containing a single haploid nucleus during its subsequent growth and division, nuclear fusion obviously is not a necessary outcome of cell fusion, nor is nuclear fusion required for viability.

It should be stressed that overall the mating process in *Saccharomyces cerevisiae* is a stately progression of events, commencing with exposure of cells to the pheromone of their opposite mating type. Although certain differentiative responses, like the induction of agglutinins, can be detected within a time as short as 30 min or so, the entire conjugation process requires about 2–3 hr before zygotes are fully formed. The situation with *Chlamydomonas* is quite the reverse: sexual cell differentiation happens quite slowly, probably because the process occurs only during nutrient limitation; however, once gametes are formed, their mating process can be completed extremely rapidly, as described below.

2. *CHLAMYDOMONAS*

2.1. Sexual Differentiation and Its Genetic Control

2.1.1. Gametogenesis. As noted above, *Chlamydomonas* cells take advantage of the sexual option in their life cycle only when faced with nitrogen starvation (109, 110) [this is true of many other unicellular eukaryotes as well, two examples being the ciliated protozoan *Tetrahymena* (111) and the cellular slime mold *Dictyostelium* (112)]. As long as NH_4^+ or nitrate are available, the algal cells carry out mitotic (vegetative) growth, and, if vegetative cultures of the two mating types (here called *plus* and *minus*, rather than *a* and α) are mixed together, no sexual interactions ensue. The gametogenesis triggered by nitrogen starvation requires 8–12 hr (110, 113); it is usually, and perhaps always, accompanied by a round of mitosis; and it results in the production of sexually competent *plus* and *minus* gametes, which, when mixed together, instantly initiate mating (109, 114).

Since hormonal signals have no apparent role in triggering sexual differentiation in *Chlamydomonas*, the signals must be generated internally: we can imagine, as a simple paradigm, that the cell responds to reduced intracellular levels of NH_4^+ by derepressing a gene whose product activates a series of cellular alterations required for mating responsiveness. A molecular understanding of this process is not likely to come easily, however, for nitrogen starvation has drastic and multifarious effects on the cell, only a fraction of which are likely to be directly involved in triggering gametogenesis. Ribosomes, membranes, and soluble proteins are selectively degraded to provide nitrogen for the gametogenic mitosis (110, 115), and the cells enter a G_0 resting state in which they can remain viable for months (110). Such resting cells can be "rescued" by one of two events: (1) nitrogen can be provided, in which case they dedifferentiate into vegetative cells and resume mitotic growth, or (2) opposite-type gametes can appear, in which case the resting cells mate and enter the diploid zygote phase of their cycle (114). Zygotes activate a unique genetic program, which includes the biosynthesis of a distinct set of wall proteins (116) that form a thick water-tight pellicle around each cell.

Not known, therefore, is at what point in the 8-hr gametogenesis the signals for differentiation are perceived, nor what the signals are, nor how they are acted upon. Because the "noise level" of general adjustments to nitrogen starvation is so high, direct analysis will probably be experimentally frustrating, and imaginative approaches will be needed. The problem is, of course, a central one in developmental biology. Differentiating cells are typically actively engaging in mitosis, yolk metabolism, migration, and so on, and the major challenge is to sort out which of the many changes one can measure are those that relate directly to the acquisition of the newly differentiated state.

A related and, to our knowledge, unanswered question is the role of cell starvation in multicellular development. Are certain genetic programs turned on when embryonic cells move away from a critical concentration of nutrient, growth factor, or inducer or where such concentrations are diluted below critical levels?

2.1.2. Genes Affecting Mating Interactions. As with yeast, a single genetic region, a mating-type locus (*mt*), exerts ultimate control over the sexual phenotype of *Chlamydomonas*. The locus segregates in a Mendelian fashion so that, of the four meiotic products from a zygote, two inherit the mt^+ locus and differentiate into *plus* gametes, and two inherit the mt^- locus and become *minus* gametes (114). The four cells that emerge from a zygote can be separated from one another with a fine glass loop on an agar surface, allowing each to form a clone. Similar manipulations with mutant strains allow one to determine whether a given mutation is linked to *mt* and/or whether it is expressed in only one or the other mating type.

Although existing mutant strains have not yet provided insight into primary differentiation mechanisms in *Chlamydomonas*, they have served to define the genes required for expression of the gametic phenotype. As with yeast, considerable rigor must be used in establishing that a nonmating strain is defective in mating-specific functions and not simply moribund or generally debilitated. A list of characterized mutant strains is presented in Table 3; their properties are considered in subsequent sections of this chapter. An important feature of all the gene loci listed (with the exception of *imp-11*, which has not yet been characterized genetically), is that they all exhibit *mt* specificity: they either map directly to the *mt* locus or else they are limited in their expression to one or the other mating type (117–122). As noted in Section 1.1.2, these same two classes of mutations have been found among mating-deficient mutants in yeast, suggesting evolutionary homologies between the two organisms despite the distinct differences in their mating strategies.

The mating-type locus of *Chlamydomonas* has two additional properties that are pertinent. First, the mt^+ locus controls the inheritance of chloroplast DNA, with mt^+ chloroplast genes normally transmitted to all four meiotic progeny from a sexual cross (123, 124). Second, the mt^- locus exerts dominant effects in diploids. Rare but stable mitotic lines of diploid mt^+/mt^- *Chlamydomonas reinhardi* can be selected following zygotic cell fusion; these can be induced to undergo gametogenesis and are found to be invariably *minus* in phenotype, mating only with mt^+ gametes (125).

Summarizing this information and the data in Table 3, we can at present state that the mt^+ locus is linked to at least one gene (*imp-1*) involved in cell fusion, controls the expression of at least two unlinked genes (*sag-1* and *sag-2*), and dictates the sexual transmission of chloroplast DNA, while the mt^- locus is linked to at least one gene (*imp-10*, *imp-12*) involved in sexual recognition, controls the expression of at least one unlinked gene (*gam-1*), and is dominant in diploids. Therefore, although the two *mt* loci have retained allelic positions on the left arm of linkage group VI (126), they appear to have diverged considerably in their properties and coding capacities. Possibly because of such divergence, the two loci give no evidence of undergoing meiotic recombination and, indeed, recombination suppression extends to contiguous loci in the chromosome (124). Cloning and molecular analysis of

TABLE 3. Properties of Mutant Strains of *Chlamydomonas* Defective in Mating

Mutation	Locus	Sex-Limited to	Phenotype	References
imp-2, *imp-5*, *imp-6*, *imp-7*, *imp-9*	*sag-1*	*plus*	No sexual agglutination, and no agglutinin activity in extracts; can bypass defect with antibody agglutination; remaining mating stages normal	117
imp-8	*sag-2*	*plus*	Same as above	117
imp-10, *imp-12*	mt^-	*minus*	Same as above	118
gam-1	*gam*-1	*minus*	Signalling abnormal; temperature-sensitive	119, 120
imp-1	mt^+	*plus*	No sexual fusion; mating structure devoid of fringe; slightly leaky	121, 122
imp-11	?	?	No sexual fusion; mating structure lacks normal fringe and membrane zone; not leaky; derives from mt^-, agglutinates and activates like mt^+	122

these loci is clearly a desired goal. J. Rochaix is presently attempting to isolate mt^- DNA using a transformation system, while D. Hourcade has generated unstable genetic elements that may carry mt^+ (127); therefore, this goal may soon be attained.

Before concluding this brief consideration of mating-type systems, it is important to raise the fundamental question: Why sex at all? Many modern eukaryotic protists appear to thrive without it (e.g., the flagellated protozoans *Euglena* and *Naegleria*), and for only a few (for example, the ciliated protozoan *Paramecium*) does it appear essential for the long-term survival of the clone. While it is true that of the various *Chlamydomonas* cell types, only the zygote can form a dehydration-resistant cell wall, other protists can form similar walls without prior sexual fusion, making this a weak rationale for sex. The textbook explanation for the evolution of sex is, of course, the desirability of intermixing and recombining genomes for selective diversity; however, many computer simulations of evolutionary situations conclude that constant gene reshuffling can work against optimal fitness. Therefore, the primordial forces that lead to the evolution of elaborate mating-type systems and meiotic mechanisms remain open to speculation (128).

2.2. Intercellular Interactions and the Mating Process

2.2.1. Flagellar Membrane Interactions. As with all fertilization systems (see Chapter 1), the *Chlamydomonas* mating reaction involves sophisticated levels of specificity. Just as sperm never fuse with sperm, nor eggs with eggs, and cross-species interactions are disallowed, so do *Chlamydomonas reinhardi* mt^+ gametes mate only with mt^- gametes, never with other mt^+ gametes or with gametes of other *Chlamydomonas* species (129). Discrimination occurs at at least two stages—once at the time of flagellar recognition and again at the time of cell–cell fusion (to be considered in Section 2.2.4).

As illustrated in Figure 8, flagellar recognition and adhesion initiate the mating reaction, and they can thus be accorded the importance of sperm-jelly or sperm-zona pellucida interactions (Chapter 1). Contacts appear to be membrane-to-membrane and are established by simple physical collision, there being no convincing evidence of any chemoattraction or comparable "lures." The simplest adnesion model to imagine is that an mt^+-specific molecular "lock" is displayed on the mt^+ flagellar surface that interacts with an mt^--specific molecular "key." Alternatively, there may exist a four-component system, with mt^+ ligand and receptor species and complementary mt^- ligands and receptors (we can discount, for discussion purposes, even more

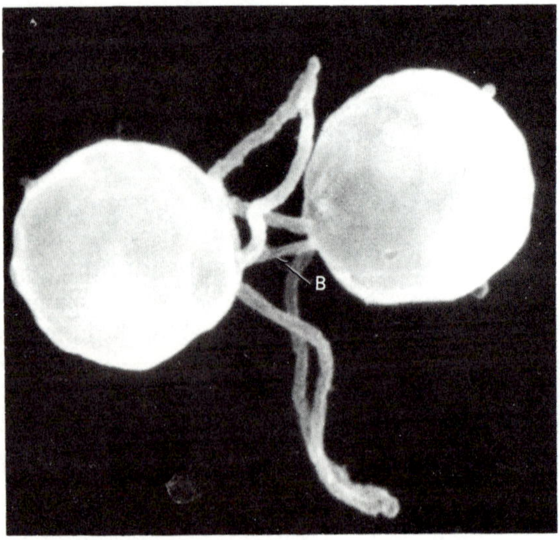

FIGURE 8. Mating reaction of *Chlamydomonas*. Two gametes of opposite mating type have adhered by their flagellar tips and have just become conjoined by a slender cytoplasmic bridge (*B*). Scanning electron micrograph by Dr. Dick Mesland. Magnification, 12,000×.

complex possibilities). While it is not yet known which model is correct, the following information is available.

Agglutinin can be isolated from gametic flagellar membranes either by detergent extraction of isolated flagella or by presenting live cells with high concentrations (12.5 m*M*) of the chelating agent EDTA (130). The detergent extracts prove to be grossly contaminated with the major intrinsic flagellar membrane protein, a large glycosylated species with an apparent molecular weight of 350,000 (350K) (131–133); therefore, the EDTA extracts are the material of choice for agglutinin purification. When EDTA is removed from these extracts and they are allowed to dry down on a glass slide, *plus* gametes stick avidly to *minus* extracts by their flagella, forming a plaque of cells (Fig. 9); an identical response is elicited in *minus* gametes by a film of *plus* extract. Using this bioassay, the extracts have been subjected to standard biochemical purification schemes (ammonium sulfate precipitation, ultrafiltration, column chromatography, and so on). The upshot is that the active material can be separated from virtually all the protein in the extract, including the 350K component, by virtue of its very large size in the native state. When this active fraction is reduced and solubilized on sodium dodecyl sulfate (SDS) and analyzed by SDS-polyacrylamide gel electrophoresis (PAGE), only a few bands are present, all of which are minor components of the original extract (130). Candidate agglutinin bands in these gels display very high apparent molecular weights, but a definitive identification of the agglutinin has not yet been made.

Extrapolating back to the flagellum, therefore, with its rich endowment of the 350K species, we can conclude that the *Chlamydomonas* agglutinin is a very minor component of the flagellar membrane, possibly represented by fewer that 100 copies per flagellum—a discouraging number for the biochemist, but apparently a reasonable number for the cell. Each flagellum (12×0.2 μm) has a surface area of about 7.5 μm^2, which means about 13 molecules per μm^2 of membrane. If these are prominently displayed on opposite-type flagellar surfaces and have high affinity for one another (as they clearly do), then recognition would readily take place.

It should be remembered that we may be dealing with a four-component system and that the large, rare molecules stripped from the *plus* and *minus* surfaces may be only two of the four. The important point is that at least one set of components has these properties, and the implications for developmental systems generally are not difficult to see: in other systems, recognition molecules may also be rare components of the membrane surface. Indeed, S. Roseman and his colleagues are finding this to be true of surface components involved in recognition between liver cells (134–136). Such an arrangment would clearly optimize the cell's flexibility, for a surface display appropriate for one developmental stage could most easily be replaced by a display appropriate for a second stage if only a few molecules are involved.

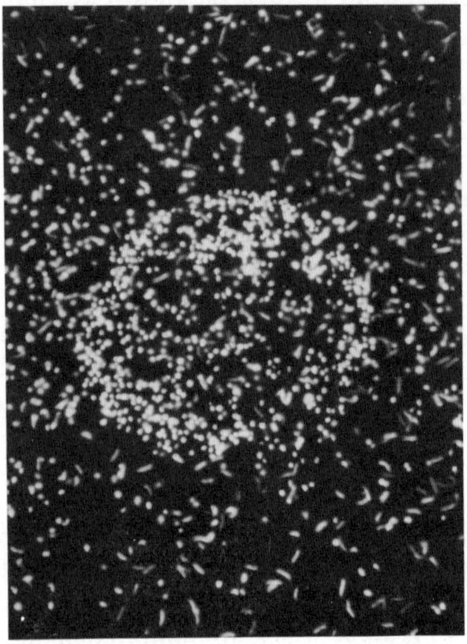

FIGURE 9. Agglutinin bioassay. A 1 μl drop of *minus* agglutinin extract was allowed to air-dry onto a microscope slide; a suspension of *plus* gametes was added, and the field was observed. Cells peripheral to the central spot appear blurred because they are swimming; cells adhered to the extract in the center appear in focus. A suspension of *minus* gametes or nonagglutinating *plus* mutants would swim right past this *minus* material. Dark-field light micrograph by Dr. W. Steven Adair and Carol Hwang.

2.2.2. The Tipping Reaction. That active adhesins are initially displayed all along the flagellar surface can be shown by fixing gametes of one mating type with the cross-linking agent glutaraldehyde: when living gametes of opposite type are presented to such "corpses," adhesions are established at all points along the flagellar length (137). The living gametes, however, soon shift the location of their active adhesins so that they are concentrated at the flagellar tips. This migration, which we term *tipping* (137), occurs as well when both mating partners are alive, the result being that after a few seconds, pairs of cells appear "locked in" (138) at their flagellar tips (Fig. 8).

That tipping entails a highly selective migration of adhesive molecules and not a grand-scale shift of membrane components is most directly demonstrated by deep-etch electron microscopy of rapidly frozen cells. Using the techniques summarized in the legend to Figure 2, it is possible to obtain dramatic views of the *Chlamydomonas* flagellar surface coat, with most of the individual protrusions presumably corresponding to individual 350K components (Fig. 2). The overall density of this coat remains unchanged throughout the mating reaction. Therefore, as we have already concluded from the biochemical studies, only a small fraction of the flagellar surface participates in adhesion, and these components are rapidly segregated to the tip. The biophysical mechanisms responsible for flagellar tipping may relate to flagellar surface motility described for *Chlamydomonas* (139–140) and/or to the patching-capping phenomena described for higher eukaryotic cells (141), but they have yet to be defined in molecular terms in any system.

An obvious consequence of agglutinin tipping [and such probes as concanavalin A and antiflagellar antiserum are tipped as well (137)] is that interacting molecules are concentrated in a small region of the cell surface. Returning to the model of the flagellum bearing 100 adhesin molecules, these would initially be dispersed and would subsequently localize in a patch. This "patching" has two apparent consequences. The first is that the cells adhere to one another much more tightly: gametes treated with agents that prevent tipping can only muster a very "loose," unstable agglutination reaction (142). The second is that the patches apparently function as signalling intermediates, for tipping of agglutinins must occur if the cells are to respond to the adhesion reaction (142). There are intriguing analogies here with "higher" eukaryotes, where the localized clustering of ligand-bound receptors appears to be the active agent in eliciting many surface-mediated responses (143–145). Why cells respond to aggregated but not to dispersed surface molecules is not yet known, but the practice is clearly an ancient one.

2.2.3. Responses to Flagellar Tipping. In *Chlamydomonas*, tipped ligands stimulate dramatic morphological responses, first in the flagellum and then in the cell body. The flagellar response has two components. First, as illustrated in Figure 10, the flagellar tip becomes filled with a dense substance termed fibrous tip material (142). Second, tip microtubules, which normally end proximal to the true end of the flagellum, grow out into the distal tip (142). How these two very unexpected responses to ligand concentration are generated is not clear, but since the agents colchicine and vinblastine inhibit both flagellar-tip activation and all subsequent mating reactions, the occurrence of tip activation appears to be an essential intermediate in adhesion-fusion coupling. Whether analogous alterations in flagellar-ciliary morphology occur in "higher" sensory cilia (e.g., those associated with mechanoreceptors or olfactory cells) has yet to be investigated.

Three cellular responses to tipped flagellar adhesions have thus far been recognized. The first response entails the release of an enzyme that causes cell walls to break down at their anterior ends so that naked protoplasts literally pop from their enclosing walls like chicks out of an egg (Fig. 11).

FIGURE 10. Flagellar tip activation. This *plus* gamete was stimulated with *minus* flagella and then fixed and critical-point dried as described (142). The flagellar tips (arrows) have dense bulbous ends and narrow stalks; unactivated tips, by contrast, are tapered and wispy. An activated fertilization tubule (F) protrudes between the flagella. Transmission electron micrograph by Dr. Dick Mesland. Magnification, 12,000×.

The second response, termed mating-structure activation, differs in the two mating types. The *plus* mating structure, located close to the basal bodies and assembled during gametogenesis, is a disc-shaped organelle; as illustrated in Figure 12, it consists of a dense membrane zone contiguous with the plasma membrane, and a doublet zone beneath. Activation proceeds in two stages. The membrane overlying the structure first forms a bud (146); the bud then rapidly fills with actin filaments (146, 147), originating from the doublet zone, which support the extension of a long narrow microvillus-like structure called the fertilization tubule (Fig. 10). In mt^- cells, on the other hand, mating structures are endowed only with membrane-zone material (Fig. 13). Flagellar adhesion triggers a doming of this structure (122) and an apparent reorganization of its intramembranous particles (148), but no actin filaments polymerize and no fertilization tubule is formed.

The third response to flagellar adhesion is most intriguing. M. Adams has discovered (149) that if polyethylene glycol (PEG) is used to fuse two different *plus* gametes that are differentially marked with mutations in their chloroplast DNA, inheritance of the two sets of chloroplast markers is random: there is no evidence of the uniparental inheritance of one set of markers observed in sexual crosses (123, 124). If, however, one mt^+ mutant strain (P_1) is first allowed to agglutinate for 15 min with flagella isolated from *minus* gametes and is then mixed with any second mt^+ (P_2) strain in the presence of

PEG, the fusion products now inherit only the plastid markers of the P_1 parent. In other words, *minus* flagellar adhesions stimulate the *mt*$^+$ locus-controlled system that causes the preferential transmission of the *plus* chloroplast genome.

Tip adhesions, then, stimulate secretion, actin polymerization, plasma membrane rearrangements, and selective gene transmission. It is not known whether all these responses are secondary to a pancellular event (such as an altered cAMP level) or whether each directly and independently perceives unidentified "tipping signals." What is known, however, is that functional agglutinins are not required for cell-wall lysis or for mating-structure activation. It is possible to take any of the *sag-1* or *sag-2 mt*$^+$ mutants (Table 3), whose flagella have no adhesin activity either in vivo or in detergent or EDTA extracts (130), and present them with antibody directed, in the extreme case, against vegetative *minus* flagella. This antiserum recognizes antigens common to both vegetative *minus* flagella and *sag-1 mt*$^+$ flagella and proceeds to agglutinate the mutant gametes to one another. The gametes respond by tipping the antibody, activating their flagellar tips, shedding their

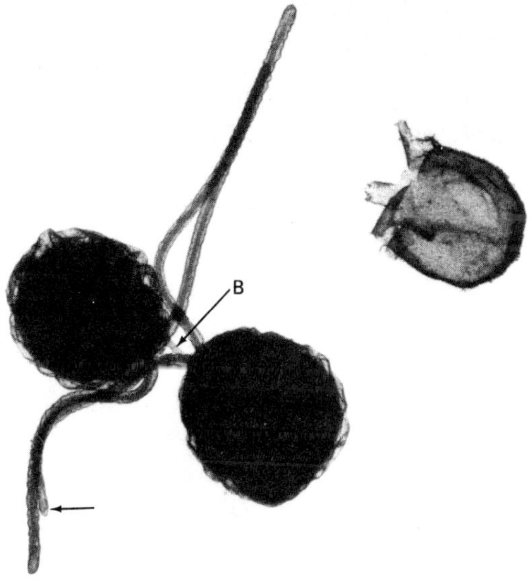

FIGURE 11. Cell fusion. The mating pair is conjoined by a cytoplasmic bridge (*B*), as in the mating reaction illustrated by scanning electron microscopy in Figure 8. The flagella have lost their tip agglutination, however, and only one (arrow) still retains its fully activated morphology. A lysed cell wall lies next to the cells. Transmission electron micrograph by Dr. Dick Mesland. Magnification, 8,000×.

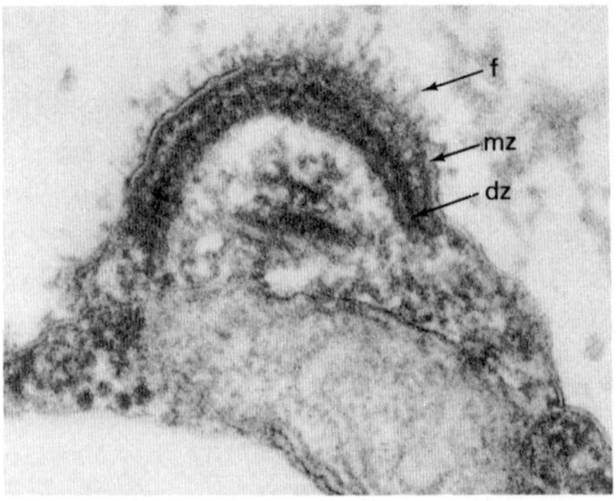

FIGURE 12. Unactivated *plus* mating structure. The cytoplasmic component of the organelle consists of a membrane zone (mz) contiguous with the plasma membrane and a bipartite doublet zone (dz) beneath it. The external surface of the mating-structure membrane carries a dense coat termed fringe (f). Magnification, 118,000×.

walls, and activating their mating structure normally. This response is not elicited if the antibody is first split into monovalent Fab fragments, even though the fragments attach to the membrane surface. Thus it appears necessary for signalling that flagellar surface components, although not necessarily active agglutinins, be cross-linked by divalent (or perhaps multivalent) interactions. We are once again reminded of observations made on "higher" eukaryotic cells: cellular responses normally triggered by the binding of insulin to its receptor, for example, can be elicited by divalent antibody directed against other surface components (150).

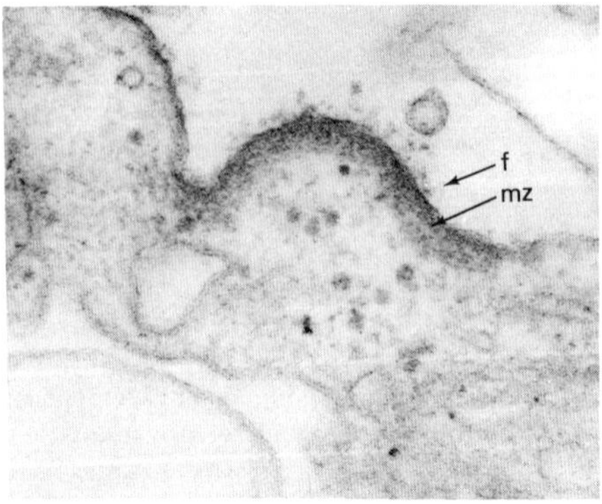

FIGURE 13. Unactivated *minus* mating structure. A membrane zone (mz) underlies the plasma membrane, and a layer of fringe (f) overlies the membrane. Magnification, 118,000×.

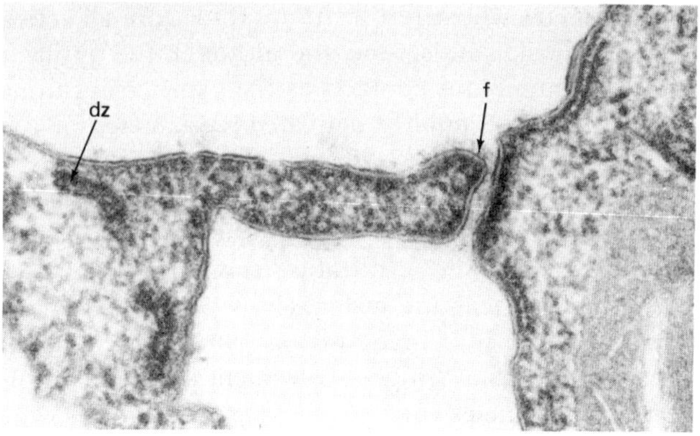

FIGURE 14. Prefusion contact. The fertilization tubule extends from the *plus* cell on the left to the *minus* cell on the right. Contact is made between the fringe (f) associated with the fertilization tubule tip and the fringe overlying the activated *minus* mating structure. Actin filaments are absent from the fertilization tubule because the gamete was treated with the drug cytochalasin D. dz, doublet zone. Micrograph by Dr. Patricia A. Detmers. Magnification, 110,000×.

2.2.4. Cell Fusion. The next step in the *Chlamydomonas* mating reaction is cell fusion itself (151). As illustrated in Figure 14, this entails membrane fusion between the tip of the mt^+ fertilization tubule and the activated mt^- mating structure (146). It is here that a second mating-type specific recognition appears to take place. One can prepare, for example, suspensions of

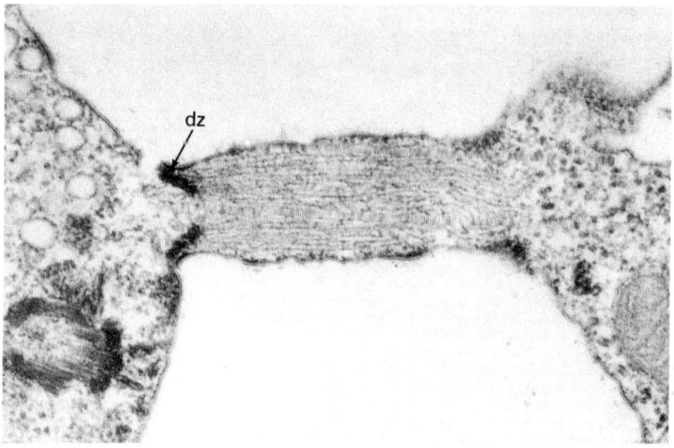

FIGURE 15. Cytoplasmic bridge formed by cell fusion. The bridge extends from the *plus* cell on the left to the *minus* cell on the right. Fusion has occurred in the region marked by the two arrows. Actin fills the bridge. dz, doublet zone. Micrograph by Dr. Patricia A. Detmers. Magnification, 60,000×.

millions of *plus* gametes whose fertilization tubules are all activated by mt^- flagella or by antiflagellar antiserum, and although the fertilization tubule tips doubtless make numerous contacts as the gametes agglutinate, cell fusion does not ensue (137). Similarly, activated mt^- gametes fail to fuse to one another.

A likely candidate for the bearer of this fusion specificity is a heavy carbohydrate coat, denoted fringe, that overlies the mt^+ mating structure (Fig. 12) and travels with the fertilization tubule tip, and a homologous fringe on the mt^- mating structure surface (Figs. 13 and 14). In the prefusion image captured in Figure 14, fringe-to-fringe contact is clearly being made. Supporting the hypothesis of fringe involvement are observations made on two mutant strains. Gametes carrying the *imp-1* mutation, which maps to

FIGURE 16. Quadriflagellate cell formed by cell fusion. Scanning electron micrograph by Dr. Dick Mesland. Magnification, 10,000×.

mt⁺ (121), and gametes carrying the *imp-11* mutation, which arose as a mating-type-switch mutation but has not yet been mapped (122), have similar phenotypes: each agglutinates normally with *minus* gametes and erects an actin-filled fertilization tubule, but neither bears the normal endowment of fringe, and neither is capable of fusing normally with *minus* gametes (122).

Once fusion has occurred between fringe-bearing membranes, the gametes are attached by a narrow cytoplasmic bridge (Figs. 8 and 11) filled with actin filaments (Fig. 15). The bridge appears to shorten and then widen until the two cells achieve side-by-side confluency (Fig. 16).

We should at this point consider the time scale of the *Chlamydomonas* mating reaction. If a mating mixture is presented with glutaraldehyde 15 sec after mixing the two mating types, the gametes already display tipped adhesion sites, activated flagellar tips, lysed cell walls, and actin-filled fertilization tubules. The first fused zygotes are found 30 sec after mixing. In other words, the events we have been describing are triggered in an extremely rapid, efficient fashion.

2.2.5. The "Off" Response. The final sequence of events in the *Chlamydomonas* mating reaction can be designated the "off" response. Following zygotic cell fusion, the flagella lose their adhesiveness and disadhere from one another, a process illustrated in Figure 11. The resultant cell, a quadriflagellated zygote (Fig. 16), is free to swim about, its four flagella beating in remarkable synchrony. These four flagella are fully nonadhesive against freshly added gametes of either *mt*, meaning that fusion somehow signals disadhesion. The experiments of Snell and his colleagues (152, 153) indicate that agglutinin activity normally undergoes a rapid loss-and-replacement cycle during the adhesion reaction, which is dependent on protein synthesis. Therefore, one hypothesis for the lack of adhesiveness of the zygotic flagella is that fusion somehow prevents the continued replacement of the adhesins, but does not block their removal.

With disadhesion, the flagellar tips return to their normal length and morphology and, within an hour after cell fusion, the doublet zone and attached actin filaments, which initially lie in the anterior cytoplasm of the zygote (146), are broken down and disappear. The zygote, meanwhile, has long since activated its new genetic program and is engaged in massive zygote wall synthesis (116, 154).

That this "off" response is triggered by cell fusion and not somehow preprogrammed can be readily demonstrated by the nonfusing mutants *imp-1* and *imp-11*. Gametes of both strains continue tip agglutination and tip activation for hours after being mixed with *mt*⁻ cells (146), but quickly disagglutinate if poisoned with the protein synthesis inhibitor cycloheximide (153).

The rapidity of the mating reaction and the subsequent "off" response has important implications in visualizing how embryonic cells might stimulate one another along diverse developmental pathways during transient contacts. One can speculate that during the course of development, cells of one germ layer may migrate past cells of a second layer and make 15-sec, signalling, reversible contacts with one another—contacts that stimulate such morphogenetic events as secretion, actin polymerization, and microtubule elongation and such genomic events as organelle DNA modification. The fact that most if not all embryonic cells bear rudimentary cilia with no known function (155, 156) gives this line of speculation additional plausibility, but no such interactions have yet been described. They would doubtless be very difficult to recognize in a multicellular eukaryotic embryo, however; therefore, if they occur at all, their detection will probably require the construction of an in vitro system analogous to the *Chalmydomonas* mating reaction.

ACKNOWLEDGMENTS

The authors gratefully acknowledge both the contributions of many students and colleagues to much of the work described in this review and the support of the NIH for these projects.

REFERENCES

1. T. R. Manney and J. Meade, Cell–cell interactions during mating in *S. cerevisiae*, in J. L. Reissig, ed., *Receptors and Recognition*, Series B, Vol. 3, John Wiley and Sons, New York, 1977, p. 281.
2. M. Crandall, R. Egel, and V. L. MacKay, *Adv. Microb. Physiol.* **15**, 307 (1977).
3. M. Crandall, *Symp. Soc. Exp. Biol.* **32**, 105 (1978).
4. J. Thorner, Intercellular interactions of the yeast *Saccharomyces cerevisiae*, in T. L. Leighton and W. A. Loomis, Jr., eds., *The Molecular Genetics of Development*, Academic Press, New York, 1980, p. 119.
5. V. L. MacKay, Mating type-specific pheromones as mediators of sexual conjugation in yeast, in J. Papaconstantinou and W. J. Rutter, eds., *Molecular Control of Proliferation and Differentiation*, Academic Press, New York, 1978, p. 243.
6. T. R. Manney, W. Duntze, and R. Betz, The isolation, characterization, and physiological effects of the *S. cerevisiae* sex pheromones, in D. H. O'Day and P. A. Horgan, eds., *Sexual Interactions in Eukaryotic Microbes*, Academic Press, New York, 1981, p. 21.
7. J. Thorner, Pheromonal regulation of development in *Saccharomyces cerevisiae*, in J. N. Strathern, E. W. Jones, and J. Broach, eds., *Molecular Biology of the Yeast Saccharomyces*, Part 1, Cold Spring Harbor Laboratory, Cold Spring Harbor, New York, 1981, p. 143.
8. D. M. Prescott, ed., *Methods in Cell Biology*, Vols. 11 and 12, Academic Press, New York, 1975.

9. D. M. Prescott, ed., *Methods in Cell Biology*, Vol. 20, Academic Press, New York, 1978.
10. R. K. Mortimer and D. C. Hawthorne, Yeast genetics, in A. H. Rose and J. S. Harrison, eds., *The Yeasts*, Vol. 1, Academic Press, New York, 1969, p. 385.
11. T. D. Petes, *Ann. Rev. Biochem.* **49**, 845 (1980).
12. D. Botstein and R. W. Davis, Principles and practices of recombinant DNA research with yeast, in J. N. Strathern, E. W. Jones, and J. Broach, eds., *Molecular Biology of the Yeast Saccharomyces*, Part 2, Cold Spring Harbor Laboratory, Cold Spring Harbor, New York, 1982, in press.
13. L. H. Hartwell, *Bacteriol. Rev.* **38**, 164 (1974).
14. J. R. Pringle and L. H. Hartwell, The *Saccharomyces cerevisiae* cell cycle, in J. N. Strathern, E. W. Jones, and J. Broach, eds., *Molecular Biology of the Yeast Saccharomyces*, Part 1, Cold Spring Harbor Laboratory, Cold Spring Harbor, New York, 1981, p. 97.
15. I. Herskowitz and Y. Oshima, Control of cell type in *Saccharomyces cerevisiae*: Mating type and mating type interconversion, in J. N. Strathern, E. W. Jones, and J. Broach, eds., *Molecular Biology of the Yeast Saccharomyces*, Part 1, Cold Spring Harbor Laboratory, Cold Spring Harbor, New York, 1981, p. 181.
16. J. B. Hicks, J. Strathern, and A. Klar, *Nature*, **282**, 478 (1979).
17. K. Nasmyth and K. Tatchell, *Cell* **19**, 753 (1980).
18. J. Strathern, E. Spatola, C. McGill, and J. B. Hicks, *Proc. Natl. Acad. Sci. USA* **77**, 2829 (1980).
19. V. L. MacKay and T. R. Manney, *Genetics* **76**, 255, 273 (1974).
20. G. F. Sprague, Jr., J. Rine, and I. Herskowitz, *J. Mol. Biol.* **153**, 323 (1981).
21. J. Rine, Ph.D. Thesis, University of Oregon, Eugene, Oregon, 1979.
22. A. J. S. Klar, S. Fogel, and K. MacLeod, *Genetics* **93**, 37 (1979).
23. A. J. S. Klar, J. N. Strathern, J. R. Broach, and J. B. Hicks, *Nature* **289**, 239 (1981).
24. K. A. Nasmyth, K. Tatchell, B. D. Hall, C. Astell, and M. Smith, *Nature* **289**, 244 (1981).
25. J. Rine, J. N. Strathern, J. B. Hicks, and I. Herskowitz, *Genetics* **93**, 877 (1979).
26. D. C. Hawthorne, *Genetics* **48**, 1727 (1963).
27. J. N. Strathern, C. S. Newlon, I. Herskowitz, and J. B. Hicks, *Cell* **18**, 309 (1979).
28. G. Kawasaki, Ph.D. Thesis, University of Washington, Seattle, Washington, 1979.
29. M. A. Fennewald, S. P. Gerrard, J. Chen, M. J. Casadaban, and N. R. Cozzarelli, *J. Biol. Chem.* **256**, 4687 (1981).
30. R. E. Malone and R. E. Esposito, *Proc. Natl. Acad. Sci. USA* **77**, 503 (1980).
31. R. Maki, A. Traunecker, H. Sakano, W. Roeder, and S. Tonegawa, *Proc. Natl. Acad. Sci. USA* **77**, 2138 (1980).
32. G. Sprague, Jr., J. Rine, and I. Herskowitz, *Nature* **289**, 250 (1981).
33. K. A. Nasmyth, K. Tatchell, B. D. Hall, C. Astell, and M. Smith, *Cold Spring Harbor Symp. Quant. Biol.* **45**, 961 (1981).
34. J. N. Strathern, J. Hicks, and I. Herskowitz, *J. Mol. Biol.* **147**, 357 (1981).
35. J. D. Levi, *Nature* **177**, 753 (1956).
36. W. Duntze, V. L. MacKay, and T. R. Manney, *Science* **168**, 1472 (1970).
37. G. Fehrenbacher, K. Perry, and J. Thorner, *J. Bacteriol.* **134**, 893 (1978).
38. B. Byers and L. Goetsch, *J. Bacteriol.* **124**, 511 (1975).
39. L. H. Hartwell, *J. Cell Biol.* **77**, 627 (1978).

40. R. Schekman and V. L. Brawley, *Proc. Natl. Acad. Sci. USA* **76**, 645 (1979).
41. D. Stötzler, H. Kiltz, and W. Duntze, *Eur. J. Biochem.* **69**, 397 (1976).
42. E. Ciejek, J. Thorner, and M. Geier, *Biochem. Biophys. Res. Commun.* **78**, 952 (1977).
43. R. Betz and W. Duntze, *Eur. J. Biochem.* **95**, 469 (1979).
44. G. Scherer, G. Haag, and W. Duntze, *J. Bacteriol.* **119**, 386 (1974).
45. F. Lipmann, *Science* **173**, 875 (1971).
46. P. Lomedico, S. J. Chan, D. F. Steiner, and G. F. Saunders, *J. Biol. Chem.* **252**, 7971 (1977).
47. J. Kurjan and I. Herskowitz, A putative α-factor precursor containing four tandem repeats of mature α-factor, in *Meeting on the Molecular Biology of Yeast*, J. B. Hicks, A. Klar, K. Nasmyth, and J. N. Strathern, eds., Cold Spring Harbor Laboratory, Cold Spring Harbor, New York, 1981, p. 242Abs.
48. L. C. Blair, A. J. Brake, D. J. Julius, J. M. Lugovoy, and J. Thorner, Synthesis and processing of yeast pheromones: Identification and characterization of mutants that produce altered α-factors, in *Meeting on the Molecular Biology of Yeast*, J. B. Hicks, A. Klar, K. Nasmyth, and J. N. Strathern, eds., Cold Spring Harbor Laboratory, Cold Spring Harbor, New York, 1981, p. 243Abs.
49. T. R. Manney and V. Woods, *Genetics* **82**, 639 (1976).
50. L. H. Hartwell, *J. Cell Biol.* **85**, 811 (1980).
51. M. M. Udden and D. B. Finkelstein, *J. Bacteriol.* **133**, 1501 (1978).
52. G. P. Samokhin, L. V. Lizlova, J. D. Bespalova, M. I. Titov, and V. N. Smirnov, *Exp. Cell Res.* **131**, 267 (1980).
53. P. F. Maness and G. M. Edelman, *Proc. Natl. Acad. Sci. USA* **75**, 1304 (1978).
54. E. Ciejek and J. Thorner, *Cell* **18**, 623 (1979).
55. D. B. Finkelstein and S. Strausberg, *J. Biol. Chem.* **254**, 796 (1979).
56. H. H. Liao and J. Thorner, *Proc. Natl. Acad. Sci. USA* **77**, 1898 (1980).
57. H. H. Liao and J. Thorner, *J. Bacteriol.* **148**, 919 (1981).
58. P. Greengard, *Science* **199**, 146 (1978).
59. M. M. Gottesman, *Cell* **22**, 329 (1980).
60. R. L. Erickson, M. S. Collett, E. Erickson, and A. F. Purchio, *Proc. Natl. Acad. Sci. USA* **76**, 6260 (1979).
61. R. Tjian and A. Robbins, *Proc. Natl. Acad. Sci. USA* **76**, 610 (1979).
62. K. Radke, T. Gilmore, and G. S. Martin, *Cell* **21**, 821 (1980).
63. T. Hunter and B. M. Sefton, *Proc. Natl. Acad. Sci. USA* **77**, 1311 (1980).
64. T. Hunter, *Cell* **22**, 647 (1980).
65. D. L. Brautigan, P. Bornstein, and B. Gallis, *J. Biol. Chem.* **256**, 6519 (1981).
66. K. Varimo and J. Londesborough, *Biochem. J.* **159**, 363 (1976).
67. F. Fujimoto, A. Tchikawa, and K. Tomita, *Arch. Biochem. Biophys.* **161**, 54 (1974).
68. J. Londesborough, *Biochem. J.* **163**, 467 (1977).
69. Y. Takai, H. Yamahura, and Y. Nishizuka, *J. Biol. Chem.* **244**, 530 (1974).
70. K. Lerch, L. Muir, and E. H. Fischer, *Biochemistry* **14**, 2015 (1975).
71. W. Kudlicki, N. Grankowski, and E. Gasior, *Eur. J. Biochem.* **84**, 493.
72. C. Dery, S. Cooper, M. A. Savageau, and S. Scanlon, *Biochem. Biophys. Res. Commun.* **90**, 933 (1979).

73. P. Jaynes, J. McDonough, and H. R. Mahler, *J. Cell Biol.* **83**, 290Abs. (1979).
74. C. Hixson and E. G. Krebs, *J. Biol. Chem.* **255**, 2137 (1980).
75. S. Zinker and J. R. Warner, *J. Biol. Chem.* **251**, 1799 (1976).
76. D. Becker-Ursic and J. Davies, *Biochemistry* **15**, 2289 (1976).
77. G. I. Bell, P. Valenzuela, and W. J. Rutter, *J. Biol. Chem.* **252**, 3082 (1977).
78. B. A. Hemmings, *J. Biol. Chem.* **255**, 7925 (1980).
79. W. Y. Cheung, *Science* **207**, 19 (1980).
80. G. Jamieson, T. C. Vanaman, and J. J. Blum, *Proc. Natl. Acad. Sci. USA* **76**, 6471 (1979).
81. W. L. Bazari and M. Clarke, *J. Biol. Chem.* **256**, 3598 (1981).
82. C. Shimoda, N. Yanagishima, A. Sakurai, and S. Tamura, *Plant Cell Physiol.* **19**, 513 (1978).
83. D. Campbell, *J. Bacteriol.* **116**, 323 (1973).
84. N. Yanagishima, K. Yoshida, M. Hagiya, Y. Kawanabe, C. Shimoda, A. Sakurai, S. Tamura, and M. Osumi, Sexual cell agglutination in *Saccharomyces cerevisiae*, in I. Ishikawa, Y. Maruyama, and K. Matsumiya, eds., *Growth and Differentiation in Microorganisms*, University Park Press, Baltimore, Maryland, 1977, p. 193.
85. R. Betz, W. Duntze, and T. R. Manney, *FEMS Microb. Letts. 4*, 469 (1978).
86. C. Shimoda and N. Yanagishima, *Arch. Microbiol.* **85**, 310 (1975).
87. N. Yanagishima and K. Yoshida, Sexual interactions in *Saccharomyces cerevisiae* with special reference to the regulation of sexual agglutinability, in D. H. O'Day and P. A. Horgan, eds., *Sexual Interactions in Eukaryotic Microbes*, Academic Press, New York, 1981, p. 261.
88. S. Doi and M. Yoshimura, *Molec. Gen. Genet.* **162**, 251 (1978).
89. S. Doi and M. Yoshimura, *Biochem. Biophys. Res. Commun.* **91**, 849 (1979).
90. H. Tohoyama, M. Hagiya, K. Yoshida, and N. Yanagishima, *Molec. Gen. Genet.* **174**, 269 (1979).
91. N. Yanagishima and K. Nakagawa, *Molec. Gen. Genet.* **178**, 241 (1980).
92. P. N. Lipke, A. Taylor, and C. E. Ballou, *J. Bacteriol.* **127**, 610 (1976).
93. J. S. Tkacz and V. L. MacKay, *J. Cell Biol.* **80**, 326 (1979).
94. P. N. Lipke and C. E. Ballou, *J. Bacteriol.* **141**, 1170 (1980).
95. R. Schekman, *Ann. Rev. Microbiol.* **36** (1982), in press.
96. C. Shimoda, S. Kitano, and N. Yanagishima, *Antonie van Leeuwenhoek J. Microbiol. Serol.* **41**, 513 (1975).
97. K. Yoshida, M. Hagiya, and N. Yanagishima, *Biochem. Biophys. Res. Commun.* **71**, 1085 (1976).
98. M. Hagiya, K. Yoshida, and N. Yanagishima, *Exp. Cell Res.* **104**, 263 (1977).
99. P. H. Yen and C. E. Ballou, *J. Biol. Chem.* **248**, 8316 (1973).
100. P. H. Yen and C. E. Ballou, *Biochemistry* **13**, 2428 (1974).
101. D. Burke, L. Mendonca-Previato, and C. E. Ballou, *Proc. Natl. Acad. Sci. USA* **77**, 318 (1980).
102. D. N. Radin, Ph.D. Thesis, University of California, Berkeley, California, 1976.
103. F. DelRay, T. Santos, I. Garcia-Acha, and C. Nombela, *J. Bacteriol.* **139**, 924 (1979).
104. I. Müller, Parental age and the life span of zygotes of *Saccharomyces cerevisiae*, in A. Goffeau and J. M. Wiame, eds., *Tenth International Conference on Yeast Genetics and*

Molecular Biology, Université Catholique de Louvain, Louvain-la-Neuve, Belgium, 1980, p. 55Abs.

105. T. Santos, F. DelRay, J. Conde, J. Villenueva, and C. Nombela, *J. Bacteriol.* **139**, 333 (1979).
106. L. Gerace and G. Blobel, *Cell* **19**, 277 (1980).
107. J. Conde and G. R. Fink, *Proc. Natl. Acad. Sci. USA* **73**, 3651 (1976).
108. S. Dutcher, Ph.D. Thesis, University of Washington, Seattle, Washington, 1980.
109. R. Sager and S. Granick, *J. Gen. Physiol.* **37**, 729 (1954).
110. N. C. Martin and U. W. Goodenough, *J. Cell Biol.* **67**, 587 (1975).
111. D. L. Nanney, Cell–cell interactions in ciliates: Evolutionary and genetic constraints, in J. L. Reissig, ed., *Microbial Interactions, Receptors and Recognition*, Series B, Vol. 3., Chapman and Hall, London, 1977, p. 351.
112. W. F. Loomis, *Dictyostelium discoideum: A Developmental System*, Academic Press, New York, 1975.
113. K. S. Chiang, J. R. Kates, R. F. Jones, and N. Sueoka. *Dev. Biol.* **22**, 655 (1970).
114. G. Smith and D. G. Regnery, *Proc. Natl. Acad. Sci. USA* **36**, 246 (1950).
115. N. C. Martin, K. S. Chiang, and U. W. Goodenough, *Dev. Biol.* **51**, 190 (1976).
116. S. A. Minami and U. W. Goodenough, *J. Cell Biol.* **77**, 165 (1978).
117. U. W. Goodenough, C. Hwang, and A. J. Warren, *Genetics* **89**, 235 (1978).
118. C. Hwang, B. C. Monk, and U. W. Goodenough, *Genetics* **99**, 41 (1981).
119. C. L. Forest and R. K. Togasaki, *Proc. Natl. Acad. Sci. USA* **72**, 3652 (1975).
120. C. L. Forest, D. A. Goodenough, and U. W. Goodenough, *J. Cell Biol.* **79**, 74 (1978).
121. U. W. Goodenough, C. Hwang, and H. Martin, *Genetics* **82**, 169 (1976).
122. U. W. Goodenough, P. Detmers, and C. Hwang, *J. Cell Biol.* **92**, 378 (1982).
123. R. Sager, *Proc. Natl. Acad. Sci. USA* **40**, 356 (1954).
124. N. W. Gillham, *Am. Nat.* **103**, 355 (1969).
125. W. T. Ebersold, *Science* **157**, 447 (1967).
126. W. T. Ebersold, R. P. Levine, E. E. Levine, and M. A. Olmsted, *Genetics* **47**, 531 (1962).
127. D. Hourcade, *Genetics* **91**, s50 (1979).
128. G. C. Williams, *Sex and Evolution*, Princeton University Press, Princeton, N.J., 1975.
129. L. Wiese, Algae, in C. B. Metz and A. Monroy, eds., *Fertilization: Comparative Morphology, Biochemistry, and Immunology*, Vol. 2, Academic Press, New York, 1969, p. 135.
130. W. S. Adair, B. C. Monk, R. Cohen, C. Hwang, and U. W. Goodenough, *J. Biol. Chem.* **257**, 4593 (1982).
131. G. B. Witman, K. Carlson, J. Berliner, and J. L. Rosenbaum, *J. Cell Biol.* **54**, 507 (1972).
132. K. Bergman, U. W. Goodenough, D. A. Goodenough, J. Jawitz, and H. Martin, *J. Cell Biol.* **67**, 606 (1975).
133. W. J. Snell, *J. Cell Biol.* **68**, 48 (1976).
134. M. S. Kuhlenschmidt, E. Schmell, C. W. Slife, T. Kuhlenschmidt, F. Sieber, U. C. Lee, and S. Roseman, *J. Biol. Chem.* **257**, 3157 (1982).
135. J. Albanese, M. S. Kuhlenschmidt, E. Schmell, C. W. Slife, and S. Roseman. *J. Biol. Chem.* **257**, 3165 (1982).
136. E. Schmell, C. W. Slife, M. S. Kuhlenschmidt, and S. Roseman. *J. Biol. Chem.* **257**, 3171 (1982).

137. U. W. Goodenough and D. Jurivich, *J. Cell. Biol.* **79**, 680 (1978).

138. U. W. Goodenough, W. S. Adair, E. Caligor, C. L. Forest, J. L. Hoffman, D. A. M. Mesland, and S. Spath, Membrane–membrane and membrane–ligand interactions in *Chlamydomonas* mating, in N. B. Gilula, ed., *Membrane–Membrane Interactions*, Raven Press, New York, 1980, p. 131.

139. R. A. Bloodgood, *J. Cell Biol.* **75**, 983 (1977).

140. J. L. Hoffman and U. W. Goodenough, *J. Cell Biol.* **86**, 656 (1980).

141. G. Schreiner and E. R. Unanue, *Adv. Immunol.* **24**, 37 (1976).

142. D. A. M. Mesland, J. L. Hoffman, E. Caligor, and U. W. Goodenough, *J. Cell Biol.* **84**, 599 (1980).

143. H. Rauvala, W. G. Carter, and S.-I. Hakomori, *J. Cell Biol.* **88**, 149 (1981).

144. Y. Shechter, L. Hernaez, J. Schlessinger, and P. Cuatrescasas, *Nature* **278**, 835 (1979).

145. F. A. Karlsson, E. Van Obberghen, C. Grunfeld, and C. Kahn, *Proc. Natl. Acad. Sci. USA* **76**, 809 (1979).

146. U. W. Goodenough and R. L. Weiss, *J. Cell Biol.* **76**, 430 (1978).

147. P. A. Detmers and U. W. Goodenough, *J. Cell Biol.* **87**, 220a (1980).

148. R. L. Weiss, D. A. Goodenough, and U. W. Goodenough, *J. Cell Biol.* **72**, 144 (1977).

149. Adams, G. M. W., *Genetics* **97**, s1 (1981).

150. D. J. Pillion, C. A. Carter-Su, P. F. Pilch, and M. P. Czech, *J. Biol. Chem.* **255**, 9168 (1980).

151. I. Friedmann, A. L. Colwin, and L. H. Colwin, *J. Cell Sci.* **3**, 115 (1968).

152. W. J. Snell and S. Roseman, *J. Biol. Chem.* **254**, 10820 (1979).

153. W. J. Snell and W. S. Moore, *J. Cell Biol.* **84**, 203 (1980).

154. D. Weeks and P. Collis, *Dev. Biol.* **69**, 400 (1979).

155. S. P. Sorokin, *J. Cell Sci.* **3**, 207 (1968).

156. J. E. Rash, J. W. Shay, and J. J. Biesle, *J. Ultrastruct. Res.* **29**, 470 (1969).

157. J. E. Heuser and S. Salpeter, *J. Cell Biol.* **82**, 150 (1979).

158. R. K. Mortimer and D. Schild, *Microbiol. Rev.* **44**, 519 (1980).

159. J. E. Haber and J. P. George, *Genetics* **93**, 1332 (1979).

160. R. B. Wickner, *J. Bacteriol.* **117**, 252 (1974).

161. R. J. Rothstein and F. Sherman, *Genetics* **94**, 871 (1980).

162. J. F. Lemontt, D. R. Fugit, and V. L. MacKay, *Genetics* **94**, 899 (1980).

163. H. C. Stark, D. Fugit, and D. B. Moshowitz, *Genetics* **94**, 921 (1980).

164. L. F. Bisson. Ph.D. Thesis, University of California, Berkeley, California, 1980.

3

The Biochemistry of Cell Adhesion in Vertebrates

STEPHEN ROTH
Department of Biology
University of Pennsylvania
Philadelphia, Pennsylvania 19104

At a time when knowledge of gene anatomy and function is increasing explosively, there are remarkably few ideas about how genes cause cells to construct tissues, organs, and organisms. One major reason for this relative lack of progress is the ubiquity and obviousness of biological form itself. All higher organisms and all of their component parts (for example, a man, his ribosomes, and all in between) are recognized and classified according to their forms and locations relative to other forms. Despite the overwhelming dependence on form in biology and medicine, it is not yet possible to describe quantitatively any of the morphogenetic phenomena shown by higher organisms. In addition to a lack of quantitation, it is almost certain that biochemical analysis of morphogenesis will eventually be forced to deal with molecules that exist in trace amounts or that differ from one another in only subtle ways. This conclusion follows from experiments that show no significant differences in protein profiles when tissues from different animal species are compared. Although the untrained observer can distinguish instantly between the embryos of the chick and quail, and the human and monkey, descriptive biochemical methods can detect no obvious differences between these.

Within the past 20 years, three factors have had important and positive effects on progress in research into morphogenetic mechanisms:

1. The realization that cell–cell or cell–substrate adhesive specificities must control the direction and extent of cell migration.
2. The development of quantitative, in vitro assays for cell adhesion.
3. The use of increasingly sophisticated methods of biochemistry and immunochemistry.

This chapter discusses the work that has led to a modern approach to the problem of morphogenesis. Additionally, it will summarize the more recent achievements and attempt to predict what successes and problems may be expected in the future.

1. HISTORICAL NOTES ON INTERCELLULAR ADHESIVE SPECIFICITY

The most acute, if not the first, observer of morphogenesis was Aristotle, who predicted that the gradual appearance of a chick embryo from amorphous yolk would be the most difficult of all biological problems. When the cell theory finally emerged some 2000 years later, it became clear that morphogenesis was the product of cell movement, either as cell sheets or less frequently as individual cells. The problem now is to understand how cells migrate with such precision. Ultimately, of course, the control of cell migration in embryos resides in the genetic apparatus. But how do the genes exert their effects? What sorts of molecules do they produce that can cause cells to undergo almost unfailingly the movements by which we come to know the organism itself? Where are these molecules located within the cell and do they have any function after morphogenesis is completed?

Modern experimentation on the mechanisms of morphogenesis began with the work of Johannes Holtfreter (1), who showed that the germ layers of early amphibian embryos could be dissociated into single cells by elevating the pH of the surrounding fluid. Holtfreter and his students systematically examined the results of pair-wise combinations of ectodermal, mesodermal, and endodermal cells that had first been dissociated. When the cells were combined and the pH lowered to normal levels, the cells reassociated and, more interesting, sorted out to form reproducible patterns. Two important results emerged: (1) cells from one germ layer tended to adhere to one another and to recreate the type of germ layer from which they were originally derived, and (2) the re-created germ layers assumed predictable positions with regard to the germ layers with which they were combined. These observations allowed the important conclusion that at least some morphogenetic properties were intrinsic to cells and did not depend on the integrity of the entire organism. Holtfreter's investigations opened the way to the study of the cell as the unit of morphogenesis.

Crucial contributions were later made by Moscona (2), who perfected methods for the dissociation and reaggregation of chick embryo tissues, and by Steinberg (3), who showed that combinations of two, dissociated chick embryo tissues formed particular patterns when allowed to reassociate together in vitro.

As recently as 1967, however, there was disagreement concerning how pattern formation occurred when dissociated cells reaggregated in vitro. Kinetic consideration led both Moscona (4) and Steinberg (3) to conclude that adhesive recognition was the mechanism for pattern formation. Others pointed out that differential migratory abilities could account for the patterns even if there were no intercellular adhesive differences among the cells (5). The last 10–15 years have seen the introduction of assays that measure the rates of cell adhesion to cells and to substrates and that, equally important, separate cell-adhesive phenomena from cell-migratory phenomena. It is clear now that adhesive recognition among cells does exist and is an important factor in morphogenetic movements. It follows that the first act of differentiation, in the sense of differential gene activity, is the elaboration of those molecules that allow the first cells to begin to gastrulate.

2. ASSAYS FOR CELL ADHESION

2.1. Dissociative Methods

Coman (6) measured the force required to separate two adhering cells using physical methods. He concluded that tumor cells were less adherent than normal cells when examined with this methodology, but others (7) have suggested alternate explanations for these data. Coman's work, however, has led to more quantitative methods in which the rates of dissociation of cells from substrates are measured when cells are exposed to physical forces (8), enzymes, and antibodies. Aside from the obviously dissociative proteases, two other classes of enzymes have been reported to detach cells from substrates and from other cells. These are ribonucleases (9) and neuraminidases (10). Antisera capable of removing cells from substrates have been produced also and these techniques will be discussed in Section 3.3.7.

More recently, a very sensitive dissociative assay for cell adhesion has been described (11). The new technique allows cells to settle on monolayers of other cells; the freshly attached cells are then inverted and subjected to centrifugal forces. Because the centrifugal forces can be so widely varied and sensitively controlled, cell adhesive stabilities that vary by five orders of magnitude can be measured and compared.

2.2. Associative Methods

The most commonly used methods for measuring cell adhesion are those that measure the rates with which cells adhere to other cells or to substrates. The first and simplest of these methods is to measure the size or number of aggregates that form when previously dissociated cells are caused to collide and aggregate in cultured suspensions (4). Although it seems intuitively obvious that cells capable of forming large aggregates are more adhesive than cells capable of forming only smaller aggregates, experiments done by Ede (12) suggest that just the opposite may be true. The most prudent path includes the use of other, more quantitative assays that make use of light scattering (13), for example, or direct cell counts (14) to determine the degree of single cell adhesion to cell aggregates. In a related strategy, viscometry (15) can be used to find the shear force that will just balance the tendency of cells to adhere to one another in suspension.

2.3. Collection Methods

For the past 10 years, collection methods for the measurement of cell adhesion have been among the most frequently used assays. Their popularity is

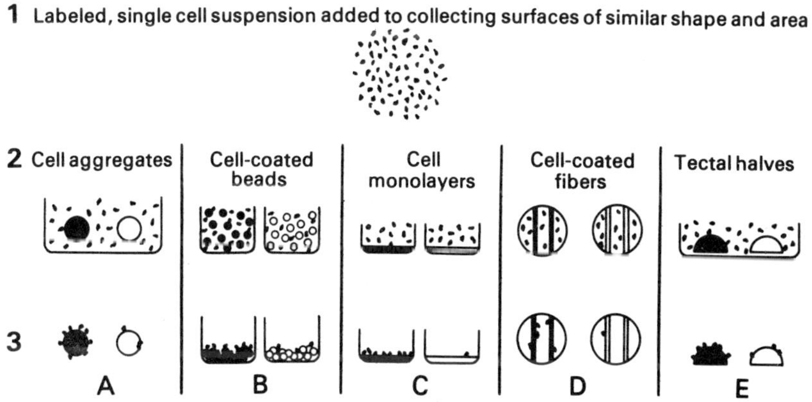

FIGURE 1. A schematic representation of collection methods for the measurement of intercellular adhesive specificity. In step 1, labeled tissues are dissociated into single cells and, in step 2, are exposed to geometrically identical, but biologically different, surfaces as shown in *A–E*. Finally, in step 3, the surfaces are washed and the adherent, labeled cells are counted and compared. In this diagram, the cell suspension is pictured as if it showed an adhesive preference for the black surfaces compared with the white in each of the five experimental designs.

due to their ability to measure adhesive specificity or recognition. In all collection methods, a suspension of labeled cells is exposed to several surfaces that are geometrically identical but biologically different. That is, the surfaces may be composed of different cell types or biochemically different substrates. When aliquots of the cell suspension are allowed to interact identically with the surfaces, the numbers of cells that adhere to the surfaces is proportional to the adhesive stabilities of the cells for those different surfaces. Some variations of this technique are illustrated in Figure 1 and described briefly in the legend. For a more complete listing and evaluation of adhesive assays, see reference 16.

3. BIOCHEMISTRY OF VERTEBRATE CELL ADHESION

3.1. General Factors

Despite the variety of methods, cells, and substrates used for the in vitro measurement of cell–cell and cell–substrate adhesion, there are some observations on which there are almost universal agreement. For virtually all measurements of intercellular adhesion, chelators of calcium and magnesium, proteases, and detergents at even very low concentrations all have detrimental effects. These additions also inhibit the adhesion of cells to substrates. Also, cells tend to adhere to substrates in proportion to the wettability of the substrate.

The simplest and most conservative conclusion that may be drawn from the vast amount of data on the role of general factors in cell adhesion is that cell adhesion depends, first, on membrane integrity. Any reagent that exerts a disruptive effect on the plasma membrane or any of its associated parts will no doubt have a similarly disruptive effect on cell adhesion and any other plasma membrane functions. There are no data, for example, that suggest a direct role for calcium in intercellular adhesion, although calcium is strictly required for cell adhesion in vitro and in vivo and calcium also has significant effects on cell surface charge measurements (17).

3.2. Serum Factors

One of earliest descriptions of a serum factor that could increase the rate of cell attachment to glass was that of Fischer et al. (18). They partly purified a glycoprotein from fetal calf serum and showed that fractions containing this agent, called fetuin, could be adsorbed to glass surfaces where it facilitated subsequent cell adhesion. Later reports questioned whether the active factor was fetuin but corroborated the existence of an attachment factor in serum.

Somewhat later, Taylor (8) systematically studied the effects of serum proteins adsorbed to glass on the attachment and spreading rates of conjunctival cells. He found, in general, that serum and adsorbed serum components retarded cell attachment to clean glass substrates. The following decade saw many studies in which serum factors either inhibited or stimulated cell adhesion (see reference 19 for a current and thoughtful review). In very few of these studies were the particular components purified and in none was the molecule identified or linked with any biologically significant class of macromolecules.

3.3. Effects of Known Compounds

Pessac and Defendi (20) showed that hyaluronic acid could itself stimulate intercellular adhesion among chick embryo neural retina cells and suggested that this effect was relatively nonspecific and due for the most part to the large, polyanionic nature of the compound.

Turley has covalently attached hyaluronic acid, chondroitin sulfate, and other glycosaminoglycans to plastic culture dishes and then measured cell attachment and migration rates to these derivatized surfaces. She found (21) that both chondroitin sulfate and hyaluronate promote attachment but that, for cultured fibroblasts, only hyaluronate promotes migration. Very recently, Hakomori's laboratory has coated cell-culture surfaces with lectins and with glycosidases (22) and shown that these coatings stimulate the rates of attachment of cultured fibroblasts. Proteins not known to interact with cell-surface oligosaccharides do not show any such stimulation.

The results of experiments by Turley and Hakomori, outlined above, represent a new level of sophistication in substrate-coating methods. At least for some fibroblastic cell lines, these data make it clear that cell surfaces contain oligosaccharides that will interact with proteins (lectins and glycosidases) sufficiently strongly to cause cell–substrate attachment. The cell surfaces must also contain proteins that will interact with hyaluronate and chondroitin sulfate.

3.4. Effects of Metabolites and Metabolic Factors

It is slowly becoming obvious that progress toward an understanding of the biochemistry of cell adhesion will require sophisticated biology and sophisticated biochemistry. Either one, without the other, will yield little useful information. In experiments that investigated the metabolic factors crucial to cell adhesion, this realization has become more apparent and, consequently, the experiments have become much more useful.

Many laboratories have found that cell adhesion is inhibited by conditions that interfere with active metabolism. For example, low temperatures (23), inhibitors of protein synthesis (24), and anesthetics and cardiac glycosides (25) all have negative effects on intercellular adhesion in a variety of assays. Trypsin and other proteases, of course, interfere generally with cell adhesion, and mild periodate oxidation can have similar effects (26). These findings led to the early idea that protein–carbohydrate interactions at the cell surface were one of the main mechanisms behind intercellular adhesion.

This hypothesis was substantiated by experiments (27) with glutamine, the amino donor for the synthesis of glucosamine. Mouse teratoma cells fail to aggregate in a simple medium unless glutamine is present. The glutamine analog diazo-norleucine overcomes the glutamine-induced stimulation of adhesion and may be, in turn, counteracted by glucosamine itself. These data strongly suggest a role for carbohydrates in cell–cell adhesion.

3.5. Factors from Cells

3.5.1. Effects on Cell–Substrate Adhesion. A prominent tactic in the battle to understand intercellular adhesion has been the characterization of materials that are produced by cells themselves and that stimulate cell adhesion. Work of this type focuses on two phenomena: (1) cell adhesion to substrates and (2) cell adhesion to other cells. When cell–substrate adhesion is under study, the interactions of cells with defined substrates are examined and compared. Alternatively, it is possible to remove cells from plastic or glass substrates and to analyze the material left behind. These investigations have been largely descriptive and have been recently reviewed (28).

To the extent that generalizations are possible, it appears that most cultured cells adhere not to the naked substrates themselves but to macromolecular components that are tightly adsorbed to the glass or plastic. Among these components are fibronectin, collagen, and glycosaminoglycans like hyaluronic acid and chondroitin sulfate (see Chapters 8 and 9). The major unanswered question here is: What is the degree of similarity between cell–substrate adhesions and the intercellular adhesions that occur during embryonic morphogenesis? Ultimately, it will be necessary to describe how these matrix materials (see Table 1) form stable complexes with cell membrane moieties and to show how the complexes could account for the specific associations that are the hallmarks of cell migration in the embryo. When hypotheses have been formulated, they should be tested in vivo. Interference with a particular molecular configuration should cause predictable perturbations in morphogenesis and these, in turn, should cause specific defects.

Currently, however, we know only that many of the compounds found in between cells and in basement membranes are capable of interacting with

TABLE 1. Glycoconjugate Structures Likely to Be Involved in Cellular Adhesion

Class	Name	Description or Structure
Glycosaminoglycans	Hyaluronic acid	Large (4–8,000,000 dalton), repeating polymer of N-acetylglucosamine and glucuronic acid; no sulfate; no covalently attached protein.
	Chondroitin sulfates	Smaller (5–50,000 dalton), repeating polymer of *N*-acetylgalactosamine and glucuronic acid; sulfated on either the 4 (CS A) or 6 (CS C) positions of the hexosamines; covalently attached to protein.
	Dermatan sulfate	Chondroitin sulfate with mixed 4 and 6 sulfate and with interspersed iduronic acid that may also be sulfated.
	Heparin and heparan sulfate	Repeating copolymers of unacetylated and acetylated glucosamine with either glucuronic or iduronic acids; the iduronic acid and the glucosamine may be sulfated; the most highly sulfated of the GAGs (6–25,000 daltons)
Serum-type glycoprotein	Ceruloplasmin, orosomucoid, and many others	Refers more to the nature of the sugar-protein linkage than to the location of the glycoprotein; the sugar is linked *N*-glycosidically to asparagines and the linkage is stable to alkaline hydrolysis; core region is usually mannose-rich and the last four sugars are often, proximal to distal, mannose, *N*-acetylglucosamine, galactose, sialic acid
Mucin-type glycoprotein	Ovine submaxillary mucin and many others	Sugar-protein linkage is via an *O*-glycosidic bond to serine or threonine; this linkage is labile to alkaline hydrolysis; oligosaccharide chains are usually shorter than are those on the serum-type glycoproteins; mucin-type proximal sugar is often *N*-acetylgalactosamine and the terminal sugar is often sialic acid
Other glycoproteins	Collagens, fibronectin, glycophorin, rhodopsin, acetyl choline receptor, and many others	Any protein with a covalently attached carbohydrate component that may be barely detectable, as in collagen, or massive, as in chondroitin sulfate
Glycolipids	Cerebroside, Forssmann antigen, blood group substances, and many others	Any molecule with lipid and carbohydrate components that are covalently attached; the lipid is often a ceramide and the proximal sugar is often glucose
Gangliosides	Tay-Sachs ganglioside (GM_2) and many others	Any glycolipid that contains sialic acid covalently attached to its carbohydrate chain; gangliosides with polymeric sialic acid residues have been described; more complex gangliosides can be reasonably soluble in aqueous solvents

one another, sometimes with a fairly high level of specifity. For example, fibronectin and collagen associate readily and to a much greater extent than either protein does with other proteins. Similarly, the carbohydrate polymers of hyaluronic acid and chondroitin sulfate show strong associative tendencies, although neither of these glycosaminoglycans shows any self-association nor any appreciable attraction for other glycosaminoglycans (29). Other interactions occur between fibronectin and glycosaminoglycans and will be covered in detail in Chapter 9.

3.5.2. Effects on Cell-Cell Adhesion. Different kinds of information come from studying the effects of shed molecules and membrane components on intercellular adhesion. Among the earliest of these observations were those of Moscona and his colleagues who have identified a component or components that seems to be shed from chick embryo neural retina cells when they are cultured for even short times in medium without serum (30). The factor is a glycoprotein of about 50,000 daltons, but its most interesting aspect is its ability to promote the aggregation of neural retina cells specifically. Moscona has suggested that similarly tissue-specific molecules reside in the cerebellum of the chick embryo. These compounds are known to exist on their respective cell surfaces, and it is reasonable to assume that they are among the molecules used by the cells to recognize one another and maintain tissue integrity.

In a closely related series of experiments, Lilien and co-workers (31) have also identified glycoproteins that are released in vitro from neural retina and cerebrum cells of the chick embryo. Each compound is able to bind specifically to its appropriate source tissue as well as accelerate the aggregation of the appropriate cells.

Glaser's laboratory has attempted to purify cell membrane components that are important in cell adhesion by beginning with an acetone powder extract of the cell membrane itself. Glaser and his colleagues searched for extractable compounds that would inhibit aggregation of neural retina cells, since that tissue was the source of their purified cell membranes. Such a compound was identified but it had a low molecular weight and could not be detected in gels either by its activity or with standard staining methods (32). A fascinating aspect of the membrane-derived component found by Glaser is its ability to inhibit the adhesion of tectal cells as well as retinal cells. This is important because the tectum is the sole target of the retinal ganglion cells as they produce the retinal projection in the midbrain during chick embryo development.

Another small molecule has been detected on the neural retina cell plasma membranes and this molecule, too, can inhibit the adhesion of neural retina cells (33). Marchase and Jakoi have called this compound ligatin and note that it is present in large amounts in a variety of tissues, and interestingly enough, that it diffuses out of gels during the staining process. The similarities between ligatin and the membrane component found by the Glaser group are intriguing and remain to be fully explored.

Mouse teratoma cells in culture release into their surrounding medium a factor that stimulates the adhesion of other teratoma cells (34). This compound, although not completely purified, is also a glycoprotein and requires terminal, β-linked galactosides for its aggregation-promoting activity.

Another class of molecules strongly implicated in the adhesion process among vertebrate cells are the lectins. Lectins are proteins with at least two binding sites for particular carbohydrate groups. Because of these properties, lectins are usually detected initially by their ability to agglutinate cells. Although genetic experiments show clearly that lectins have an adhesive function in slime molds, similar conclusions are not yet valid for any vertebrate lectin (see Chapters 7 and 10 for a more detailed treatment of lectins).

3.6. Disrupting Cell–Cell Adhesion with Enzymes

The first enzyme shown to have a detrimental effect on intercellular adhesion was crude trypsin, the preparation used by Moscona to dissociate embryonic tissues into single cells. There have been surprisingly few subsequent studies making use of enzymes as specific perturbing agents in cell-adhesion assays.

Glycosidases were used (35) to disrupt adhesive recognition among neural retina cells. The enzymes were partly purified from supernatants of pathogenic bacterial cultures and only one enzyme—β-galactosidase—had a consistent effect on adhesion. In this sytem, galactosidase stimulated neural retina cell adhesion to neural retina cell aggregates but not to other aggregate types. Neuraminidase, which removes *N*-acetylneuraminic acid from cell-surface glycoproteins and glycolipids (see Table 1 for a brief description of the glycolipids), also stimulates adhesion among some, but not all, cell types (for a review of neuraminidase effects, see reference 10).

Most recently, Marchase and his colleagues (36) have used an extensive battery of glycosidases to investigate the mechanism of adhesive recognition in vitro between chick embryo retinal cells and optic tecta. On the basis of the effects of two galactosidases and two *N*-acetylgalactosaminidases, Marchase concluded that one cell surface moiety active in recognition was the ganglioside GM_2, also known as the Tay-Sachs ganglioside (Table 1 gives a brief description of gangliosides). This conclusion was supported by experiments in which only GM_2 oligosaccharide sequences inhibited in vitro recognition between these two cell types (37).

In the retino-tectal system, as in most others, trypsin can eradicate the adhesive recognition manifested in vitro. This result implies that there must also be a proteinaceous receptor for the GM_2. If so, this receptor remains to be purified.

In a recent and important experimental series, Gottlieb (38) has shown that an adhesive gradient exists across the chick embryo neural retina such that ventrally located cells show a high adhesive stability toward dorsally located cells and a gradually diminishing preference for cells located more ventrally. This gradient is consistent with the biochemical data obtained in the same system (35, 36), but an actual gradient of GM_2 has not been detected across the retina or the tectum.

Nirenberg and his colleagues (56), however, have found a gradient. They immunized mice with dorsal retinal fragments and have produced a monoclonal antibody that shows a gradient-type binding to the chick embryo retina. In terms of pmoles of antibody per unit area of retina, the gradient varies by a factor of 20–30 with the dorsal-most parts being higher than the ventral-most. The gradient is already present in the four-day embryo and the antibody appears to bind to most of the cells in the appropriate retinal areas. The antigen against which this antibody is directed, therefore, shows many of the properties of the ganglioside GM_2 except that, using biochemical methods, Marchase was not able to detect any GM_2 gradient. Direct testing should establish if the monoclonal antibody can recognize the Tay-Sachs ganglioside or oligosaccharides with similar sequences.

3.7. Surface Enzymes as Recognition Molecules

When galactosidases were seen to stimulate neural retina cell adhesion specifically (35), one conclusion was that retinal cell-surface galactosides were masking another glycoside that was involved in the recognition process. Early attempts were made to replace the removed galactoside by incubating galactosidase-treated cells with a crude preparation of galactosyltransferases and with the galactose donor used by these enzymes, uridine diphosphate galactose. Control experiments showed that galactose was incorporated onto the cell surfaces, even when the crude enzyme preparation was omitted from the incubations. This preliminary indication of a cell-surface glycosyltransferase has been substantiated and generalized to include a great many cell types (see reference 39 for a recent review).

The evidence that surface glycosyltransferases may actually be recognition molecules for vertebrate cells is entirely circumstantial, although for the cariogenic bacterium *Streptococcus mutans*, similar enzymes located in the cell wall mediate the adhesion of the bacteria to dental placques that consist of carbohydrate polymers (40). The advantage of a hypothesis that proposes transferases as recognition molecules is its ability to deal with de-adhesion as well as adhesion. An enzyme can let go of its substrate and modify it at the same time. Proteins like antibodies and lectins that are merely complementary to other molecules have no obvious way, at least, to allow deadhesion.

Most recently, Balsamo and Lilien (57) showed that an N-acetylgalactosaminyltransferase activity at the neural retina cell surface correlated with cell adhesion among these cells. In this case, the transferase activity in plasma membranes was dependent on the presence of calcium. To generalize from transferases to other, likely adhesion components, it is possible, therefore, that the effects of calcium may be due to its ability to keep adhesion proteins in the membrane when cells are dissociated. Calcium-independent adhesive processes may result from surface components that do not require calcium in order to remain in the membrane.

For the present, suffice it to say that surface transferases have been implicated in cell migration in the early chick embryo, sponge cell adhesion, algal mating interactions, lymphocyte-target cell interactions, in vitro growth control, murine T-locus effects, and cell adhesion in many systems (39). Although the transferases remain attractive candidates for mediating intercellular adhesions in a variety of situations, unequivocal data on this point for any given system have not yet been accumulated.

3.8. Immunological Definitions of Adhesion Molecules

An increasingly popular strategy for the identification of adhesion molecules makes use of antisera raised initially against whole cells. If the antisera affect adhesion among the cells used originally, then two avenues are open to the investigator. First, it is possible to absorb the antisera with other, related cell types so that unwanted antibodies are lost but the adhesion effect is retained. When an antiserum enriched in the desired antibody is obtained, immunoprecipitation of cell homogenates or cell membranes may yield a small number of compounds that can be tested for their effects in the adhesion assay being used. Alternatively, an investigator may use a blocking assay in which cell membranes and homogenates may be fractionated and the fractions tested for their ability to inhibit the original antiserum effect. Both these methods, and numerous variations, are being pursued extensively, but the following discussion will concentrate on the data from only a few, representative laboratories.

Buck, Damsky, and their colleagues have prepared antisera that affect cell adhesion to substrates and cell morphologies in culture. They have described an antiserum that causes fibroblasts to detach from culture dishes, another that causes epithelial cells to detach, and a third that causes epithelial cells to resemble fibroblasts. Figure 2 shows examples of the morphological effects that can be caused by these antisera. Recent work from this laboratory (41)

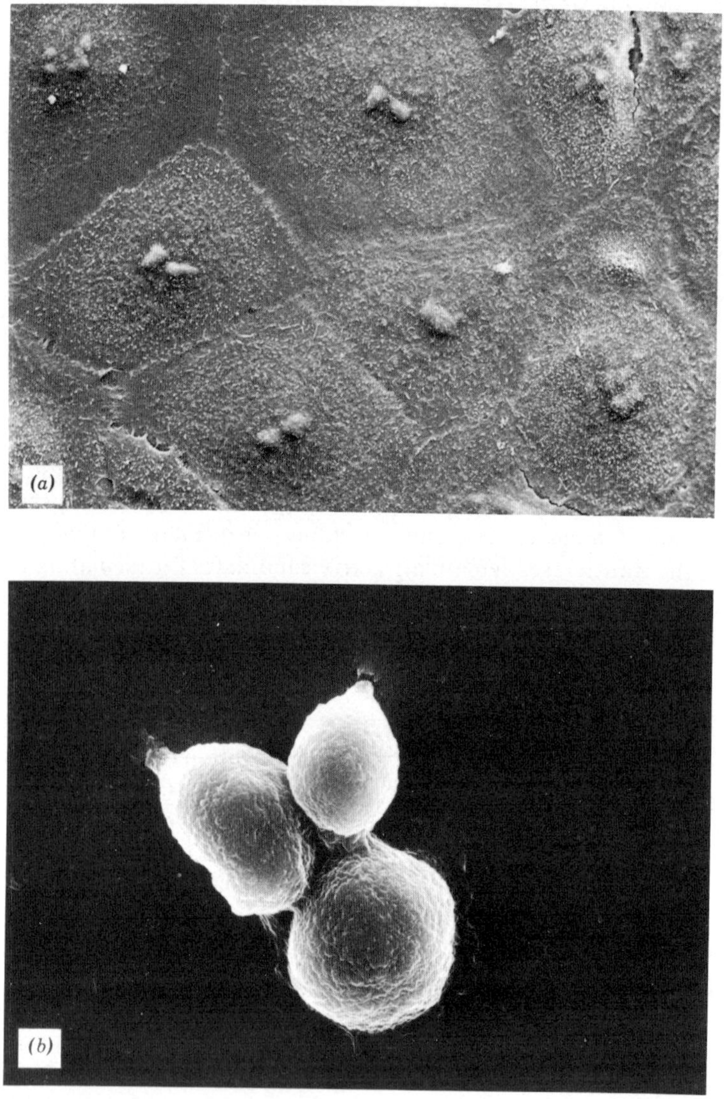

FIGURE 2. Scanning electron micrographs of mouse mammary tumor epithelial cells. Untreated cells (*A*) are flat and closely adherent. In *B*, the same cells have been exposed for 2 hr to one antiserum directed against surface components. The cells round up and, eventually, detach from the culture dish. After exposure to a different antiserum for longer periods of time (*C*), the same epithelial cells do not round up or detach but, instead, change their morphology and assume a more fibroblastic appearance. Magnifications, about 2000×. [These micrographs were the kind gift of Dr. C. Damsky (41).]

FIGURE 2. (*continued*)

has focused on a search for membrane components that will inhibit the detachment of epithelial cells by the appropriate antiserum. Affinity chromatography of detergent-extracted membranes shows that a small number of glycoproteins (with molecular weights between 130,000 and 140,000 daltons) are effective antagonists to the antiserum. Similar experiments with the other two antisera indicate that the antagonists to each of the three antisera are largely different from one another. These data are important for two reasons. First, they suggest that a small number of membrane components may affect cell adhesion to plastic substrates and that these components may differ according to cell type. Perhaps more important, however, the strategy used by Buck and Damsky is a general one that should be applicable to virtually any cell-adhesion system. Furthermore, the introduction of monoclonal antibodies to this general protocol should vastly increase its power, since the specificity of the antisera will be made much greater.

In a closely related project, Takeichi et al. (42) have prepared antisera against components released from hamster fibroblasts by gentle treatment with trypsin. The resultant antiserum inhibited intercellular adhesion among the cells and studies with the fluoresceinated antiserum show that the antigens involved are distributed over the entire surface of fibroblasts but are not present on hamster epithelial cells. Immunoprecipitation studies suggest that the antiserum is reacting with a single glycoprotein with a molecular weight of 125,000.

Edelman, Rutishauser, and co-workers have applied these techniques to developing systems and, more specifically, to the adhesive recognition known to occur among cells of different organs from the chick embryo. By inoculating rabbits with whole retinal cells, they produced an antiserum that inhibits the adhesion of retinal cells. This antiserum effect is blocked by compounds that are released into the medium when retinal cells are maintained in culture. These blocking compounds were subsequently used to raise another antiserum that was capable of precipitating a small number of proteins from detergent extracts of retinal cell membranes. The major protein has a monomer molecular weight of 140,000 daltons and it shares antigenic determinants with the proteins released by the retinal cell culture (43). The 140,000-dalton protein does not, however, inhibit retinal cell adhesion. Using a very similar protocol, the Edelman group has recently identified another membrane component (with a monomer molecular weight of 68,000) that seems to be involved in the adhesion of chick embryo liver cells (44). The McGuire research group has also used these techniques and has found a similar, if not identical, protein that blocks the inhibition of liver cell adhesion by a specific antiserum (45). McGuire has made the interesting observation, as well, that this liver antigen is no longer detectable in adult livers. This is, of course, consistent with the idea that the protein is involved in the organogenesis of the liver and that after organogenesis occurs the component is no longer needed.

4. TEMPORAL CONSIDERATIONS IN CELL ADHESION

Just as time is the single most important dimension in developmental biology, time must be considered when designing or discussing cell-adhesion experiments. The tenacity and specificity of cell adhesions change in the minutes and hours after tissues are dissociated into cells. On a larger time scale, tenacity and specificity also change over the life span of the individual.

Holtfreter (1) noticed that when he combined dissociated cells from different germ layers of the amphibian embryo, the cells adhered to each other initially without regard to type. Only after some time passed, did the cells begin to sort out into groups according to their origins. When more quantitative assays came into use, they showed (46) that adhesive recognition among cells of different tissues occurred only after a variable lag period, during which adhesions tended to be relatively nonspecific.

A number of subsequent studies (47, 48) have indicated that in vitro adhesive stabilities among cells increase in stages. Depending on the cell type under study and the conditions of the experiment and assay, increments in tenacity may be dependent on temperature, metabolism, cations, and serum proteins, or any combination of these. It no longer seems likely, however, that any of these stages correspond with nonspecific, intercellular adhesion. The laboratories of McGuire (48) and McClay (11) have shown, independently, that if care is taken with the dissociation of chick embryo tissues, and if sensitive assays for cell adhesion are used, then specificity among cells of different tissues can be detected almost immediately.

Adhesive recognition among embryonic cells can change also as a function of developmental time. Glaser and his colleagues (49) found that membranes from 7-day chick embryo retinas can inhibit the adhesion of 7-day retinal cells but not the adhesion of 8-day retinal cells. The reciprocal experiment, using 8-day retina membranes, showed a similar temporal specificity. Barbera (50) described an abrupt change in chick embryo retinal cells between days 6 and 7 of incubation. After this time, ventral retinal cells adhere to dorsal surfaces of the optic tectum in accord with the ultimate map of the retinal ganglion cell fibers on the tectum. On or before day 6, however, ventral retinal cells adhere to ventral tectal surfaces.

5. IN VIVO EXPERIMENTS WITH INTERCELLULAR ADHESION

In vivo experiments on cell adhesion in embryos are rarely attempted because of the small size and limited accessibility of most embryos. In adults, morphogenetic processes are, with two exceptions, completed. These exceptions are, first, the constant migration of the small lymphocytes between the blood and the lymphatic system (see Fig. 3) and, second, malignancy, which is characterized by the dissemination throughout the body of the tumor cells. In both cases, a relatively discrete population of cells has the ability to recognize some aspect of the vascular endothelium and leave the blood. Small lymphocytes can be injected into animals and will populate lymph nodes selectively and quickly. Similarly, tumor cells often tend to show predictable, metastatic pathways, although the degree of their organ specificity has never approached that of the lymphocytes.

In an extensive and elegant series of experiments, Gesner and his colleagues (51) showed that the ability of lymphocytes to "home" to lymph nodes after injection into an animal could be affected by proteases and glycosidases. The effects of the proteases could be reversed by allowing the cells to metabolize in the absence of the enzyme for 12 hr. The effects of the glycosidases could be partly reversed by including L-fucose and *N*-acetylgalactosamine with the enzyme mixture. It appears, therefore, that cell-surface proteins or glycoproteins are involved in endothelial recognition by the small lymphocytes and that the proteins, once removed or destroyed, can be regenerated in about 12 hr.

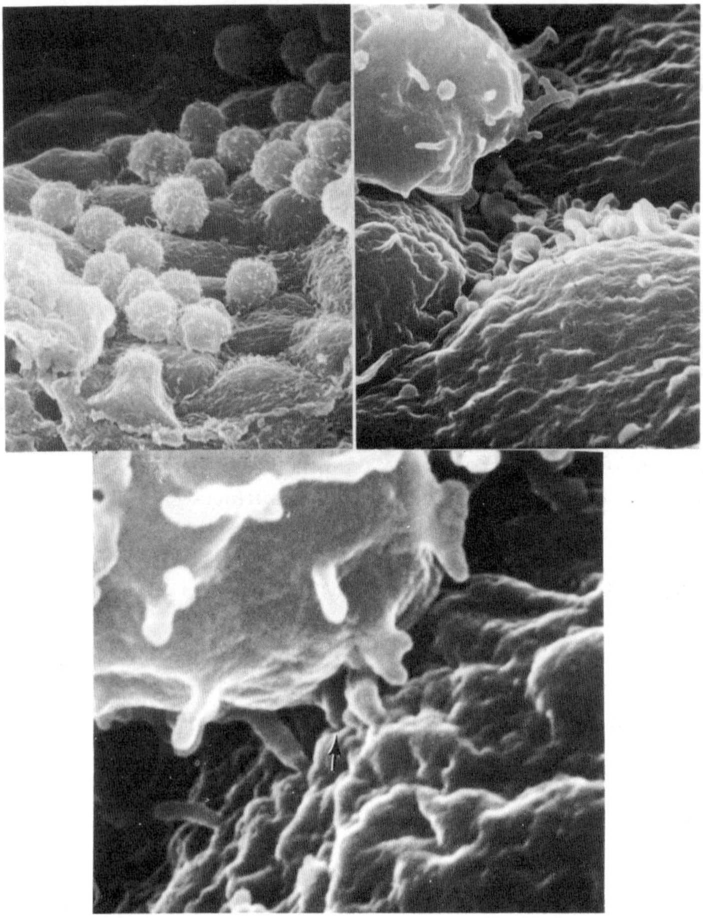

FIGURE 3. Scanning electron micrographs of increasingly higher magnifications showing the attachment of lymphocytes to venule endothelial cells in the mouse. (*Upper left*) lymphocytes studding the blood vessel linings. At a higher magnification (*upper right*), a single lymphocyte begins its migration through the intercellular clefts of the endothelia. At a still higher magnification (*lower*), the lymphocyte extensions are seen interacting with the endothelial pits. (These micrographs were the kind gift of Dr. A. Anderson.)

The inhibitory effects of the glycosidases, coupled with the antagonism by only two monosaccharides, suggest that specific glycoproteins or glycolipids may also be involved in the recognition process and that fucose and *N*-acetylgalactosamine residues may be especially significant. Similar studies have shown, for example, that neuraminidase as well as cytochalasin A and colchicine (52) can interfere with the migration of lymphocytes to lymph nodes.

Curtis and de Sousa have combined in vitro with in vivo methods, using mouse, rat, and human lymphocytes (53). They describe small (3000–9000 dalton), heat-sensitive factors that can be isolated from cultures of different lymphocyte types. The factors do not affect the homologous lymphocytes (the cells that produce the factors) but do affect in vitro adhesion and in vivo "homing" among heterologous lymphocytes.

Very similar experimental strategies have been employed using tumor cells. Gasic and Gasic (54) found years ago that neuraminidase could affect the fate of malignant cells when they were treated with the enzyme and then injected into animals. More recently, it has been found that melanoma cells, selected for their ability to metastasize to lung, show a definite adhesive preference for lung cells in adhesive assays in vitro (55). Because the melanoma cells were injected intravenously, their rapid and relatively selective lodging in the lung must mean that they recognize and adhere to lung endothelial cells preferentially when compared to other endothelia. Because the in vitro experiments were carried out with whole lung tissue, as opposed to just endothelia, the demonstrated preference of the melanoma cell strain for lung cells must also mean that "lungness," whatever this means in biochemical terms, is expressed not merely on many lung cells but also on endothelial cells of blood vessels that permeate the lung.

6. PATTERNS AND PREDICTIONS

Do any patterns emerge from the studies outlined above? In fact, there are several, overall conclusions from research in cell adhesion over the last three decades. First, there is the happy, unmistakable trend toward quantitation and function. Subjective and intuitive assays for cell adhesion are disappearing and are being replaced by assays that result in numbers. This is important for results from one laboratory to be tested in many other laboratories. Function, too, is becoming a touchstone of cell adhesion research. Controls of increasing biological subtlety coupled with increasingly sophisticated biochemical methods are yielding more and more information that is not only reliable technically but relevant from developmental, pathological, and physiological standpoints.

Second, there is a clear, biochemical picture emerging when the totality of the data is considered, although the details are admittedly still quite vague. More serious, there is as yet no indication of the master pattern, if one exists. Most cell-adhesive mechanisms seem to employ proteins and carbohydrates but the questions now become: Precisely which proteins and carbohydrates interact in any particular system? How do they interact? Is there an overall relationship among all or most adhesive molecules?

There are three common paradigms for protein-carbohydrate interactions, and all three make use of molecular complementarity: antibody–antigen, lectin–oligosaccharide, and enzyme–substrate. All three may play a role in morphogenesis and, if so, they may all be related evolutionarily. Intercellular recognition, at least at the grossest levels, is a highly conserved phenomenon in that organotypic affinities exist even when animals as disparate as birds and mice are compared. The molecular apparatus responsible for this recognition must similarly be conserved.

In theory, adhesive components at the cell surface should have a lineage that can be traced to the time of the acquisition of multicellularity. It is simpler to imagine that a protein and its oligosaccharide complement in protozoans were put to a new use in metazoans, rather than to postulate the simultaneous and spontaneous appearance of a previously nonexistent molecular species. In either case, however, the novel moieties or their novel use allowed protozoans to remain attached after mitosis or, alternatively, to form stable adhesions with one another. As the complexity of the metazoans increased, the most credible hypothesis is that the original adhesive compounds diversified to fulfill ever more diverse recognition functions. It is reasonable to hope, therefore, that a pattern of structures will emerge as more and more adhesive molecules become defined and subsequently characterized. If this occurs, searches for homologies in other proteins and carbohydrates might be able to shed light on the nature of the ancestral, and hypothetical, adhesion molecules. In addition to the intellectual value of information of this type, significant, practical applications will be inevitable. If a general theory of adhesion and morphogenesis is possible, malignant metastases should be comprehensible in biochemical terms and, therefore, rational and specific therapies should be forthcoming. Finally, of course, teratology could become a predictive science with spontaneous birth defects being detected and, perhaps, even treated in utero.

REFERENCES

1. P. L. Townes and J. Holtfreter, *J. Exp. Zool.* **128**, 53 (1955).
2. A. Moscona and H. Moscona, *J. Anat.* **86**, 287 (1952).
3. M. Steinberg, *Proc. Natl. Acad. Sci. USA* **48**, 1769 (1964).
4. R. E. Hausman and A. Moscona, *Proc. Natl. Acad. Sci. USA* **72**, 916 (1975).
5. A. S. G. Curtis, *The Cell Surface: Its Molecular Role in Morphogenesis*, Logos/Academic Press, London, 1967.
6. D. R. Coman, *Cancer Res.* **4**, 625 (1944).
7. L. Weiss, *The Cell Periphery, Metastasis and Other Contact Phenomena*, North Holland, Amsterdam, 1967.

8. A. C. Taylor, *Exp. Cell Res. Suppl.* **8**, 154 (1961).
9. L. Weiss and E. Mayhew, *J. Cell Physiol.* **68**, 345 (1966).
10. C. W. Lloyd, *Biol. Rev.* **50**, 325 (1975).
11. D. R. McClay, G. M. Wessel, and R. B. Marchase, *Proc. Natl. Acad. Sci. USA* **78**, 4975 (1981).
12. D. A. Ede and G. S. Agerbak, *J. Embryol. Exp. Morphol.* **20**, 81 (1968).
13. R. B. Kemp, B. M. Jones, T. Cunningham, and M. C. M. James, *J. Cell Sci.* **2**, 323 (1967).
14. A. S. G. Curtis and M. F. Greaves, *J. Embryol. Exp. Morphol.* **13**, 309 (1965).
15. A. S. G. Curtis, *J. Embryol. Exp. Morphol.* **22**, 305 (1969).
16. R. B. Marchase, K. Vosbeck, and S. Roth, *Biochim. Biophys. Acta* **457**, 385 (1976).
17. M. Collins, *J. Exp. Zool.* **163**, 23 (1966).
18. H. W. Fischer, T. T. Puck, and G. Sato, *Proc. Natl. Acad. Sci. USA* **44**, 4 (1958).
19. W. Frazier and L. Glaser, Surface components and cell recognition, in E. E. Snell, P. Boyer, A. Meister, and C. C. Richardson, eds., *Annual Reviews of Biochemistry*, Annual Reviews Inc., 1979, p. 491.
20. B. Pessac and V. Defendi, *Science* **175**, 898 (1972).
21. E. A. Turley and S. Roth, *Cell* **17**, 109 (1979).
22. H. Ranvala and S.-I. Hakomori, *J. Cell Biol.* **88**, 149 (1981).
23. S. Roth, *Dev. Biol.* **18**, 602 (1968).
24. E. J. McGuire, *J. Cell Biol.* **68**, 90 (1976).
25. G. Ungar and A. S. Keats, *Anesthesiology* **39**, 362 (1973).
26. A. A. Moscona, *J. Cell. Comp. Physiol. Suppl.* **1**, 65 (1962).
27. S. B. Oppenheimer, M. Edidin, C. W. Orr, and S. Roseman, *Proc. Natl. Acad. Sci. USA* **63**, 1385 (1969).
28. F. Grinnell, *Int. Rev. Cytol.* **53**, 65 (1978).
29. E. A. Turley and S. Roth, *Nature* **283**, 268 (1980).
30. R. E. Hausman and A. A. Moscona, *Proc. Natl. Acad. Sci. USA* **73**, 3594 (1976).
31. J. Balsamo and J. Lilien, *Biochemistry* **14**, 167 (1975).
32. R. Merrell, D. I. Gottlieb, and L. Glaser, *J. Biol. Chem.* **250**, 5655 (1975).
33. E. R. Jakoi and R. B. Marchase, *J. Cell Biol.* **80**, 642 (1979).
34. J. T. Meyer and S. B. Oppenheimer, *Exp. Cell Res.* **102**, 359 (1976).
35. S. Roth, E. J. McGuire, and S. Roseman, *J. Cell Biol.* **51**, 525 (1971).
36. R. B. Marchase, *J. Cell Biol.* **75**, 237 (1977).
37. J. M. Pierce, in preparation.
38. D. I. Gottlieb, K. Rock, and L. Glaser, *Proc. Natl. Acad. Sci. USA* **73**, 410 (1976).
39. J. M. Pierce, E. A. Turley, and S. Roth, *Int. Rev. Cytol.* **65**, 1 (1980).
40. H. Mukasa and H. D. Slade, *Infect. Immun.* **10**, 1135 (1973).
41. C. H. Damsky, K. A. Knudsen, R. J. Dorio, and C. A. Buck, *J. Cell Biol.* **89**, 173 (1981).
42. H. Urishihara and M. Takeichi, *Cell* **20**, 363 (1980).
43. J.-P. Thiery, R. Brackenbury, U. Rutishauser, and G. M. Edelman, *J. Biol. Chem.* **252**, 6841 (1977).
44. R. Bertolotti, U. Rutishauser, and G. M. Edelman, *Proc. Natl. Acad. Sci. USA* **77**, 4831 (1980).

45. E. J. McGuire, in preparation.

46. S. Roth, *Dev. Biol.* **18**, 602 (1968).

47. J. Umbreit and S. Roseman, *J. Biol. Chem.* **250**, 9360 (1975).

48. E. J. McGuire, *J. Cell Biol.* **68**, 90 (1976).

49. D. I. Gottlieb, R. Merrell, and L. Glaser, *Proc. Natl. Acad. Sci. USA* **71**, 1800 (1974).

50. A. J. Barbera, *Dev. Biol.* **46** 167 (1975).

51. B. M. Gesner and V. Ginsburg, *Proc. Natl. Acad. Sci. USA* **52**, 750 (1964).

52. A. O. Anderson, N. D. Anderson, and J. D. White, in W. Muller-Ruchholtz and H. K. Muller-Hermelink, eds. *Advances in Experimental Medicine and Biology,* Vol. 114, *Function and Structure of the Immune System,* Plenum, New York (1979).

53. A. S. G. Curtis and M. A. B. de Sousa, *Cell. Immunol.* **19**, 282 (1975).

54. G. Gasic and T. Gasic, *Proc. Natl. Acad. Sci. USA* **48**, 1172 (1962).

55. J. L. Winkelhake and G. L. Nicolson, *J. Natl. Cancer Inst.* **56**, 285 (1976).

56. G. D. Trisler, M. D. Schneider and M. Nirenberg, *Proc. Natl. Acad. Sci. USA* **78**, 2145 (1981).

57. J. Balsamo and J. Lilien, *J. Biol. Chem.* **257**, 349 (1982).

4

Embryonic Induction

KLAUS KRATOCHWIL
Institut für Molekularbiologie der Österreichischen Akademie der Wissenschaften
Salzburg, Austria

The phenomenon of *embryonic induction*, in which a short-range interaction of two dissimilar tissues determines the further developmental course of at least one of them, was discovered 80 years ago (1). Of all types of cellular interactions, it brings about the most dramatic developmental consequences, yet in molecular terms it is certainly the least understood. Therefore, this chapter will not describe the phenomenology of the great variety of known inductive systems. Many of them are dealt with in embryology textbooks, and in addition there are a number of excellent reviews available, which in this field have the further advantage of not becoming obsolete rapidly. Instead, an attempt is made to examine our present concepts of induction and the prospects for its analysis. Some characteristic features of a few typical inductive systems are presented in Section 2 to provide a basis for this discussion.

1. ROLE OF TISSUE INTERACTION IN VERTEBRATE DEVELOPMENT

Diversification of cell types within the progeny of the fertilized egg can be achieved by two basically different mechanisms. In certain invertebrates, for instance, the cytoplasm of the egg cell is already inhomogeneous due to a prior segregation of molecules capable of regulating gene activity. The unequal distribution of these still unknown substances (called *morphogenetic determinants*) is stabilized by division of the egg cytoplasm during cleavage, and this accounts for the differential gene expression of nuclei that originally were totipotent. In the so-called "mosaic egg," cell diversification is therefore basically due to cytoplasmic predetermination and unequal cell division with respect to cytoplasmic factors and is consequently characterized by relatively strict cell lineage. Differentiation is governed by purely intracellular regulation with little or no developmental interaction taking place between cells. Embryogenesis of nematodes (e.g., *Caenorhabditis elegans*) is a classical example of this type of development (2, 3). In less extreme form, cytoplasmic predetermination is also typical of the eggs of other spiralia, of ascidia, and to some extent also of insects.

Differentiation of cell types in other embryos, notably those of vertebrates, follows a distinctly different pattern: the developmental fate of a cell and its progeny is not determined by its lineage relationship but rather by its position within the embryo or within a blastema. In the development of these embryos, diversification occurs within relatively large and homogeneous populations of cells having the same developmental potential. From this follows that the number of cells possessing any specific developmental potential is always greater than the number of cells eventually expressing it in differentiation.

When groups of cells within such a homogeneous population are to enter a developmental pathway different from their surroundings, they require special cues. According to the hypothesis of *positional information* (4) cells may receive cues from their position within some sort of gradient extending over a usually two-dimensional population. Frequently, however, it is an external influence that determines which part of the population is to express a particular developmental potential. In most cases, the decisive signal comes from another tissue that comes into contact with it by morphogenetic movements, and the process is then called *embryonic induction*. For example, the potential to form a lens extends over wide areas of the ectoderm in most vertebrate embryos, but it becomes realized only where the ectoderm comes into close contact with the optic vesicle (the precursor of the retina). The optic vesicle, therefore, is said to act as the inductor* of lens differentiation.

The fact that the inductive event selects a small group of cells out of a much larger competent area (population) guarantees that the inductor can determine precisely where the induced development will occur and therefore where the induced structure will form. This can be considered to constitute

* Throughout this chapter, the term *inductor* refers to the tissue and *inducer* to the hypothetical substances involved in induction.

the principal benefit for embryos relying on this mechanism for cell diversification. The organs of vertebrates, for example, are usually formed by two or more tissue components. These tissues derive from different blastemas that frequently make contact only immediately before organ development. Nevertheless, a high degree of architectural coordination between the tissues is often required for the function of the organ that can best be achieved if at least one of the tissues develops under the control of the other. As a consequence, inductor and responding tissues are related in function rather than by developmental history or cytodifferentiation.

This case can again be illustrated in the development of the eye. The precursor of the retina, the optic cup, is an evagination of the brain, whereas the lens develops from the overlying head ectoderm. As mentioned, most of the embryo's ectoderm is capable of lens differentiation, but its formation anywhere else than in precise alignment with the optic cup would make very little sense. The dependence on an inductive signal from this optic cup would seem to guarantee this alignment much more reliably than cytoplasmic predetermination within the ectoderm. The same argument applies to other structures, especially those where small-scale coordination is required within rather large populations of cells (e.g., in the kidney where two duct systems of different origin must be connected; see Section 2.2.1).

In contrast to predetermination, inductive cell interaction is able to correct for the variability that is necessarily introduced by the involvement of large numbers of cells and of complicated morphogenetic movements in the vertebrate embryo. It ensures maximal architectural precision even when cell diversification continues into later stages of embryogenesis and therefore allows greater developmental complexity. It is certainly no coincidence that embryos relying most heavily on the alternative system of cytoplasmic predetermination and cell lineage for cell diversification (e.g., nematodes) have only small and even constant numbers of cells.

2. BIOLOGY OF SOME INDUCTIVE SYSTEMS

A great number of inductive systems have been described in vertebrate embryogenesis; not all of them have received the same attention. With the development of organ culture methods, emphasis has shifted from amphibians to bird and mammalian embryos and, simultaneously, from "early" to "late" events. Nevertheless, the same inductive systems have been found in lower and higher vertebrates, and the basic generality of the processes seems also established by the fact that inductive interaction does occur in experimental combinations between tissues taken from different species and even from different classes.

During embryogenesis, tissue interactions not only affect the decision between developmental alternatives, but they continue beyond the time of (presumed) final commitment of the tissues, regulating growth, morphogenesis, and cyto-differentiation of one or both interactants. This latter type of interaction is especially important in organogenesis (5), but this chapter will deal only with embryonic induction in the more narrow sense of regulating developmental choices.

2.1. Induction of Ectodermal Differentiation

It is remarkable that the most typical induction systems have the ectoderm or its derivatives as the responding tissue (induction of the neural plate, nose, lens, and ear; in later periods, of the cornea, skin, and tooth). The ectoderm of the early amphibian gastrula seems to be its least committed cell population, which can be induced even to mesodermal and endodermal differentiation. In later stages, the development of the epidermis and of related epithelia (cornea, enamel epithelium, skin glands) can still be modified by experimental combination with other tissues. The ectoderm is the tissue most susceptible to, and most dependent on, external cues for its differentiation.

2.1.1. Induction of the Lens. Although the action of the optic cup on head ectoderm is generally regarded as the classical example of embryonic induction, the retina anlage is neither the only nor the first inductor of the lens. Lens formation independent of the optic vesicle has been obtained in some amphibian species (for reviews see references 6 and 7) and in the chick (8, 9) and mouse embryo (10). Such lenses synthesize lens-specific products, the crystallins (11), although they develop poorly and eventually degenerate.

This strong developmental "bias" of the head ectoderm toward lens differentiation is apparently acquired in earlier inductive events: before its association with the optic cup, the presumptive lens ectoderm of amphibians is first underlain by foregut endoderm, then by heart mesoderm. In the newt, both tissues have been shown to act as inductors (12). In the chick, where lens differentiation could be obtained from primitive streak head ectoderm, a comparable action of the hypoblast and possibly some mesoblast cells on the epiblast has been postulated (9). After this event, the presumptive lens ectoderm presumably requires only nonspecific triggers or support (e.g., combination with various unrelated tissues, such as dermis, meso- or metanephos, salivary or lung mesenchyme, etc., or even merely improvement of the culture medium) to express its developmental bias (9, 10). In contrast, body ectoderm of 2-day chick embryos specifically requires association with an optic cup for lens differentiation (10).

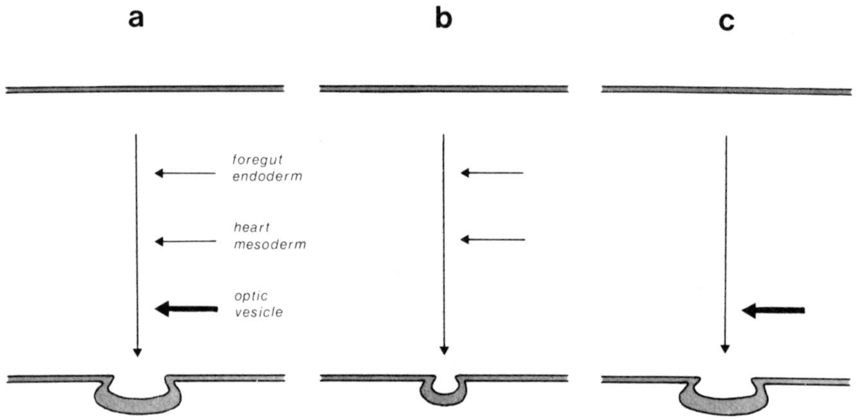

FIGURE 1. Induction of the lens. (*a*) Three different tissues are assumed to act sequentially and synergistically on head ectoderm, causing development of the lens. (*b*) In some species, the effect of the early inductors is strong enough to allow lens formation even without optic vesicle. (*c*) The optic vesicle can induce a lens in ectoderm that has not been exposed to the earlier inductors (e.g., trunk ectoderm). The experiments shown in (*a*) and (*b*), and others, indicate that the three inductors cause the same qualitative effect in head ectoderm, rather than "preparing" the tissue for the next step in an obligatory sequence of inductive events.

Lens induction thus appears to be a step-wise process with at least three different tissues acting synergistically as natural inductors (12) (Fig. 1*a*). The importance of the earlier inductions may vary to the degree of even allowing "spontaneous" lens formation without optic vesicle (Fig. 1*b*). On the other hand, the optic cup does not necessarily require previous action of the other tissues, since it alone can induce a lens in body ectoderm (Fig. 1*c*). At early stages, the ectoderm of the entire dorsal body is capable of responding to the lens-inducing stimulus; in the chicken, even extraembryonic ectoderm can respond (13). This competence gradually becomes restricted to the prospective lens-forming area, and this restriction occurs faster in species where early inductions play a more important role; this process leads to the well-known inverse relationship between spontaneous lens formation and inducibility of the body ectoderm in amphibian species.

2.1.2. Induction in Skin Development. The skin develops regionally diverse structures, such as feathers or scales in birds. Induction not only decides where these structures are formed but it also specifies their type. The inductive events preceding these decisions are analyzed by the use of interclass, interspecies, and heterotopic dermal-epidermal combinations (14–16).

In bird skin, the first inductive event affects the demarcation of future feather-forming areas from featherless regions. Experimental recombination of dermis and epidermis from different sites has established that the dermis is responsible for this decision and that epidermis from normally featherless areas is competent to respond with feather formation. Interestingly, even mouse dermis can induce feather formation not only in chick epidermis (17) but also in 5-day chick cornea (18). Conversely, mouse epidermis initiates hair development in association with chick dermis but not with mouse dermis from the hairless sole of the foot (17).

It is therefore the dermis of specific regions, irrespective of its taxonomic origin, that determines where differentiated skin structures are formed. The epidermis responds according to its genetic potential and basic developmental program, which is "feathers" in birds and "hair" in mammals.

Both feather and hair formation in these interclass combinations do not proceed far. Full development is achieved by the addition of small fractions (about 10%) of species-specific dermal cells. Most importantly, these cells need not come from an area where skin appendages are formed. For example, chick epidermis forms no structures with chick ventral dermis and only arrested feather cones with mouse dorsal dermis, but it gives rise to fully developed feathers when combined with a mixture of the two. After induction by a region-specific, but not class-specific, signal the epidermis now requires class-specific, but no longer region-specific, dermis to participate in the development of appendage structures (16, 19).

This lack of dependence on the regional origin of the dermis, however, exists only for the "basic" epidermal program (feathers in bird skin, hair in mammals). Other developmental options (e.g., scales, beak, spur, comb) are only expressed with dermis taken from the appropriate region (Fig. 2). For instance, scales are formed in combinations of chick dorsal epidermis with tarso-metatarsal dermis, and only when the dermis is of a relatively advanced stage (just before the appearance of scale ridges). With younger dermis, the graft will still form feathers. It appears that a property of the dermis, unique to certain regions and developmental stages, can override the basic feather program of chick epidermis (14).

On the other hand, epidermis reveals unexpected developmental potentials in response to nondermal mesenchymes. On gizzard mesenchyme, 5-day chick epidermis forms a mucus-secreting, even ciliated epithelium (20), and 14-day mouse foot epidermis combined with a mesodermal tooth papilla was reported to develop an enamel organ (21). The existence of a basic developmental program in the epidermis does not seem to be in conflict with a high plasticity and susceptibility to external influences.

2.2. Induction of Mesodermal Differentiation

There are two better-studied inductive systems that affect the development of mesodermal cells: formation of kidney tubules and differentiation of carti-

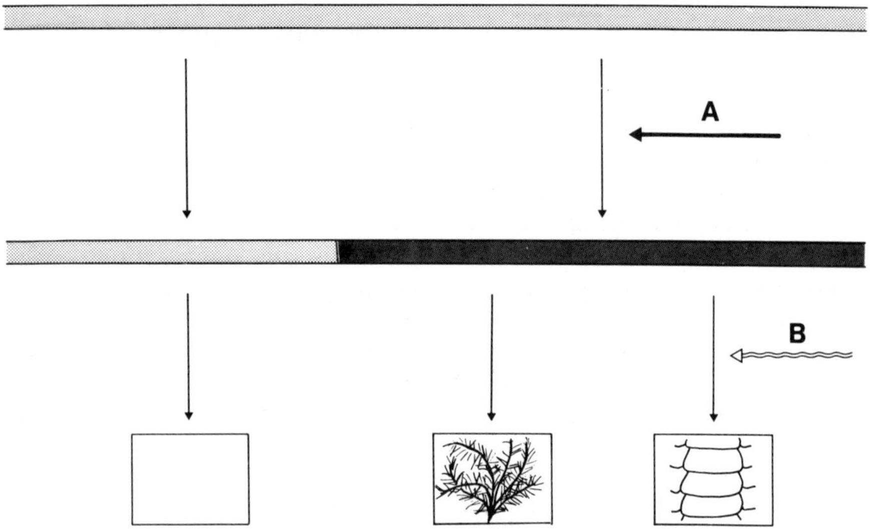

FIGURE 2. Inductive events in (bird) skin development. In order to develop cutaneous appendages, embryonic epidermis requires an inductive stimulus coming from the underlying dermis. This signal (*A*) is provided by dermis of certain regions (of future feather tracts, but also of the scale-forming foot), and it can even be provided by mammalian dermis. It is therefore not considered to represent an order specifying "make feathers," but rather: "make appendages (any kind)" (15). In the absence of further modifying directives, the epidermis will proceed to form feathers. In contrast, for the development of other cutaneous appendages, such as the scales shown in this figure, the epidermis requires a specific inductive signal (*B*), again provided by regionally specific dermis (in this case of the 12-day or older foot). As Sengel and Dhouailly put it: "bird epidermis forms feathers, unless told otherwise" (16).

lage. Both processes are late events and in both the responding tissue has already developed far in the direction of its final "induced" differentiation before coming under the influence of the inductor.

2.2.1. Induction of Kidney Tubules. Development of the amniote kidney begins when the ureteric bud grows into the metanephrogenic mesenchyme. There, the ureter branches and gives rise to the pelvis and the collecting tubules of the kidney. In contrast, the proximal kidney tubules and glomeruli are derived from the metanephric mesenchyme. Proximal and collecting tubules eventually connect to produce the integrated duct system required for the excretory function of the organ.

Proximal tubules develop precisely at the tips of the ureter branches, and experiments have shown that tubule formation and subsequent development is in fact induced by the ureter (22–24) (see also Fig. 5). This tubule induction is characterized by a remarkable lack of specificity: a great variety of embryonic tissues were found to be active as inductors (24–26). In fact, spinal cord has now virtually replaced the natural inductor in experiments done in this system.

Although formal experimental proof is not available (and may be difficult to obtain), it can be assumed that all metanephric mesenchymal cells are competent to form tubules. The mesenchyme also shows a strong "bias" for this differentiative pathway by its ready response to a variety of heterotypic and rather unrelated tissues. Nevertheless, unlike the situation found in lens differentiation in some species, no "spontaneous" tubule development has been reported and the requirement for an inductor persists despite this strong bias. This lack of specificity can be tolerated, since no tissue other than the ureter normally invades metanephric mesenchyme. It is more important that induced tubuli form in close proximity to the ureter-derived collecting ducts, and it therefore makes sense that the final step in mesenchymal differentiation is delayed until the cells come in contact with the ureter.

2.2.2. Induction of Cartilage. Cartilage induction is equally regarded as an event that promotes and stabilizes a preexisting developmental pathway rather than evoking a new one. "Preinduced" prospective cartilage cells possess the enzymes specifically required for chondroitin sulfate synthesis (27, 28), and inductor-independent chondrogenesis can be obtained by improving the culture medium (29) or by in ovo explantation (30).

Cartilage of apparently identical biochemical properties develops in various sites in the embryo. Chondrogenesis of the somitic sclerotome is induced by the notochord and spinal cord (31–33), and in limbs, the apical ectoderm has been identified as the inductor (34, 35). With this similarity in the induced cellular phenotype and the preexistence of the characteristic biosynthetic pathways, one might expect that cartilage inductors are interchangeable between different systems. They are not: Whereas limb ectoderm is capable of inducing somite chondrogenesis, notochord—the most potent and reliable inductor in somites—is ineffective in limb mesoderm (35, 36). In another such pair, notochord induces cartilage in ear mesoderm, whereas in the reciprocal combination, the otic vesicle fails to induce somites (37). The surprising conclusion emerges that different types of chondrogenic mesenchymes, although appearing equally biased for cartilage formation before induction, nevertheless require very specific and different conditions (inductors) to achieve the same type of cytodifferentiation (Fig. 3).

2.3. Induction of Endodermal Differentiation

Although the primitive gut gives rise to a variety of different organs (derivatives of the pharyngeal pouches, lung, liver, pancreas), it is not known in any single case whether inductive events select the cell population for the endo-

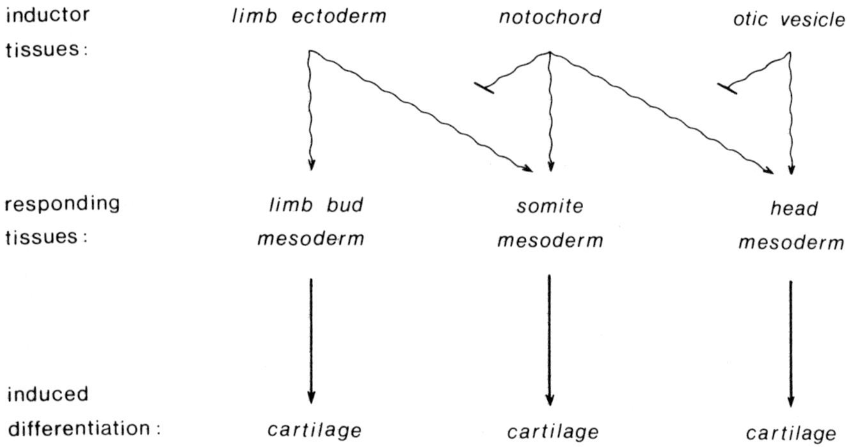

FIGURE 3. Test for interchangeability of cartilage inductors of various chondrogenic systems. Shown are three types of mesenchyme capable of forming cartilage under the influence of their respective inductors: limb mesoderm (forms skeletal cartilage), somite mesoderm (vertebrae), and head mesoderm (ear capsule). Although the induced differentiation appears the same in all three cases ("cartilage"), the inductors are not fully interchangeable: notochord fails on limb mesoderm and the otic vesicle cannot induce somite mesoderm. This situation argues against a specific "cartilage" signal in induction, but suggests that different tissues require different conditions to express their developmental potential.

dermal component of such an organ. Initial regionalization and diversification of the endoderm apparently occurs very early in the embryo and perhaps independently of inductive tissue interactions. In contrast, the importance of endodermal-mesodermal interactions in later organogenesis is well recognized (for review see reference 38).

3. NATURE OF INDUCTIVE SIGNALS AND THEIR TRANSMISSION

3.1. Attempts to Isolate Inducer Substances

Most of the work on attempting to determine the molecular nature of inducers has been performed on amphibian primary induction (for review see references 39–41). Formation of the medullary plate by the ectoderm and regionalization of the central nervous system seems to result from a two-step induction by the underlying archenteron roof (42–44). Although this conclusion is mainly based on results obtained with nonnatural (*heterotypic*) inductors, it is considered to reflect the events taking place in the gastrula (Fig. 4). In the first step, the ectoderm becomes "neuralized" and, without further modification, this induction yields forebrain (*archencephalic*) structures only.

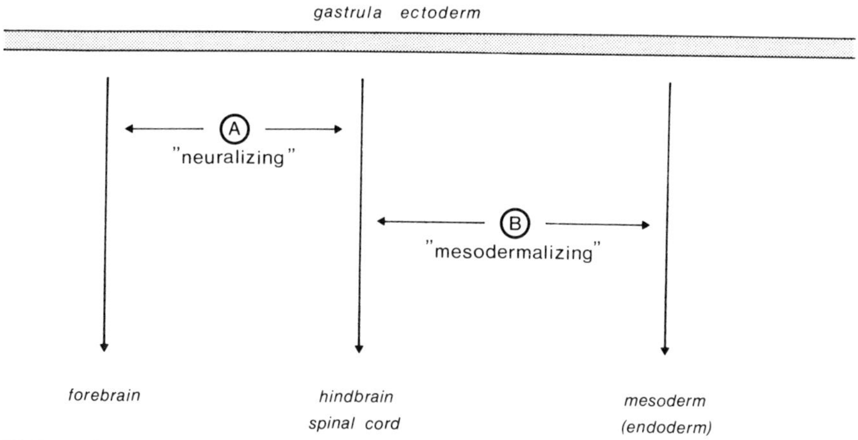

FIGURE 4. Highly simplified representation of the two-step hypothesis of neural plate induction. Two inductive influences are postulated, the first (*A*) causing uniform neuralization of the ectoderm. Cranio-caudal regionalization of the central nervous system is then brought about by an anterior-posterior gradient of a second inductive influence (*B*), which by itself causes mesodermal differentiation.

Development of hindbrain and spinal cord is due to the additional action of a second inductive influence, the strength of which determines increasing caudal characteristics of the CNS. In the absence of the first "neuralizing" influence, this second induction evokes mesodermal and even endodermal differentiation in the ectoderm. It is therefore referred to as either a "mesodermalizing" (45), or a "vegetalizing" (46) factor.

A neuralizing activity (A in Fig. 4) was detected in virtually all tissues tested, but so far attempts to purify it have not been successful. The activity withstands short heat treatment [e.g., 70°C for HeLa cells (43)] but is susceptible to proteases (46). A mesodermalizing activity was purified from 11-day to 13-day chick embryos; the active fraction has a molecular weight of about 30,000 daltons and its activity is destroyed by proteases and heating (46). The mesodermalizing effect of HeLa cells is also abolished by heating (43).

It is obvious that attempts to isolate and purify potential inducer substances require quantities of tissue that can only be obtained from such "heterotypic" sources, but it may be questioned how much the active fractions have in common with molecules actually operating in primary induction. It is certainly not impossible that molecules regulating embryonic development are also produced by adult tissues [such a situation, for instance, has allowed the isolation of the nerve growth factor (47)], but then the neuralizing agent would have to be a rather ubiquitous substance. It seems that this question will need to await antibodies against these fractions from heterotypic sources, which can then be tested for their ability to interfere with the process of neural plate induction. To date, not a single active cell-free fraction has been obtained from any "natural" inductor.

3.2. Transmission Characteristics

Different transmission characteristics have been described for various inductive events. In primary induction, the "neuralizing" activity appears to be due to diffusible substances that can travel considerable distances, pass through Nuclepore filters of 0.1 and 0.05 μm pore size (excluding cell contacts, see below), and even through dialysis membranes that prevent the passage of molecules over 12,000 daltons (48, 49). These experiments were performed with natural inductors. A possible reason for concern, however, is the extreme readiness of gastrula ectoderm to become neuralized, especially by treatment causing slight tissue damage. This is reflected in occasionally high control values (i.e., neuralization without inductor; cf. reference 49). In contrast, the mesodermalizing activity does not pass dialysis membranes or Nuclepore filters that prevent cell contact, and therefore this process is considered to be mediated by direct cell-to-cell interaction (50).

Passage through dialysis membranes was also reported for the lens-inducing activity of the optic vesicle in chick embryos (51). Karkinen-Jääskeläinen was careful to use trunk ectoderm as the responding tissue to avoid ambiguities caused by spontaneous lens differentiation independent of the optic vesicle.

Interposing mechanical barriers between inductor and responding tissue has become the classical test for the need of cell contacts after Grobstein (52, 53) introduced the use of thin Millipore filters with a nominal pore size of 0.35 or 0.45 μm (Fig. 5). He, and later others, found that any known interactive effect would pass the filter (54). The transmission characteristics, however, could not be reconciled with free diffusion (see also reference 55). There was a strict limitation in the range of interaction between 50 and 100 μm, depending on the system. Since the Millipore filter was originally thought to prevent cell contacts, Grobstein proposed a third mechanism of tissue interaction, mediated by cell-surface-associated matrix material (5). The question of transfilter cell contacts was later reopened by Saxén and his group using Nuclepore filters with straight channels that allowed easier observation of their penetration by cell processes (56). In the spinal cord/kidney mesenchyme combination, transfilter cell contacts were detectable with pore sizes of 0.2 μm and larger, and these filters also allowed tubule induction. With smaller pores, no cell contacts and no induction were found. This result suggests that kidney tubule development depends on actual cell contact with

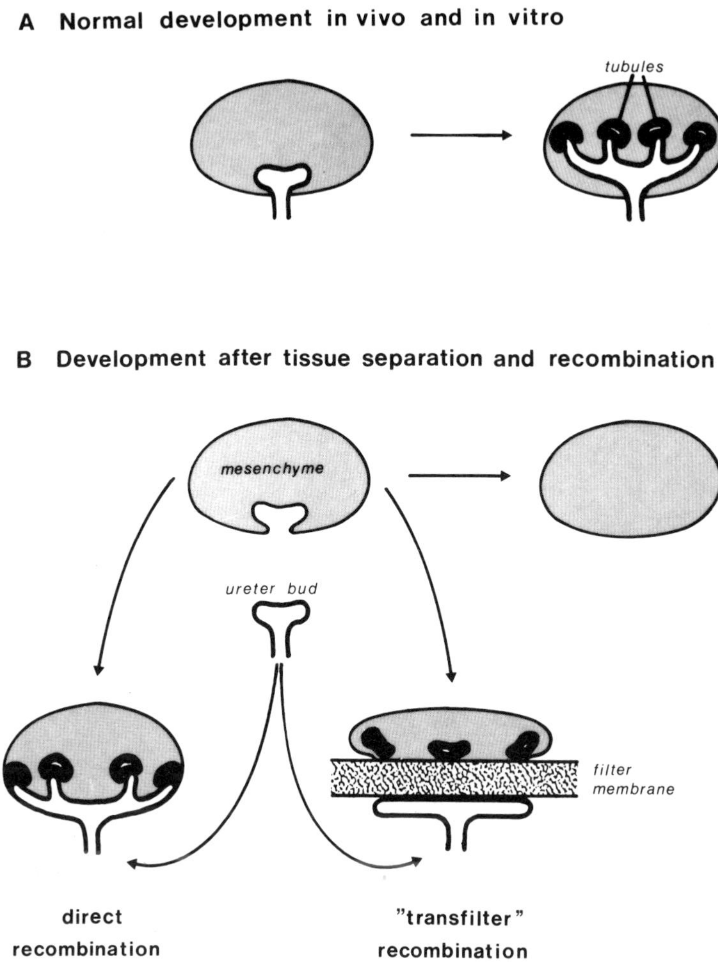

FIGURE 5. Direct and "transfilter" association of inductor and responding tissue, illustrated in one of the best analyzed systems, the kidney. (*A*) In normal development, the ureter bud grows into the metanephrogenic mesenchyme, branches, and induces the formation of proximal tubules in the mesenchyme. (*B*) Ureter bud and kidney mesenchyme can be separated, usually with the aid of trypsin, and then recombined again. The two tissues can be brought into direct contact (*left*), or they are glued opposite each other to a porous filter membrane. In successful recombinations, tubules will appear after two days in culture. The Millipore filter used in the original experiments was about 25 μm thick and had a nominal pore size of 0.35 or 0.45 μm. Because of its foamy structure, it was impossible to establish whether cell processes penetrating from either side made contact within the filter matrix. In contrast, the thinner (15 μm) Nuclepore filter is perforated by straight channels, which allow easier observation of cell processes. Using such filters of varying pore size, a strong correlation between penetration of the filter by cell processes and positive transfilter induction has been found in the kidney and also in the tooth system. (The drawing is schematic and not to scale. In the actual experiments, the ureter was replaced by a "heterotypic" inductor, the spinal cord.)

the inductor (see also reference 57). Later, cell contacts were also found in the developing tooth rudiment between odontoblasts and ameloblasts, via cell processes extending across a rather thick layer of matrix material (58). As the recent work employing Nuclepore filters and dialysis membranes in neural plate, lens, and kidney induction was all done in the same laboratory, it can be assumed that the different results obtained reflect genuine differences among these inductive systems.

The best evidence for the role of extracellular matrix material in the regulation of the developmental behavior of cells comes from "late" systems of organogenesis, involving the interaction of already committed tissues, and here the most intensive research is done in epithelio-mesenchymal organs where the extracellular material separating the interacting tissues is present in the form of a highly regular structure, the basal lamina. The role of this lamina in epithelial morphogenesis is discussed by Bernfield and Banerjee (59).

In summary, there is probably no single, general mechanism for the transmission of inductive signals. (For a more detailed discussion on the transmission problem see reference 50.)

4. CONCEPTS

4.1. Attempts to Classify Inductive Events

Even more important than the question of the nature of inductive signals is the central problem of the nature of the effect of induction on the responding cell. Clearly, by selecting part of a competent population of cells for a given developmental pathway, the inductor also affects the choice of a developmental alternative within these cells. The basic question is: To what extent is this induced development an expression of the cell's intrinsic program (or even bias)—with the inductor operating a sort of developmental switch—and to what extent is this particular development imposed on a receptive and malleable tissue by an "instructive" action of the inductor?

The dramatic results obtained in Spemann's transplantation experiments (60) initially favored the instructive hypotheses of induction. When the dorsal lip of an early gastrula was grafted to another site, it invaginated there to give rise to all mesodermal dorsal structures—notochord, somites, nephric elements—and in addition it also induced the overlying ectoderm to form a neural plate. Since transplant-derived and induced structures formed in perfect spatial and temporal coordination, the experiment eventually resulted in the development of an almost complete secondary embryo. The grafted piece was thought to hold the instructions for an entire embryo (the "organizer" concept) and this idea led to a long and disappointing search for responsible substances and molecules. The apparent lack of specificity of neuralizing factors or manipulations, together with data obtained from other inductive systems, has since discredited such naive instructive hypotheses.

Several attempts have been made to classify inductive systems into *instructive* and *permissive* events, but all of them remain unsatisfactory because of inadequate criteria (see discussion in Chapter 4 of reference 61). We speak of induction when a specific development of one tissue occurs only in the presence of another tissue (the inductor) but not in its absence. This criterion certainly applies to permissive as well as instructive types of tissue interaction. The term instructive was proposed for inductions that fulfill two more criteria: (1) The responding tissue does not develop in the specific way in response to nonspecific stimuli. (2) The inductor must be able to induce that specific development in another tissue that is normally destined for an alternative differentiative pathway. The first criterion is unsatisfactory, since it is easily imaginable that for certain tissues, only very specific conditions allow the expression of a preexisting developmental program (cf., cartilage induction). It is also obvious that a test for the second of these criteria depends on the competence of other tissues (i.e., on the presence of the required developmental potential). The main objection to the term instructive induction is that it implies the transfer of informational molecules from inductor to responding tissue, for which there is absolutely no experimental evidence available.

More recently, Saxén has proposed two other categories of inductive events: *directive* and *permissive.* He defines directive events as inductions that affect the choice between two (or more) developmental options of the responding tissue. In contrast, in a permissive event "the inductor acts upon a tissue that is already determined toward its final fate, but which still requires an exogenous stimulus for the expression of its phenotype" (62). In this form, the criteria for directive versus permissive induction are not mutually exclusive and therefore do not seem useful for a classification: even a cell that needs the inductive signal merely for the expression of a predetermined developmental program does something else in its absence. Inevitably then, even a permissive event effects a developmental choice. (According to the proposed definition, permissive inductions could at best be detected in artificial experimental situations where one tissue has been deprived of its normal partner and therefore fails to develop at all. This nondevelopment must be without counterpart in embryogenesis, because then it would already constitute an alternative development.) The difficulties stemming from a strict application of Saxén's criteria are perhaps best seen in the fact that they are not used in his own classification of kidney tubule induction. Although the event clearly involves a developmental choice (tubules/nontubule mesenchyme), it is regarded as a typical example of a permissive induction on different grounds: low inductor specificity and failure of the ureter to induce tubules in nonkidney mesenchyme (62, 63).

It would seem that, given our present state of knowledge, attempts to define and classify inductive events are somewhat premature and of little heuristic value. In the absence of any molecular understanding of the process of developmental decision, and of the basis of developmental option or determination, our thinking about induction must be based on purely phenomenological information and it essentially relies on two testable properties of inductive systems: (1) the competence of the responding tissue and (2) the requirement for inductor specificity.

4.2. Competence

Competence describes the ability of a tissue to respond to induction. It can be tested by exposure to the relevant inductor, but its molecular basis is far from understood. We have seen that the competent area (or cell population) is larger than the region (cell number) that will actually undergo this particular development. Competence, however, is never unlimited, neither topographically nor in time. Not all tissues of a given embryonic stage are competent for a specific inductor, and those that are can respond only at a certain stage. Competence must be acquired before the time of induction, and it is lost again, usually soon thereafter. Blastomeres, even those that will give rise to the ectoderm, cannot be induced directly to form a medullary plate, and neither can body ectoderm after neurulation. Cells at any stage of development have a restricted developmental potential, and models calling for a gradual restriction of developmental potentials during embryogenesis are misleading if they are applied to cells rather than their (diversifying) progeny (see 64). The presence of the relevant developmental potential is therefore a prerequisite for induction, but the need for the inductor shows that the cell also has the potential for at least one alternative pathway of development. The minimum that the inductor must contribute is to influence a pending decision between these alternatives in a yes/no manner, and, in fact, there are only few indications that it ever does more than that. In the terminology of the preceding paragraphs, we would say that the inductor permits the expression of one developmental option to the exclusion of the other. Only if a cell has more than two options will the inductor have to play a more specific role.

4.3. Inductor Specificity

The main body of information on the requirement for inductor specificity is also compatible with a minimalistic view of induction. Of course, even the strictest requirement for unique inductor tissues would not contradict the concept, since it can be argued (and in certain cases has been shown) that even committed cells can express their phenotypic differentiation only under very specific extracellular conditions. On the other hand, the frequent lack of inductor specificity constitutes a strong case against an instructive influence, in the sense that the inductor specifies the developmental course of a naive target cell.

Among all the instances of inductive tissue interaction affecting a developmental decision in the responding tissue, inductor specificity is the exception rather than the rule. It is hard to imagine that all the various tissues, extracts, and manipulations that lead to neuralization do so by a message whose specific function is instructing the ectoderm to form a neural plate. It also does not seem likely that pharynx endoderm, heart mesoderm, and optic vesicle, three tissues derived from all three different germ layers, produce the same inductive signal specifying lens differentiation. Spinal cord and the great variety of other tissues probably do not release a signal specifically calling for kidney tubule formation. Finally, in association with chicken epidermis, mouse dermis cannot be assumed to specify the induced structure, since the combination develops feathers.

In addition, the finding that cartilage inductors are only partly interchangeable among various chondrogenic systems—though causing the same type of differentiation in tissues already predisposed—also argues against the idea that inductive signals represent specific orders for the development induced.

A crucial finding that will require a modification of this extreme permissive view of induction would be a tissue that can be shown to have more than two developmental options that are expressed under the influence of different inductors. Such a situation may possibly be found in the development of the epidermis, with its wide repertoire of differentiative potentials. As far as feather (or hair) formation is concerned, induction of the epidermis could affect a simple yes/no decision for a preexisting developmental program, as in other systems. However, in the induction of nonfeather (nonhair) structures, it is difficult to avoid the conclusion that the dermis specifies rather than permits epidermal development (Fig. 2). It is therefore very important to know whether an epidermal cell has, in fact, more than two options, or whether these multiple potentialities of the tissue can only be achieved by a series of binary decisions by cells with intervening cell division. Most theoretical models are based on strictly binary decisions (cf. references 64 and 65), but it should be remembered that this is still a theoretical, though appealing, postulate.

Certainly, the greatest difficulties for the permissive concept are created by the very rare reports of inductor-specific effects in tissues not even expected to possess the corresponding developmental potential: the induction of mucous secretion in the epidermis by stomach mesenchyme (20), the formation of an enamel organ by foot epidermis in response to dental mesenchyme (21), or the induction of cartilage in dental mesenchyme by limb ectoderm (66). Clearly, such cases deserve further investigation as they seem to argue for a much more specific role of the inductor than could be deduced from most classical systems.

4.4. Developmental Decisions

It is obvious that our present phenomenological knowledge of the variety of inductive systems does not yet allow a consensus on a unifying concept. In general, however, we have become increasingly aware of the importance of the developmental status of the responding cell, its developmental potential, and the need to make developmental decisions. A number of observations, not only in induction, have contributed to the impression that early embryonic cells rush through a rather rigid temporal developmental program until they have reached a state of final commitment. The times at which decisions between developmental alternatives must be made [*epigenetic crises* (65)] seem to be determined rather precisely. Such decisions probably cannot be delayed, postponed, or reversed [except for the nucleus alone in certain transplantation experiments (67)]. It may be quite significant that pluripotential embryonic cells, that is, cells whose progeny give rise to various differentiated cell types (e.g., blastomeres), cannot be maintained in this state as a cell line in vitro. "Undifferentiated" normal cells propagated in culture are actually cells committed to a particular differentiation without expressing its phenotype. (Exceptions are the persistence of pluripotential stem cells in hematopoiesis, and embryonal teratocarcinoma cells; see Section 5.1.)

It is the characteristic of the regulative type of development that embryonic cells, in making these decisions between differentiative alternatives, are susceptible to external cues. The period of responsiveness to external influences may vary in length (cf., lens), but the signal must be received at the proper time. The pending decision will even be made in the absence of the inductor and this will then be reflected in the loss of competence. In other words, inductions do not bring about decisions, they influence them. This function can be fulfilled without obvious need for a high specificity of such external influences, particularly not in binary decisions. To secure the necessary architectural precision in development, presumably the principal function of inductive tissue interactions (Section 1), a short range of inductive influences would seem to be much more relevant than any specificity.

5. PROSPECTS

5.1. Effect of Induction on the Target Cell

From the previous discussion, it is apparent that our most important area of ignorance concerns the molecular basis of a cell's developmental potential, bias toward a particular phenotype (if that exists), and finally commitment to a given differentiative pathway. There are still too many possibilities open for speculation but progress made in recent years has increased the hope that some models, especially those based on differential transcription and/or hnRNA processing (e.g., reference 68) can be put to test within reasonable time. This would, of course, require work on homogeneous populations of embryonic cells, rather than on entire embryos, and a particular cell type would have to be studied at consecutive stages of its development. Since we must assume that the same developmental decision is made only in a small group of cells, it seems that tremendous technical difficulties still remain to be overcome. Nevertheless, there is no other obvious way in which to learn how a cell makes developmental decisions, how these decisions are related to the division cycle, and how they can be influenced by extracellular factors. At the moment we can only say that we have no evidence for specific development program selector genes in vertebrates, as suggested in insects by the phenomena of homoeotic mutation and of transdetermination (69).

Even with more molecular understanding of the gradual progress of cell differentiation in embryogenesis, work on embryonic induction will probably still be difficult due to the small numbers of cells affected and the lack of adequate alternative experimental systems. The ideal system in which to study cellular events in response to induction would be bi- or multipotential cell lines receptive to inducing signals. With one possible exception, no such material is available now, and since this fact may be a direct consequence of the characteristics of the process of cell differentiation per se (as discussed in Section 4.4), there is little hope that one will ever be.

The one exception are teratocarcinoma cells, and recent work suggests that they might indeed be useful for the study of inductive tissue interaction. These malignant transplantable mouse tumors contain stem cells (*embryonal carcinoma cells*) whose virtually unlimited developmental repertoire makes them comparable to very early embryonic cells. They also share antigenic properties with epiblast cells of 5½-day blastocysts (70, 71) and most recently such cell lines were established directly from in vitro cultures of mouse blastocysts (72). Embryonal carcinoma cells may therefore provide a material to study primarily the earliest developmental cell interactions in the mammalian embryo. One such cell line was induced in vitro to synthesize marker proteins for visceral or parietal endoderm, respectively (73). This is a promising result in view of the finding that the differentiation of the primitive endoderm in vivo is regulated by tissue interaction (74, 75).

If embryonal carcinoma cells indeed correspond to early embryonic cells in terms of their developmental status and potential, they cannot be expected to respond directly to classical inductors in vitro, as all of them act only at later stages. When, however, these cells are implanted into blastocysts, they give rise to differentiated progeny in almost all tissues of the body (76, 77).

The most likely explanation for this result is that embryonal carcinoma cells develop in accordance with neighboring cells in the embryo, thus showing that they do respond to extracellular signals, at least when received in proper sequence between the stages of the blastocyst and of organ formation.

An obvious and consistent early phenomenon accompanying induction is a change in the shape and/or intercellular relationship of the induced cells. Epithelial cells usually elongate to form a high columnar epithelium (neural plate, lens, nose, ear, enamel organ), or at least form distinctive placodes (skin); mesenchymal cells also change their mutual adhesiveness (kidney tubules) or arrange themselves in characteristic patterns (frequently seen in cartilage). Changes in cell membrane properties correlated with differentiation have been noted for years, but only more recently has cellular interaction via membrane-bound molecules (the H-Y antigen) and membrane receptors been directly implied as the mechanism for the morphogenetic and differentiative decision in gonad development (78). The notion that a cell's shape and cytoskeletal organization could affect its differentiative state is also relatively new. It is emphasized that the changes in cell morphology occur early, at the time of commitment, and well before full phenotypic biochemical differentiation. They could therefore be part of the mechanism by which an extracellular signal is transmitted to the nucleus, and may provide a promising area of investigation.

5.2. Inductive Signals and Intercellular Transmission Mechanism

To evaluate our chances of success in this area of research, it may be worthwhile to compare induction to better-known systems of tissue interaction relying on hormones. There is no finding that forbids such a comparison in principle. In fact, the old model of complementary surface molecules can be regarded as a hormone-receptor interaction, with both molecules linked to the cell membrane. Hormones, too, act only on competent target cells, and cells respond according to their developmental or physiological potential (resulting in different responses of different cells to the same hormone). Hormones have even found a way for transmembrane signalling of their message, although the need for such a mechanism is not as apparent for small and diffusible molecules as it would seem to be for cell interactions.

Inductive tissue interaction is distinct from classical hormone-mediated interaction mainly by its short range and relative lack of specificity. The short range might be explained by a tighter linkage of the signal molecules to the inductor tissue (like peptide hormones bound to Sepharose beads) and a lack of specificity in induction can be afforded just because of this short range of its influence. Only when a variety of signal molecules diffuse widely within the organism, as is the case with hormones, are specific receptors required, first to make a cell their target, and second, in case this cell responds to more than one hormone, to allow it to distinguish between different messages. After specific receptor-mediated hormone recognition, cells then utilize a transmembrane and intracellular signal transfer mechanism common to many different hormones (79). In contrast, a cell responding to induction is a target because of its proximity to the inductor, and in most cases it does not have to distinguish between different inductive signals. It is therefore imaginable that it can do without the highly specific receptor in the signal interpretation mechanism.

The isolation and characterization of hormones has been aided by the fact that they can be extracted from adult-sized, enriched sources and from blood, that they are water-soluble, and that suitable assay systems are available both in vivo and in vitro. Even precise knowledge of their molecular structure has not per se elucidated their mechanism of action, but the availability of purified and labeled hormones has permitted the detection of their receptors. Work on cell-surface receptors has progressed more slowly, since they are present only in low concentrations and not found in blood. In addition, functional receptors must be embedded in the plasma membrane. Still, the mechanism of hormone action can be studied in target organs of comfortable size and, in addition, the research has greatly benefited from the use of homogeneous cell lines in vitro (cf., 80).

None of these favorable conditions prevails in the search for inductive signals and in the study of their receptors. Not only is the number of cells producing and receiving inductive signals exceedingly small (even by comparison with the hypothalamus—and what effort had to be invested to isolate its releasing hormones!), but both signal and receptor will probably be bound to, or at least associated with, the cell membrane. If the apparent lack of specificity in induction is a characteristic of the signals and/or their receptors, the difficulties in detecting either one will be further enormously increased. We must even be prepared to find that, in a strict sense, neither inducer molecules nor receptors exist, but that a cell is influenced by variable compositions of its environment (e.g., Chapters 6, 8, and 9).

This comparison shows that even in the most favorable case, when the mechanism of induction holds no surprise and we can apply experience and knowledge from the study of hormone action, a molecular understanding of embryonic induction is not within easy reach. Its analysis will require a great amount of imagination.

REFERENCES

1. H. Spemann, *Verh. Anat. Ges.* **15**, 61 (1901).
2. U. Deppe, E. Schierenberg, T. Cole, C. Krieg, D. Schmitt, B. Yoder, and G. von Ehrenstein, *Proc. Natl. Acad. Sci. USA* **75**, 376 (1978).

3. J. E. Sulston and H. R. Horvitz, *Dev. Biol.* **56**, 110 (1977).
4. L. Wolpert, Positional information and pattern formation, in A. A. Moscona and A. Monroy, eds., *Current Topics in Developmental Biology*, Vol. 6, Academic Press, New York, 1971, p. 183.
5. C. Grobstein, *Natl. Cancer Inst. Monogr.* **26**, 279 (1967).
6. V. Twitty, Eye, in B. H. Willier, P. A. Weiss, and V. Hamburger, eds., *Analysis of Development*, W. B. Saunders, Philadelphia, 1955, p. 402.
7. J. W. McAvoy, *Differentiation* **17**, 137 (1980).
8. T. Mizuno, *C. R. Hebd. Séanc. Acad. Sci. Sér. D* **271**, 2190 (1970).
9. T. Mizuno, *J. Embryol. Exp. Morphol.* **28**, 117 (1972).
10. M. Karkinen-Jääskeläinen, *J. Embryol. Exp. Morphol.* **44**, 167 (1978).
11. T. Mizuno and Y. Katoh, *Proc. Jpn. Acad.* **48**, 522 (1972).
12. A. C. Jacobson, *Science* **152**, 25 (1966).
13. M. S. McKeehan, *J. Exp. Zool.* **117**, 31 (1951).
14. M. E. Rawles, *J. Embryol. Exp. Morphol.* **11**, 765 (1963).
15. P. Sengel, *Morphogenesis of Skin*, Cambridge University Press, London, 1976.
16. P. Sengel and D. Dhouailly, Tissue interactions in amniote skin development, in M. Karkinen-Jääskeläinen, L. Saxén, and L. Weiss, eds., *Cell Interactions in Differentiation*, Academic Press, London, 1977, p. 153.
17. D. Dhouailly, *J. Embryol. Exp. Morphol.* **30**, 387 (1973).
18. J. L. Coulombre and A. J. Coulombre, *Dev. Biol.* **25**, 464 (1971).
19. D. Dhouailly and P. Sengel, *C. R. Hebd. Séanc. Acad. Sci. Sér. D* **281**, 1007 (1975).
20. C. B. McLoughlin, *J. Embryol. Exp. Morphol.* **9**, 385 (1961).
21. E. J. Kollar and G. R. Baird, *J. Embryol. Exp. Morphol.* **24**, 173 (1970).
22. C. Grobstein, *J. Exp. Zool.* **130**, 319 (1955).
23. L. Saxén, O. Koskimies, A. Lahtii, H. Miettinen, J. Rapola, and J. Wartiovaara, Differentiation of kidney mesenchyme in an experimental model system, in M. Abercrombie, J. Brachet, and T. King, eds., *Advances Morphogenesis*, Vol. 7, Academic Press, New York, 1968, p. 251.
24. L. Saxén, P. Ekblom, and E. Lehtonen, The kidney as a model system for determination and differentiation, in M. Ritzen and A. Larsson, eds., *Biology of Normal Human Growth*, Raven Press, New York, 1981, p. 117.
25. M. Lombard and C. Grobstein, *Dev. Biol.* **19**, 41 (1969).
26. B. Unsworth and C. Grobstein, *Dev. Biol.* **21**, 547 (1970).
27. J. W. Lash, Somitic mesenchyme and its response to cartilage induction, in R. Fleischmajer and R. E. Billingham, eds., *Epithelial-Mesenchymal Interactions,* Williams and Wilkins, Baltimore, 1968, p. 165.
28. J. Medoff, *Dev. Biol.* **16**, 118 (1967).
29. M. L. Ellison, E. J. Ambrose, and G. C. Easty, *J. Embryol. Exp. Morphol.* **21**, 331 (1969).
30. B. Christ, *Z. Anat. Entwicklungsgesch.* **129**, 177 (1969).
31. C. Grobstein and G. Parker, *Proc. Soc. Exp. Biol. Med.* **85**, 477 (1954).
32. G. Avery, M. Chow, and H. Holtzer, *J. Exp. Zool.* **132**, 409 (1956).
33. J. W. Lash, S. Holtzer, and H. Holtzer, *Exp. Cell Res.* **13**, 292 (1957).
34. J. Milaire and M. Mulnard, *J. Embryol. Exp. Morphol.* **20**, 215 (1968).

35. M. Gumpel-Pinot, *C. R. Hebd. Séanc. Acad. Sci. Sér. D* **274**, 2786 (1972).

36. O. Luger, Ph.D. Thesis, Vienna (1970).

37. J. A. A. Bénoit, *J. Embryol. Exp. Morphol.* **8**, 39 (1960).

38. K. Kratochwil, Tissue interaction during embryonic development: general properties, in D. Tarin, ed., *Tissue Interactions in Carcinogenesis*, Academic Press, London, 1972, p. 1.

39. L. Saxén and S. Toivonen, *Primary Embryonic Induction*, Academic Press, New York, 1962.

40. L. Saxén and J. Kohonen, Inductive tissue interactions in vertebrate morphogenesis, in G. W. Richter and W. A. Epstein, eds., *International Review of Experimental Pathology*, Vol. 8, Academic Press, New York, 1969, p. 57.

41. L. Saxén, Neural induction: past, present, and future, in A. A. Moscona and A. Monroy, eds., *Current Topics in Developmental Biology*, Vol. 15, Part I, Academic Press, New York, 1980, p. 409.

42. S. Toivonen and L. Saxén, *Exp. Cell Res. Suppl.* **3**, 346 (1955).

43. L. Saxén and S. Toivonen, *J. Embryol. Exp. Morphol.* **9**, 514 (1961).

44. L. Saxén, *Med. Biol.* **56**, 293 (1978).

45. H. Tiedemann, The molecular basis of differentiation in early development of amphibian embryos, in A. A. Moscona and A. Monroy, eds., *Current Topics in Developmental Biology*, Vol. 1, Academic Press, New York, 1966, p. 85.

46. H. Tiedemann, *J. Embryol. Exp. Morphol.* **3**, 437 (1976).

47. R. Levi-Montalcini and P. Calissano, *Sci. Am.* **240**, 44 (1979).

48. S. Toivonen, D. Tarin, L. Saxén, P. J. Tarin, and J. Wartiovaara, *Differentiation* **4**, 1 (1975).

49. S. Toivonen, *Differentiation* **15**, 177 (1979).

50. L. Saxén, P. Ekblom, and J. Thesleff, Mechanisms of morphogenetic cell interactions, in M. H. Johnson, ed., *Development in Mammals*, Vol. 4, Elsevier/North-Holland, Amsterdam, 1980, p. 161.

51. M. Karkinen-Jääskeläinen, *Differentiation* **12**, 31 (1978).

52. C. Grobstein, *Nature (London)* **172**, 869 (1953).

53. C. Grobstein, *Exp. Cell Res.* **10**, 424 (1956).

54. L. Saxén, Interactive mechanisms in morphogenesis, in D. Tarin, ed., *Tissue Interaction in Carcinogenesis*, Academic Press, London, 1972, p. 49.

55. S. Nordling, H. Miettinen, J. Wartiovaara, and L. Saxén, *J. Embryol. Exp. Morphol.* **26**, 231 (1971).

56. J. Wartiovaara, S. Nordling, E. Lehtonen, and L. Saxén, *J. Embryol. Exp. Morphol.* **31**, 667 (1974).

57. L. Saxén and E. Lehtonen, *J. Embryol. Exp. Morphol.* **47**, 97 (1978).

58. H. C. Slavkin and P. Bringas, *Dev. Biol.* **50**, 428 (1976).

59. M. R. Bernfield and S. D. Banerjee, Basal lamina in epithelial–mesenchymal morphogenetic interactions, in N. A. Kefalides, ed., *Biology and Chemistry of Basement Membranes*, Academic Press, New York, 1978, p. 137.

60. H. Spemann and H. Mangold, *Arch. Entwmech. Org.* **100**, 599 (1924).

61. N. K. Wessells, *Tissue Interactions and Development*, W. A. Benjamin, Menlo Park, California, 1977.

62. L. Saxén, Directive versus permissive induction: a working hypothesis, in J. W. Lash and M. M. Burger, eds., *Cell and Tissue Interactions*, Raven Press, New York, 1977, p. 1.

63. L. Saxén, Morphogenetic tissue interactions: an introduction, in M. Karkinen-Jääskeläinen, L. Saxén, and L. Weiss, eds., *Cell Interactions in Differentiation*, Academic Press, London, 1977, p. 145.

64. H. Holtzer, Cell lineages, stem cells and the "quantal" cell cycle concept, in B. J. Lord, C. S. Potten, and R. J. Cole, eds., *Stem Cells and Tissue Homeostasis*, Cambridge University Press, London, 1978, p. 1.

65. W. D. Stein, *J. Theor. Biol.* **82**, 663 (1980).

66. R. J. Hata and H. C. Slavkin, *Proc. Natl. Acad. Sci. USA* **75**, 2790 (1978).

67. J. B. Gurdon and H. R. Woodland, On the long-term control of nuclear activity during cell differentiation, in A. A. Moscona and A. Monroy, eds., *Current Topics in Developmental Biology*, Vol. 5, Academic Press, New York, 1970, p. 39.

68. E. H. Davidson and R. J. Britten, *Science* **204**, 1052 (1979).

69. W. Gehring, The stability of the determined state in cultures of imaginal disks in Drosophila, in W. Beermann, J. Reinert, and H. Ursprung, eds., *Results and Problems in Cell Differentiation*, Vol. 5, Springer, Berlin, 1972, p. 35.

70. M. J. Evans, R. H. Lovell-Badge, P. L. Stern, and M. G. Stinnakre, Cell lineages of the mouse embryo and embryonal carcinoma cells; Forssman antigen distribution and patterns of protein synthesis, in N. LeDouarin, ed., *Cell Lineage, Stem Cells and Cell Determination*, INSERM Symposium 10, Elsevier/North-Holland, Amsterdam, 1979, p. 115.

71. A. Kapadia, T. Feizi, and M. J. Evans, *Exp. Cell Res.* **131**, 185 (1981).

72. M. J. Evans and M. H. Kaufman, *Nature* **292**, 154 (1981).

73. S. Strickland, *Cell* **24**, 277 (1981).

74. M. Dziadek, *J. Embryol. Exp. Morphol.* **46**, 135 (1978).

75. B. L. M. Hogan and R. Tilly, *J. Embryol. Exp. Morphol.* **62**, 379 (1981).

76. K. Illmensee and B. Mintz, *Proc. Natl. Acad. Sci. USA* **73**, 549 (1976).

77. M. J. Dewey and B. Mintz, *Dev. Biol.* **66**, 550 (1978).

78. S. Ohno, *Cell* **7**, 315 (1976).

79. R. W. Butcher, G. A. Robison, and E. W. Sutherland, Cyclic AMP and hormone action, in G. Litwack, ed., *Biochemical Actions of Hormones*, Vol. 2, Academic Press, New York, 1972, p. 21.

80. G. H. Sato and R. Ross, eds., *Hormones and Cell Culture, Cold Spring Harbor Conference Cell Proliferation*, Vol. 6, Cold Spring Harbor Laboratory, Cold Spring Harbor, N.Y., 1979.

5

Cell Migration and Its Directional Guidance

ALBERT K. HARRIS

Department of Zoology
University of North Carolina
Chapel Hill, North Carolina 27514

The active migration of the body's component cells serves many important functions, ranging from the morphogenetic movements of early embryonic development to the migration of leukocytes to areas of inflammation in the adult body. Not only the location of cells but also their mutual arrangement in tissues (and in many cases even their shape) all result from cell motility. For example, the paths of nerve fibers are laid down by active migration of their tips, called *growth cones*, but which are really akin to pseudopodia.

The locomotion of tissue cells is analogous but not identical to that of amoebae (Fig. 1). The propulsive force in either case is provided by cytoplasmic actin and myosin. There is no single mechanism of amoeboid locomotion, however, and there may be even more different ways to propel cells with actin and myosin than there are different ways to propel multicellular animals with muscles (1). Tissue cells of all differentiated types and metazoan species, from *Drosophila* hemocytes to human fibroblasts, all seem to employ the same basic propulsive mechanism. This mechanism differs great-

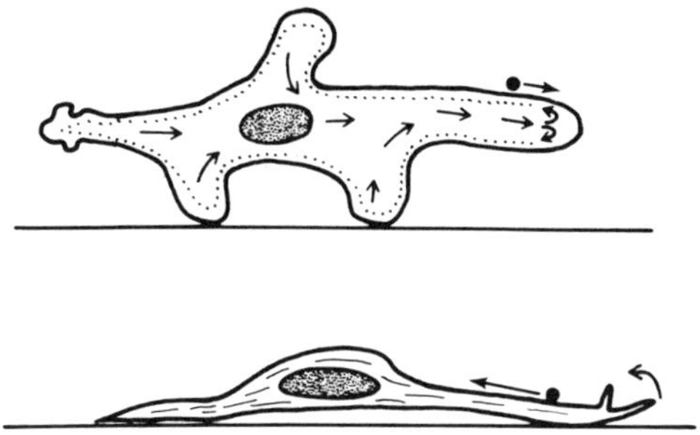

FIGURE 1. Diagrammatic comparison between the locomotion of *Amoeba proteus* (*top*) and a fibroblast (*bottom*). Both cells are moving toward the right. The amoeba raises itself well above the substratum to which it is only weakly adherent, and it can extend pseudopods through the fluid medium for a considerable fraction of its total length. The fibroblast adheres firmly to the substratum upon which it spreads, and it can extend processes only a few micrometers beyond its adhesions. The amoeba cytoplasm flows forward and gels at the advancing tip (fountain zone) to form an exoplasmic tube down which the flow continues. Cytoplasmic flow is not seen in fibroblasts. A particle attached to the upper surface of an amoeba pseudopod is carried forward at a constant distance behind the tip. On the fibroblast surface, attached particles are carried rearward relative to the substratum, even when the cell is not actually advancing. Complex rearward movements of membrane protrusions such as the lamellipodia shown at the right of the cell are characteristic of tissue cells but are not seen in amoeba. The rate of migration of *Amoeba proteus* is much faster than that of any tissue cell; it advances at several micrometers per second, compared to a fraction of a micrometer per minute for a fibroblast.

ly from that of the familiar *Amoeba proteus*, whose locomotion has been studied intensively (2). For example, the body's cells lack the fountain-zone cytoplasmic streaming characteristic of this category of amoebae, and only a few cell types show any cytoplasmic flow at all. Instead, the most distinctive features of tissue cell locomotion are a series of rearward movements of parts of the cell surface, as will be described in Section 1.4.

Because their active, directed migration serves to move body cells to certain locations rather than at random, the mechanisms that guide or steer migration are as important as the actual propulsive mechanisms. We shall consider in detail the two best understood guidance mechanisms: chemotaxis (guidance by diffusing chemicals) and contact guidance (guidance by the physical contours of the substratum).

1. GENERAL DESCRIPTION OF MIGRATION

1.1. Methods of Observation

Most information about cell migration is derived from tissue culture studies, because these afford optimal visibility as well as unlimited experimental control over the chemical and physical environment of cells. The best-studied cells are fibroblasts (generalized mesenchymal cells), which emigrate readily from tissue fragments of nearly any part of the body (usually of embryos) explanted into a suitable medium on a glass, plastic, fibrin, collagen, or other substrate. There have also been many successful observations of cell migrations in vivo, as recently reviewed by Trinkaus (3). For example, the early embryos of some echinoderms are sufficiently transparent to allow observations of the morphogenetic cell migrations within (4, 5), and tadpole tails afford similar opportunities (6). Considerable information has also been gained using special observation chambers in which cover slips are sealed across holes punched through the ears of rabbits (7).

1.2. Speed and Polarity of Migration

Tissue cells migrate much more slowly than amoebae, and must be studied by time-lapse photography. From 0.5 to 1 μm/min is considered rapid motility for fibroblasts or most other cell types. Nerve growth cones can advance at twice this speed, while lymphocytes and granulocytes (i.e., polymorphonuclear leukocytes such as neutrophils) move 30–40 μm/min.

One reason why granulocytes and lymphocytes migrate much more rapidly than other cell types may be the polar morphology they adopt while migrating. They develop a front and a definite rear, called a uropod or uroid, which is a bulblike protrusion from which trail approximately a dozen very thin (0.1 μm) cytoplasmic strands. These strands are called retraction fibers, and they connect the cell to residual adhesions to substrates that are gradually broken as the cell moves forward. Granulocytes and lymphocytes are also peculiar in possessing cytoplasmic flow. This flow occurs as intermittent forward surges, but does not follow the fountain zone pattern seen in *Amoeba proteus* (8).

In contrast, fibroblasts typically have several competing fronts that stretch the cell body between them. These flattened organelles of locomotion, by which the cell pulls itself along, have been termed *leading lamellae* (9), which is a more inclusive synonym for the older term *ruffled membrane*. Although fibroblasts can be temporarily unipolar, the typical spindly, polygonal shape of these cells results from two, three, or more leading lamellae crawling away in opposing directions. If any net translocation results, it will be in the direction of the broadest and strongest of the competing lamellae of a cell. The leading lamellae on the losing side often become inactive and are dissected into a series of retraction fibers. On flat surfaces, fibroblasts tend to move in a given direction for about 2–3 hr, interrupted by sudden random turns: this pattern has been analyzed mathematically (10). In general, the straight paths reflect the period of dominance of a particular leading lamella that pulls the remainder of the cell after it. Sharp turns occur when a temporary reduction in the cells' stretched tension permits a new leading lamella to become dominant and to drag the cell after it. Although it has been claimed that the directions of these turns are correlated in sister cells after mitosis (11), the required statistical analysis has yet to be reported.

1.3. Adhesion to a Substrate

In order to migrate in a tissue, cells need to attach to some surface that is sufficiently adhesive and rigid to withstand the cell's contractility. Thus, weakly contractile cell types such as macrophages and neurons can spread and move on weakly adhesive substrates or even on the protein film at the surface of culture medium. More strongly contractile cells require substrates that are more adhesive and solid. The earliest attempts at tissue culture were not successful until Harrison (12) provided his cells with a suitably rigid substrate of clotted lymph, and fibrin clots long remained a standard culture substrate until displaced in recent decades by glass and polystyrene. Reprecipitated collagen provides a particularly good substrate (13). Cell migratory behavior is strongly influenced by the adhesiveness of substrates, as has been shown using inert metals in various stencilled patterns as substrates, among which cells show marked preferences (14).

The pulling forces that cells exert through their adhesions can be visualized by using flexible rubber culture substrates which wrinkle visibly where stressed (Fig. 2, reference 15). This method has revealed that fibroblast traction is several orders of magnitude stronger than the traction of several other cell types and is apparently greater than necessary to move fibroblasts from place to place. This excessively strong traction can serve as a very effective mechanism for remodeling collagen fibers into certain anatomical patterns, which suggests that this remodeling could be the real purpose of the phenomenon studied for so long as fibroblast "locomotion" (15).

The adhesive sites through which these pulling forces are exerted can be directly observed beneath migrating cells by the optical technique called interference reflection microscopy (Fig. 3; references 16 and 17). Cells cultured on glass are illuminated from below using light projected through the microscope objective (epiillumination), and the reflected rays are observed. Destructive interference between the rays reflected from the lower surface of the cell and those reflected from the surface of the adjacent substratum results in an optical image in which close contacts between these two surfaces appear dark. The general rule is that the darker the image, the closer the plasma membrane to the glass surface. There are complicating factors, however, such as variations in the fraction of light reflected at each surface depending on the local increments in refractive index. Thus, an area where

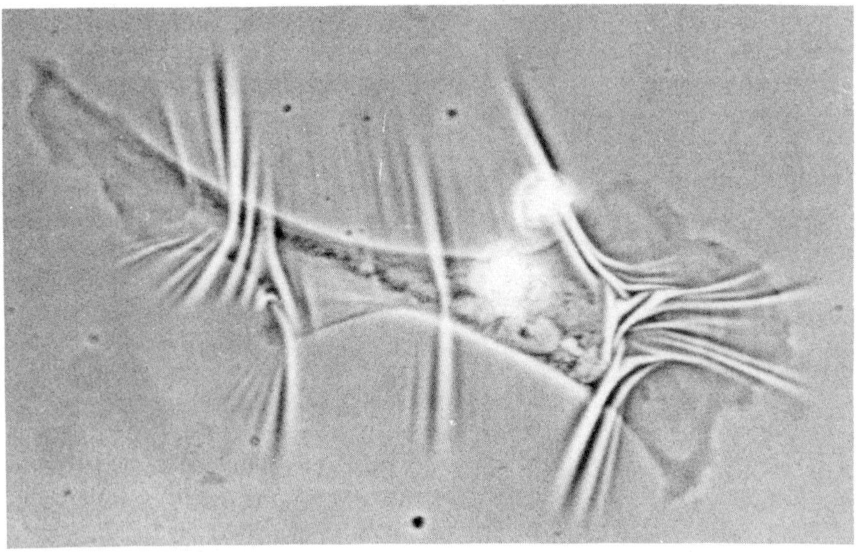

FIGURE 2. Photograph of an individual embryonic chick fibroblast cultured on a thin sheet of silicone rubber. The contractile forces by which the cell spreads and migrates are made visible by the wrinkling pattern produced in the rubber sheet.

the density of cytoplasm adjacent to the membrane was increased as a result of higher protein concentrations could appear darker even though the contact distance was unchanged.

The cell-to-substrate contacts observed by this method fall into two categories: small, black, focal contacts with a cell-to-substrate distance of 10–15 nm and broad, grey contacts with a gap distance of about 30 nm (Fig. 4). The "black" contacts are usually elongated, 1–10 μm long and 0.25–1 μm wide, and located where the actin stress fibers attach to the inner surface of the plasma membrane. The black contacts form within the much broader grey contacts, which are often tens of micrometers across. A typical fibroblast would have two, three, or more broad grey contacts: one such contact would be adjacent and internal to each part of the periphery where ruffling or blebbing is occurring. Each grey contact contains 10–20 black, focal contacts, with the long axes of these focal contacts oriented parallel to the cell axis or radial relative to the center of the cell (17). Trailing parts of the margin often have only the black contacts. The black contacts seem to form only within preexisting grey ones, but some cells form only the grey adhesions without black foci (e.g., macrophages and some lines of transformed cells) and this absence is associated with the absence of stress fibers in the cytoplasm.

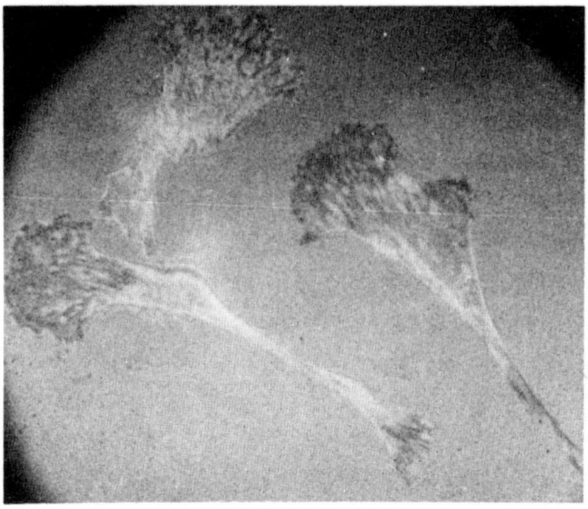

FIGURE 3. Three chick heart fibroblasts as seen by interference reflection microscopy. The darker areas in each cell are the contacts of the cell with the glass substratum.

1.4. Cell-Surface Movements

1.4.1. Their Relation to Migration. Instead of the cytoplasmic flow characteristics of the more familiar kinds of amoebae, complex movements of the cell surface are the most evident feature of propulsion in tissue cells. These cell surface movements include the forward extension of several classes of surface protrusions as well as the rearward transport of markers attached to

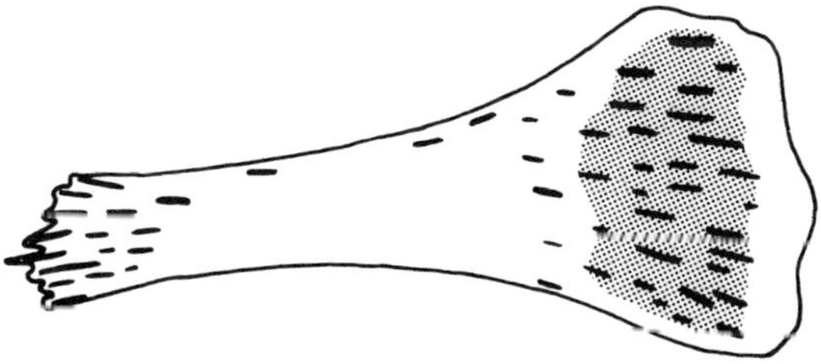

FIGURE 4. Diagram of a typical fibroblast as seen by interference reflection microscopy. The closer the plasma membrane approaches to the glass substratum, the darker the image at that point, so that the closest contacts appear black and the less close contacts appear grey (the boundary of this cell is indicated by a black line, although this would not actually be seen). The advancing end of this cell has a broad grey adhesion, within which many of the black focal contacts (or adhesion plaques) have formed with their long axes parallel to that of the cell itself. The retreating "tail end" of this cell is adhering to the glass by another group of the black focal adhesion plaques which are gradually being broken loose (or broken off). Note that the broad grey adhesions are more characteristic of the leading margin and are often absent where the margin is retreating. Note also that the cell margin extends outward 5–10 nm in advance of either class of adhesions along the advancing margin but not along the trailing margin.

the plasma membrane. If we consider cellular locomotion in terms of a "forward-reaching component" together with a "backward-pulling component," then the surface protrusions appear to provide the former element while the rearward surface transport reflects the latter. Taken together, these surface movements represent the transfer through the plasma membrane (to the outside) of the cytoplasmic forces developed by the actomyosin network.

1.4.2. Forms of Cell-Surface Protrusion. Lamellipodia ("ruffles") are the flattened, sheet- or veillike protrusions of plasma membrane that are so characteristic of fibroblasts, granulocytes, macrophages, and other cell types, particularly when they are cultured on flat substrates (18). However, on collagen and fibrin, fibroblasts rarely show lamellipodia, and in vivo other classes of protrusion are also more common.

Lamellipodia are only 0.1–0.2 μm thick but are 5–10 μm or more in length. They extend tens of micrometers in width, sometimes across the entire advancing margin of a leading lamella. These structures typically form by forward extension from the margin of a leading lamella and are protruded out over the substrate (parallel to, but not in contact with it). From this initial horizontal position, lamellipodia fold gradually upward and backward, requiring 0.5–1 min to reach the vertical, and they usually continue to bend backward before being resorbed into the cell. Sometimes they flop back down onto the substrate or undergo curious spiraling movements. Because adjacent parts of a lamellipodium will almost always fold backward at slightly different angular velocities, the outer edge develops a sinuous or lacy curve. The older term "ruffle" is highly descriptive. In phase contrast microscopy, the outer margins of upfolded lamellipodia are highly refractile, so that the backfolding process is easy to mistake for the rearward propagation of a peristaltic wave. The bases of lamellipodia do sometimes move rearward across the cell surface and lamellipodia can be protruded from part of the cell other than the margin (19).

Lamellipodia are optically clear and exclude organelles such as mitochondria. Electron microscopy shows them to be full of filaments the size of actin microfilaments, but they never contain microtubules. Their contents could reasonably be described as all cell cortex and no cytoplasm.

Filopodia (*microspikes, microvilli, spicules*) are thin but stiff, needlelike protrusions about 5–10 μm long, though occasionally much longer. They are pushed out rigidly from the cell surface and should not be confused with the superficially very similar retraction fibers, which are flexible strands pulled out from the cell surface where it retracts away from residual adhesions, for example in mitosis. Filopodia seem to be equivalent to very thin lamellipodia, and the two are frequently intermingled. Like lamellipodia, filopodia are protruded along the advancing margin, folding upward and rearward. They are not flexible, however, and bend rigidly at angles rather than curving. They have a core of actin filaments (20).

Blebs are hemispherical or bulbous protrusions from the cell surface. Larger protrusions of this shape have also been called lobopodia. Mechanically, blebs appear to be blisters in the plasma membrane where it has pulled away from the underlying actin-containing cortex and has been pushed outward by hydrostatic pressure. Blebs protrude suddenly (2–5 sec), but are resorbed more slowly over a time span of 20 sec to 1 min. When forming, a bleb lacks a microfilamentous cortex, but a new one quickly assembles just within the protruded hemisphere of membrane (21).

Blebs can be symptomatic of sick and dying cells, a fact that has discouraged their study, even though they also occur on healthy cells and are more prevalent in vivo than are lamellipodia or filopodia. Cells can bleb over their whole surface, for example, in mitosis or after trypsinization; but blebbing is most frequent along the advancing margins where it merges into lamellipodium formation as cells become more completely spread (19).

1.5. Rearward Transport of Materials Attached to the Cell Surface

Migrating tissue cells transport attached particles, lectins, and antibodies tangentially across their plasma membranes. The direction of this transport is rearward in the same direction as the traction forces being exerted on the substratum by that part of the cell, as if equivalent forces were also being exerted on the attached particles and molecules. This surface transport therefore seems to provide important information concerning the mechanism by which the forces of cytoplasmic contraction are transmitted through the plasma membrane to the substratum.

Particles dropped onto the cell surface or picked up from the substrate are carried rearward at 2–4 μm/min and accumulate in a ring around the nucleus in multipolar cells such as fibroblasts and on the uropod in granulocytes (Fig. 5; references 19 and 22). Test particles of plastic, glass, and metal in sizes from 0.2 to 2 μm in diameter are all transported in the same patterns, so long as they adhere to the cell surface and are small in size relative to the cells used. Lectins specific for cell-surface polysaccharides and antibodies against surface antigens are transported in the same pattern, except that these molecular markers first aggregate together into small patches (*patching*) before being transported to a cap of accumulated marker (*capping*) (23, 24).

The molecular mechanism that moves these attached markers laterally remains uncertain. Though most agree that the force is supplied by an actomyosin network inside the plasma membrane, it is debated as to how the membrane components themselves move. The central question is whether

FIGURE 5. Particles attached to the upper surface of a migrating fibroblast. Such particles are picked up from the substratum by the advancing margin of the cell and are transported rearward on the upper, outer surface of the plasma membrane. They eventually accumulate in a ring around the nucleus (N).

parts of the plasma membrane that are not attached to a marker (particle, lectin, or antibody) also move rearward, or whether these unattached components flow forward in a countercurrent between the rearward-moving marked areas. In the latter case, it would be necessary for the rearward movement of a given area of membrane to be induced only where some marker was bound. The alternative would be for the whole membrane to flow rearward, with its components being recycled through the cytoplasm back to the leading edge.

The failure of monovalent antibodies to be capped by some cells can be interpreted to mean that cross-linking of receptors on the surface to form patches induces their transport. Cross-linking could either stimulate the assembly of actomyosin cables or their attachment to the cross-linked area of membrane, or alternatively could render these areas more susceptible to being swept along by a general rearward flow of other membrane components. This latter alternative would be analogous to bunches of leaves being caught between tines of a rake, whereas individual ones could slip through them.

It has seldom been possible to compare the movements of monovalent and multivalent markers simultaneously on the same cells. Neither has it been shown that cells ever fail to cap monovalent antibodies or lectins while at the same time either capping attached particles, or even undergoing locomotion (unfortunately, most of these studies have been performed on lymphocytes in suspension). The closest to a simultaneous observation of mono- and bivalent markers is an elegant study by Mendoza and Metzgar (25) that deserves to be better known in this context. They studied mast cells, a granulocyte whose plasma membrane contains receptors specific for the Fc end of antibody molecules. Since there is only one Fc part of each IgG molecule, bonding of these antibody molecules to the mast cell surface is equivalent to a monovalent attachment. When the corresponding antigen is added, however, the IgG molecules bind to the antigen molecules, thus becoming cross-linked to each other. Such cross-linked antibodies were capped, whereas antibodies free of antigen were not, even on the same cells at the same time.

In this and other capping studies where fluorescent tags are used, it is too seldom recognized that what one is actually observing is not movement of markers per se, but instead whether the markers accumulate somewhere or remain dispersed. Obviously, if they accumulate they have been moved, but if they do not accumulate anywhere it does not follow that they have not moved. They could have merely continued to recirculate without accumulating at a single point.

Recirculation need not involve any resynthesis, but the appearance of newly synthesized membrane components may tell us something about the spatial pattern of plasma membrane recirculation. What is observed in virus-infected cells and in the products of cell fusion is that newly synthesized plasma membrane molecules first enter the surface along the advancing margins where ruffling and blebbing occur; they subsequently spread centripetally to occupy the whole surface (26, 27). These observations are most easily explained by the hypothesis that the plasma membrane of migrating tissue cells continually flows rearward from the advancing margin (carrying particles, antibodies, and other markers rearward with it), is disassembled in some central or trailing position (leaving the surface markers and their cross-linked receptors trapped there), and recycles back to the advancing margin to be reassembled together with any newly synthesized membrane molecules, repeating the cyclic process over and over again.

2. MOLECULAR ORIGIN OF THE PROPULSIVE FORCES

In motile tissue cells and amoebae there are arrays of proteins whose molecular properties closely resemble those of actin, myosin, and the other muscle proteins tropomyosin and α-actinin (Fig. 6). It is generally believed that these proteins provide the motive forces causing cell migration, but the mechanisms by which they cause the spreading and translocation of cells are not known (reviewed in references 28–32). The presence of actin and myosin in amoebae indicates that muscle contraction evolved secondarily from a system whose primary function was cell propulsion. Nevertheless, the functions of these proteins in muscle are so much better understood that it may be best to proceed by asking first how the requirements for moving a cell

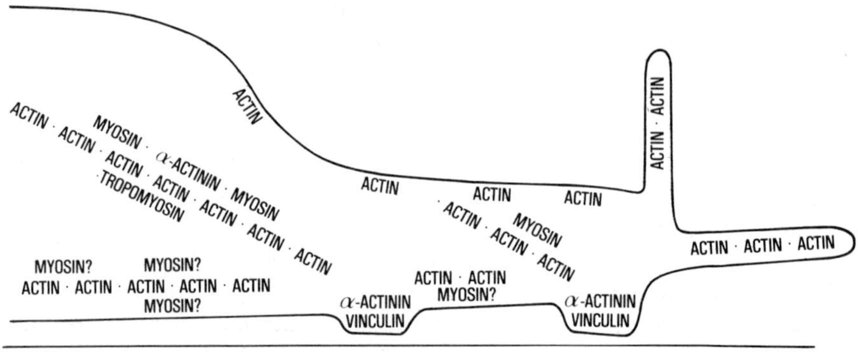

FIGURE 6. Diagram of the characteristic patterns of localization of contractile proteins.

from place to place differ from those of simply contracting it in one place as in muscle, and second how the differing molecular properties of the cell motility forms of actin, myosin, and other molecules could help to accomplish these special requirements.

Perhaps the most important difference is that muscle contraction involves a simultaneous activation of a cell's entire network of contractile proteins, whereas migration requires local development of contractile force in only certain parts of the cell at certain times. Different mechanisms of coordination would be anticipated, with the development and breakdown of cell–substrate adhesions somehow playing a controlling role so that the contractions will occur at the correct sites relative to these adhesions. Another difference is that motility must involve a continual breakdown and reformation of the contractile protein networks. Fiber assembly, and the localization and polarity of this assembly, will be as essential as contraction itself. Migration also involves the outward protrusion of parts of the plasma membrane for at least several microns.

2.1. Actin

Of the musclelike proteins, the molecular properties of cellular actin differ the least from its counterpart protein in muscle. The molecular weights are virtually identical and the amino acid sequences differ only slightly, so that it has often been difficult until recently to separate the proteins by electrophoretic or comparable techniques (29, 31). Even the actins found in amoebae and green plants differ only slightly. On the other hand, actins with slightly different amino acid sequences have been found in the tissue cells of a single species.

The motility actins are able to interact specifically with myosin from muscle. It is therefore possible to use proteolytic fragments of myosin (heavy meromyosin and subfragment-1) to identify fibers of motility actin within permeabilized cells and even to determine the directional polarity of these fibers by the direction of the "arrowheads" formed by the myosin fragments along the fiber (33). This is a highly specific interaction, but raises the question why (or whether) there are no interactions between the motility actins and myosins and their counterpart proteins in migrating muscle cells, which necessarily have both types of actin and myosin. Whether the various kinds of actin are ever able to copolymerize with each other in vivo is also not clear.

Actin is found in the highest concentration of any of these motility proteins. Concentrations of 10%, or even 20%, of dry weight have been reported (as compared with 19% in muscle). Isolated membrane fractions are often highly enriched—as if the motility actins had a special affinity for the plasma membrane. The actin content of cell types varies rather mysteriously and is not necessarily correlated with contractile strength, since it is especially high in nerve fibers, which contract only very weakly. Considerable quantities of actin are even found in erythrocytes, which are the one entirely nonmotile cell type.

The molecular property in which motility actin differs most from muscle actin is solubility, that is in the state of the equilibrium between assembled or fibrous actin (F actin) and disassembled globular or G actin. In muscles this equilibrium lies far toward assembly with few freely diffusing actin molecules, but in the case of motility actins a substantial fraction (½ or more) of the protein seems to exist in the soluble G-actin form at any given time (though this may be largely the effect of inhibition of polymerization by special actin-binding proteins rather than the properties of actin itself). The state of this equilibrium is very much a function of salt concentrations, of course, but under a given set of isolation conditions the motility actins are almost always found to be more soluble and the same difference has been extrapolated to the cell.

The functional explanation for this greater solubility seems clear; it is the need to make and unmake contractile fibers repeatedly in order to achieve translocation. Muscle fibers function by repeated contraction without the need for any reassembly of the fibers between contractions, but in most cellular propulsion mechanisms one can imagine that the cytoplasmic fibers may contract only once before being taken apart and put together again at a more anterior location. The greater tendency to convert to a freely diffusing state would permit this dynamic turnover, and there are similar differences in the properties of the motility myosins. The locations and conditions within the cell causing fiber assembly are as important a part of propulsion as the conditions of contraction itself, and where assembly occurs relative to contraction may be the major difference between the different mechanisms of amoeboid locomotion (e.g., fibroblasts versus *Amoeba proteus*).

The necessity for a dynamic disassembly and reassembly of actin fibers seems to explain the paradox that cytochalasin quickly blocks tissue cell migration, cytokinesis, and other processes attributed to actin even though muscle contraction is essentially insensitive to this drug. Cytochalasin probably acts by blocking the polymerization of actin rather than its contractile function (34). The depolymerization of actin filaments seems also to be a necessary component of cell migration, as its inhibition with phalloidin also inhibits fibroblast migration (35).

The spatial distribution of actin within fibroblasts and other migrating cells has been observed with heavy meromyosin decoration and transmission electron microscopy and through the use of fluorescent molecules bound to heavy meromyosin (36), to antibodies against (denatured) actin (37), and to certain drugs that bind specifically to actin (38).

Lamellipodia and filopodia are found to be densely packed with actin fibers. Curiously there is little or no myosin in these protrusions, suggesting that the actin there is serving a skeletal or pushing function similar to that recently discovered in the acrosomal filament and brush border microvilli (39). Meromyosin arrowheads attached to this marginal actin always point away from cell surface. Arrays of actin fibers are also found in the cell cortex associated with the inner surface of the plasma membrane (40) as well as concentrated together in cytoplasmic fiber bundles called stress fibers. These stress fibers attach to the inside surface of the plasma membrane precisely where the outside surface forms close adhesion plaques to the substratum. Myosin is also present in these stress fibers and meromyosin arrowheads point in both directions, supporting a widespread but unproven idea that these fibers represent little cytoplasmic muscles by which the cell pulls its way along. We should be aware that the most motile kinds of cells, including leukocytes and nerve growth cones, seem to lack stress fibers.

2.2. Myosin

Depending on the differentiated cell type, from 0.1 to 2% of cell protein consists of myosinlike proteins (as compared with 35% in muscle). These motility myosins differ more from muscle counterparts than is the case with the actins, but they do meet the criterion of myosin as Ca^{2+} and actin-stimulated ATPases (32). They can also be assembled tail to tail into double-ended aggregations much like muscle thick filaments.

Antibodies to myosin stain the cell cortex and stress fibers if the latter are present, as well as the cytoplasm diffusely (41, 42). In the stress fibers, the myosin staining is localized in a series of transverse bands each a fraction of a micrometer long, which apparently form a miniature version of muscle sarcomeres. Staining by antibodies against myosin has also been reported along the inner surface of the plasma membrane (43, 44).

Exactly how the actin and myosin move relative to one another in propelling the cell is one of the single most important unanswered questions in this whole phenomenon. One category of possibility is for cables of mixed actin and myosin to shorten. A rather different possibility is that cortical layers of actin and myosin slide past one another, with one of them coupled to the plasma membrane. The myosin antibody staining near the cell surface in certain studies could be interpreted as support for such cortical sliding, were it not that the polarity of the actin fibers near the advancing margin is such that meromyosin arrowheads point rearward (45). In muscle contraction, the active sliding of myosin along actin is in the direction opposite to the arrowheads. Thus, if active sliding were to occur in the cortex, it would be the actin that would move rearward with the membrane. Other studies do show actin in the cortex near the plasma membrane and myosin more toward the interior of the cells (42), localizations consistent with a model of lateral sliding of plasma membrane-attached actin. However, more direct tests of this hypothesis are needed.

2.3. α-Actinin

This is the major protein component of the Z lines of striated muscle, where the actin fibers of adjacent sarcomeres connect (46). Antibodies against muscle α-actinin cross-react with certain nonmuscle proteins that are localized at the focal adhesions to the substratum and along stress fibers. These α-actinin antibodies stain the vertices of the interlocking polygons, which some kinds of fibroblasts form between adhesion plaques (37, 47).

α-Actinin appears to act as a crosslinking molecule between actin filaments (48), and it can exist in close proximity to the plasma membrane (49). It may therefore have a function in organizing actin filaments.

2.4. Tropomyosin

This protein also has a motility analog that joins together with fibrous actin to form filaments. The motility form of the protein is both lighter in weight and shorter in length than its muscle counterpart. It is found in stress fibers of fibroblasts (37).

2.5. Vinculin

This recently discovered protein is concentrated in focal adhesive contacts and is located in very close apposition to the plasma membrane at sites where actin-containing microfilament bundles attach to the membrane. It can also cap the ends of actin filaments (50). These properties suggest that vinculin may be involved in the interaction of actin with membranes, although its mechanisms of action are as yet unknown.

3. CHEMOTAXIS

3.1. Criteria for Chemotaxis

Cell migration can be influenced by diffusible chemicals in several ways. They can accelerate or slow its movement or control its direction. When the direction of movement of a cell is altered according to differences in the concentration of a chemical, the phenomenon is called chemotaxis—positive chemotaxis when the cells move toward increasing concentrations and negative chemotaxis when they move toward decreasing concentrations. The apparent simplicity of chemotaxis conceals several rather insidious conceptual pitfalls, and although many differentiated cell types have been reported to migrate under chemotaxis (e.g., positive chemotaxis toward the gonads by primordial germ cells, mutual negative chemotaxis between melanocytes), by the most stringent criteria, chemotaxis has been demonstrated unequivocally only by granulocytes, macrophages, and (most recently) nerve growth cones.

Chemotaxis is positive in all three of these cell types. Although there is some evidence for mutual negative chemotaxis between crowded granulocytes (51), this could also be explained by local depletion of attractant substances. That is, cells that soak up an attractant will seem to be emitting a repellant and vice versa. Granulocytes as well as macrophages use chemotaxis to find bacteria and sites of inflammation (Fig. 7). Their attractants include bacterial exudates and proteolytic fragments released by the complement system. For nerve growth cones, the protein nerve growth factor has been found to be an attractant (52). It has long been supposed that such types of chemotaxis may guide nerve endings to their proper points of connection.

To achieve true chemotaxis a cell must be able to do three things: detect the concentration of the attractant at two or more locations; (2) compare the concentrations at these locations; and (3) reorient its locomotion so as to move toward (in the case of positive chemotaxis) the higher concentration of the attractant substance. Although it is an interesting fact that many chemotactic attractants do accelerate locomotion, acceleration itself is not evidence of chemotaxis. A nondirectional stimulation of migration will not cause cells to accumulate near the source of a substance, but will instead promote their dispersion. Conversely, an accumulation of cells near the source of a gradient does not prove chemotaxis. The substance may only be inhibiting their

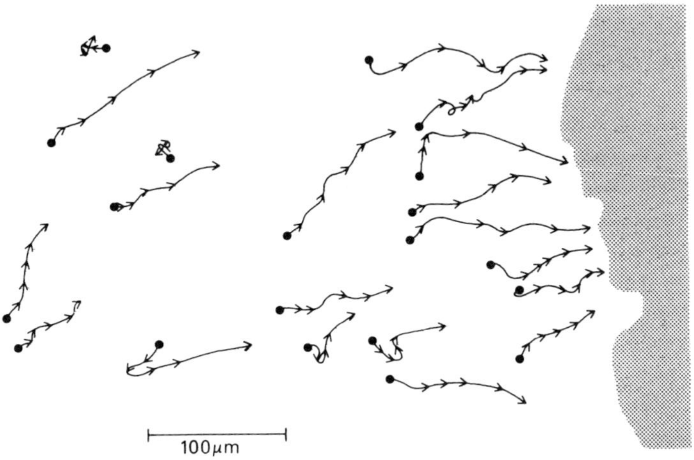

FIGURE 7. Chemotoxis by 20 human granulocytes. The dotted area along the right side indicates the edge of a large mass of bacteria whose exudate acts as an attractant for granulocytes. The initial positions of the 20 cells are marked with black circles and their paths by lines, with each arrowhead indicating the position of the leading margin of each cell at 100-sec intervals. (Tracing from a time lapse film kindly provided by Scott Ramsey.)

movements and trapping them there. Chemotaxis is a directional response, but we also must not fall into the trap of thinking that gradients exert an "attractive" force on cells. The only forces exerted are by a cell's own propulsive mechanism. Chemotaxis steers this mechanism.

3.2. Chemotactic Assays

The most widely used assay for chemotaxis in granulocytes and macrophages is the Boyden technique (68), which measures the locomotion of cells through the pores of Millipore (or equivalent) filters toward an attractant (Fig. 8). The number of cells penetrating through the filter after a set time (or sometimes the maximum distance penetrated after some shorter time period) is determined. If more cells penetrate through the filter (or penetrate farther) when a putative attractant substance is placed on the opposite side of the filter, as compared with the number penetrating when ordinary culture medium is placed on both sides, this result is usually taken as a demonstration (and measurement) of chemotaxis. Unfortunately, if a substance happens merely to stimulate the speed of locomotion, or even just makes the cells more inclined to burrow through the holes in the filter, it will seem to be a chemotactic attractant by this criterion. As an experimental control for this possibility, one should also compare the amount of penetration when the

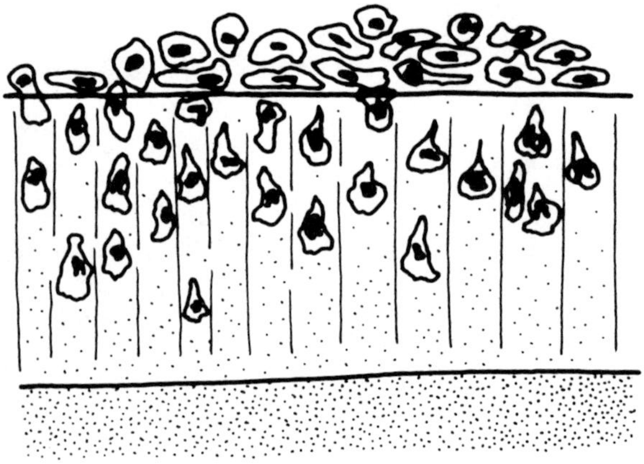

FIGURE 8. The principle of operation of the Boyden chamber. Molecules of attractant diffuse through the filter (from below) and the migration of granulocytes or other cells from above is monitored (see Section 3.2).

chemical substance is placed on both sides of the filter. As Zigmond and Hirsch (53) have pointed out, however, some substances can promote migration at low concentration (as when their source is on the opposite side of the membrane) but inhibit migration at higher concentrations, as when added to the side where the cells are. A third control is therefore desirable (the only remaining permutation, with the suspected attractant added on both sides). Even then the results can be ambiguous, especially if there is a mixture of effects. The many cell types suspected of chemotaxis need to be reexamined with these ambiguities in mind (54).

A less equivocal approach to the study and measurement of chemotaxis is simply to observe the cells through a microscope as they crawl about on a flat surface in a gradient of the substance (Fig. 9) (55). For convenience, time-lapse photography is often used, and tracing the paths of a few dozen cells (or sometimes even just a single cell) can provide information about chemotaxis that may be more definitive than any number of Boyden chamber experiments.

3.3. Alternative Mechanisms of Chemotaxis

Although chemotaxis requires that the cell compare attractant concentrations at two or more locations, these need not be two different locations on the cell's surface. Many bacteria achieve chemotaxis by comparing concentrations around themselves at sequential times as they swim along in relatively straight lines. Periodically they "tumble" about and change directions to a new and randomly chosen direction (having flipped themselves instead of a

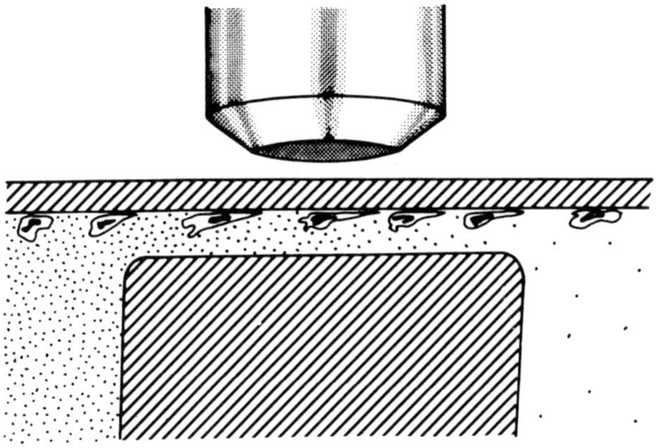

FIGURE 9. Diagram of the principle of operation of a Zigmond Chamber for observing the chemotaxis of granulocytes and other tissue cells. The cells are cultured on the lower surface of a cover slip through which they are continuously observed with the objective of a light microscope. The chamber itself consists essentially of a depression slide with two wells separated by an impermeable but transparent barrier (glass or polystyrene as indicated by diagonal cross-hatching), which rises to within a known distance of the cover slip so as to leave a narrow space (e.g., 10 or 25 μm in height). Medium containing a putative attractant is placed in one well and ordinary medium in the other. A diffusion gradient develops through the narrow intervening space and the directional locomotion (if any) of the cells is directly observed.

coin). If they then detect an increasing concentration of attractant as they move forward, turning is inhibited; but if the sequential measurements reveal a decline in concentration, they choose a new direction. The response to a repellant substance is the reverse. This kind of mechanism is called *temporal chemotaxis*, as opposed to *spatial chemotaxis*, in which the gradient direction is determined by multiple measurements of attractant concentration made more or less simultaneously at separate locations on the cell surface. This mechanism is shown by their ability to orient their movements correctly in a newly formed gradient through which they have not previously migrated (56, 57).

The study of how granulocytes detect and compare attractant concentrations has recently been greatly helped by the discovery of several tripeptides that are very effective attractants (58). With the availability of these synthetic attractants, it became possible to study cellular responses to precisely known concentrations of attractant. In one such study Zigmond and Sullivan (59) observed the responses of granulocytes to sudden increases and decreases in tripeptide concentration. In both cases, the granulocytes responded by halting forward locomotion and rounding up, protruding lamellipodia in all directions when the attractant concentration had been raised, but blebbing all over when this concentration had been lowered. Since lamellipodium protrusion is normally restricted to the leading edge of cells, the reaction to increased attractant concentration is as if the whole cell surface were trying to become the front of the cell. After the delay of from 1 to 10 min, the treated cells then resume locomotion. Significantly, the length of this delay is proportional to the change in the number of receptors occupied that had induced the stoppage. Furthermore, this delay was even additive when the concentration jump was carried out in two steps. Differences were also observed in the minimum size of the concentration increments required to stimulate this stoppage and generalized ruffling. Over some parts of the concentration range, a 50% increase was sufficient, while at other parts of this range even a doubling of concentration fails to elicit the reaction. The smallest proportional increments—about a 20% increase—are required in the concentration range nearest the binding constant (K_d) of the cell surface receptor for the tripeptide attractant. What this seems to imply is that the difference in the absolute number of receptors occupied (rather than their proportion, for example) is the determining factor in this reaction, and probably also in granulocyte chemotaxis in general (see Section 4.5).

The response of cells to these concentration-jump experiments suggests that there could be some temporal element to their spatial chemotaxis. However, it would be a mistake to conclude that their chemotaxis must be temporal, similar to that of bacteria. These responses to temporal jumps in concentration may only reflect the finite time lag inherent in whatever mechanism the cells employ to compare attractant concentrations between one side of themselves and the other. Unless these spatial comparisons could somehow be carried out instantaneously, the poor cells will be subject to being tricked by sudden changes in the overall concentration of attractant. When every part of the surface senses a large increase in concentration, naturally each part of the plasma membrane will "think" itself destined to become the new front end. The whole surface reacts appropriately by protruding lamellipodia everywhere, but this does not mean that the comparison mechanism is temporal rather than spatial, only that the spatial comparisons are not instantaneous. Although much more information is needed, these studies are providing important information about the mysterious but all-important internal mechanism by which the cells compare attractant concentrations at different points on their external surface.

3.4. Molecular Mechanism for Comparing Attractant Concentrations to Detect Gradients

At the molecular level, the control of cellular propulsion by attractant concentrations has been most closely analyzed in the case of bacteria. It may be possible to borrow some of the concepts developed to explain temporal chemotaxis and to apply those concepts in modified form to the interpretation of spatial chemotaxis in higher animal cells such as granulocytes.

To explain how bacteria are able to detect whether the attractant concentration is rising or falling, Koshland (60) has proposed a model in which a series of interlocking biochemical reactions together serve as a sort of "logic circuit." Just as the logic circuits of computers are constructed by combining capacitances, resistances, and so on, we must expect that complex cellular behaviors are products of interlocking enzymatic reactions, diffusion, patterns of cytoplasmic contractility, and the like.

The operations of this particular biochemical calculator are illustrated in Figure 10. There is good evidence that the chemical reactions happen to be methylations (e.g., reference 41), but this is incidental to the general problem of how to detect a gradient biochemically. Two enzymes are postulated, both of which are converted into an active form at rates controlled by the local concentration of the chemotactic attractant. The difference is that the activation of enzyme 1 is faster than that of enzyme 2. The activated forms of these two enzymes are supposed to catalyze two sequential steps of another reaction. The first enzyme creates substance X while the second destroys it. Thus, when the attractant concentration is rising, more of X will be created than destroyed and the reverse when the attractant concentration is falling, in either case because the activation of the creating enzyme 1 is a faster reaction than the activation of the destroying enzyme 2. If the probability of turning is an inverse function of the concentration of X, positive chemotaxis results.

As Zigmond and Sullivan (59) have pointed out, a similar biochemical logic circuit could also govern spatial chemotaxis, provided that both enzyme 1 and substance X are relatively nondiffusible, while enzyme 2 is free to diffuse. The difference in (spatial) diffusion rates takes the place of the difference in (temporal) activation rates. The concentration of activated enzyme 2 would be proportional to the average concentration of attractant over the whole cell surface, while the amount of activated enzyme 1 at each point

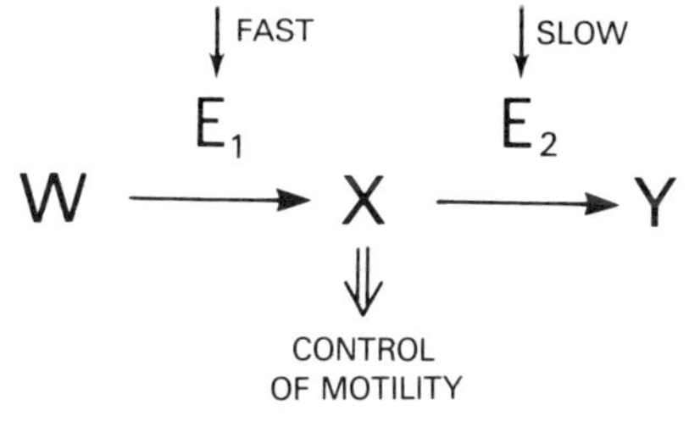

FIGURE 10. A hypothetical biochemical mechanism or "logic circuit" for controlling the frequency of direction changes in bacteria according to whether the attractant concentration is rising or falling (see Section 3.4). Modified after Koshland (60).

would reflect the attractant concentration over that particular part of the cells' surface. Substance X would rise in concentration wherever the attractant concentration is higher than anywhere else. If X promoted lamellipodial protrusion, chemotaxis would result.

Other molecular models may be imagined, but to achieve granulocyte chemotaxis over a broad range of attractant concentrations it is essential to have a means of comparing the overall average concentration of attractant with the local concentrations at different points on the surface (Fig. 11). The attractant itself cannot simply promote lamellipodial protrusion in proportion to its concentration, particularly not in cells such as granulocytes that maintain a consistent morphology (and a highly monopolar one at that)

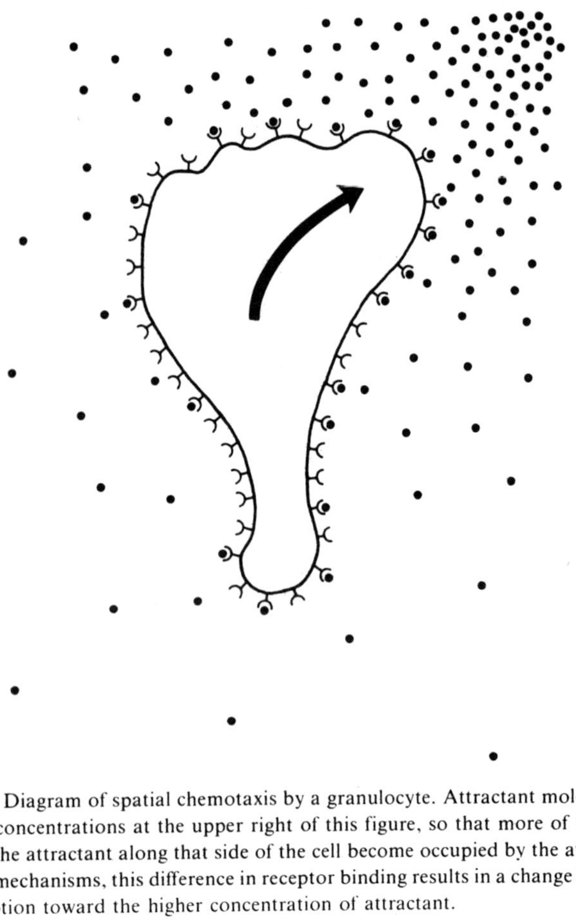

FIGURE 11. Diagram of spatial chemotaxis by a granulocyte. Attractant molecules are present in higher concentrations at the upper right of this figure, so that more of the cell-surface receptors for the attractant along that side of the cell become occupied by the attractant. By as yet unknown mechanisms, this difference in receptor binding results in a change in the direction of cell locomotion toward the higher concentration of attractant.

over a broad range of attractant concentrations. So little is known about the linkage between the cells' propulsive machinery itself and this system for comparing attractant concentrations that it would be premature to discuss it here, except to suggest the likelihood that some of these activated enzymes and X substance, which it is necessary to hypothesize, may in fact correspond to some of the molecular species responsible for propulsion (e.g., actin, myosin, or other motility proteins).

3.5. Granulocyte Attractants and Their Receptors

Quite a variety of different chemicals have been found to be effective chemotactic attractants for granulocytes. Among these are normal components of serum, bacterial exudates, denatured or otherwise especially hydrophobic proteins, partially oxidized lipids, the C5a fragment (or complement factor 5), and a series of tripeptides having a formyl group at their aminoterminal end and a phenylalanine at their carboxyl end. For example, the tripeptide *N*-formylnorleucyl-leucyl-phenylalanine is now commercially available and being widely used. In addition to acting as chemotactic attractants, these substances produce other effects in granulocytes, including an acceleration of locomotion. Such an accleration effect is called *chemokinesis* as opposed to chemotaxis, which refers specifically to the directed orientation of locomotion. Until recently, it was accepted that chemotactic attractants do not affect the speed of locomotion but only steer its direction. It now appears, however, that this was misinterpretation based on studies in which enough serum proteins were already present to stimulate locomotion maximally before the (intended) attractants were added (61). Granulocytes actually remain quiescent in saline lacking any attractant (62). This chemokinetic effect of attractants has important implications, however unclear, for the nature of the linkage between attractant detection and cellular propulsion. It would be valuable to know whether any substances can stimulate and accelerate locomotion in granulocytes without also being attractants (directing locomotion when present as a gradient), as would be expected in principle.

By comparing the chemotactic effectiveness and mutual competitive interaction of a series of different synthetic polypeptides, it has been possible to learn a good deal about the cell-surface receptors that mediate chemotaxis. There appear to be several different classes of receptors. For example, the strong attractant C5a does not interfere with the binding of f Met–Leu–Phe. Using tritium-labeled attractants, it has been found that the bound tripeptides are cleaved at the cell surface, apparently by an enzymatic action of the receptor molecule itself. Furthermore, the differing proteolytic susceptibilities of the various tripeptides are paralleled by their differing potencies as attractants, and several inhibitors of proteolysis also inhibit chemotaxis.

Even though granulocytes can orient to chemotactic gradients over a broad range of absolute concentrations, their ability to follow gradients of a given proportional steepness (a tenfold concentration change within 1 mm, for example), does vary with concentration. Granulocytes can orient best in concentration ranges nearest the dissociation content of the tripeptide-receptor complex (55). This constant (about 10^{-8} M) corresponds to the molar concentration of tripeptide at which half the receptors are occupied. The finding that the cells are most efficient at comparing concentrations at this level seems to imply that the cells are actually sensing the difference in the actual number of receptors occupied at opposite sides of the cell. One might have expected that they would sense differences in the proportion of receptors bound, but in that case the granulocytes should have been most sensitive to shallow gradients in the range one or two orders of magnitude lower than the K_d.

The number of receptors on the granulocyte surface declines markedly after exposure to attractant (59). An average granulocyte might begin with about 50,000 receptors scattered over its surface, but exposure to 2×10^{-8} M Met–Norleu–Phe will cause about half of these to be internalized in a phenomenon reminiscent of the "down-regulation" of hormone receptors in other cell types. At any given level of attractant, the number of receptors will reach a plateau level that depends inversely on the attractant concentration. The internalization process has been observed using fluorescently tagged peptides, but curiously no capping of the attractant was observed. It remains unclear how (or whether) the internalization of receptor-attractant complexes is related to gradient detection, or why these complexes do not participate in the rearward surface flow which caps other materials.

4. CONTACT GUIDANCE AND ITS MOLECULAR MECHANISM

Tissue culture cells direct their locomotion so as to orient themselves parallel to fibers, ridges, grooves, or more generally along the axis of minimum curvature of their substratum. This phenomenon is called contact guidance (63) and is an extremely effective means of orienting cell migration (and axonal extension, in particular) over long distances. So nearly universal is contact guidance among cultured cells, and so ubiquitous are tracts of oriented fibers within the body, that it would be remarkable if contact guidance did not guide many or most cell migrations within the body. Some specific cases have been documented (64) through the role of this phenomenon in guiding neuronal connectivity (65) is still debatable. Localized cellular traction can orient collagen fibers over distances of several centimeters (15), and these aligned tracts can serve almost like railway tracks for migrating cells.

Weiss explained contact guidance as essentially a form of capillarity in which parts of the cell margin are pulled outward by their adhesion to fibers and grooves, much as water is drawn up narrow tubes of a wettable material such as glass. However, such an explanation is metaphorical at best, especially considering that cellular migration is an active process of locomotion, rather than a passive wetting of the substratum. This capillarity hypothesis also cannot explain why cells cultured on glass fibers many times their own diameter should orient their own elongation along the fiber axis (66).

Recently, a new hypothesis has been proposed that explains contact guidance in terms of the cytoplasmic contractile proteins, which provide the force for cell migration (66, 67). This theory holds that bending stress fibers or other arrays of actomyosin interfere with their contractile function. Thus, when a cell spreads on a curved substratum and itself follows the curvature, this bending effect will differentially inhibit cell spreading in directions of greatest curvature (across a groove or around a fiber). Fibroblasts cultured on glass cylinders of varying diameters fail to orient longitudinally on the thicker cylinders, and the threshold degrees of surface curvature required to elicit contact guidance correspond closely to the curvature required to bend stress fibers extending from a cell center to the margin around the curve (Fig. 12) (66).

Fibroblasts were also cultured on a series of specially made prisms shaped something like the pitched roof of a house but with different angles across the ridge line ranging from 1° (nearly flat) up to 32° (approximately an ordinary roof). The occasional cells that happened to straddle these ridge lines were necessarily bent at this angle also. When the angle exceeded 4°, the cytoplasmic actomyosin cables refused to cross the ridge at all and separate independent cytoskeletal patterns would develop in the parts of each cell on either side of the ridges (66). Apparently these cytoplasmic fibers are mechanically incapable of bending around even such slight corners, which supports this theory of contact guidance (Fig. 13). If there were cells whose cytoplasmic fibers were more flexible (or perhaps just shorter relative to the

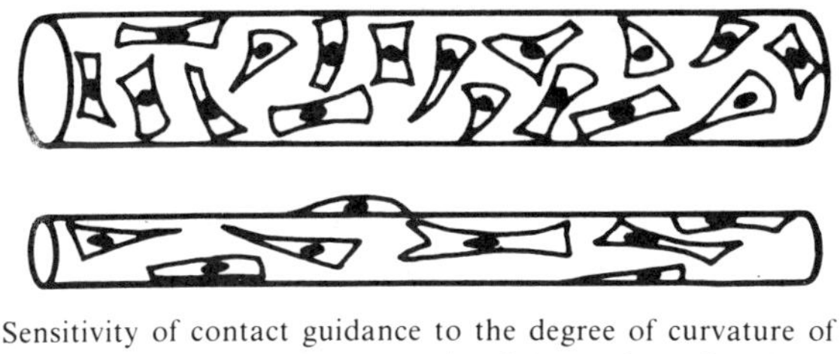

FIGURE 12. Sensitivity of contact guidance to the degree of curvature of the substratum. Fibroblasts align themselves along the long axis of the surface of a glass cylinder of small diameter, but show much less alignment on another cylinder with twice the diameter. [After Dunn and Heath (66).]

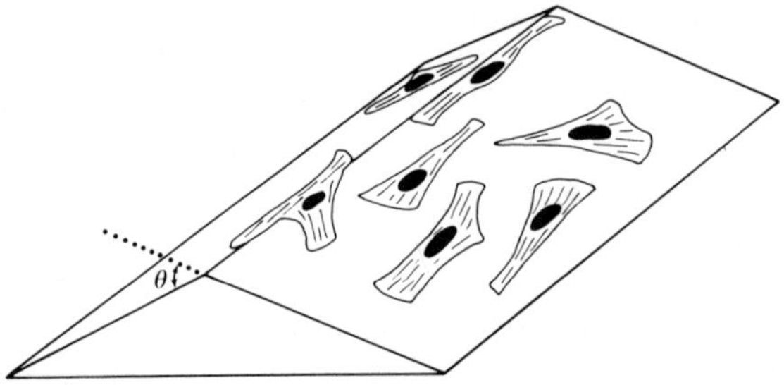

FIGURE 13. Diagram of the prism experiment of Dunn and Heath (66). Fibroblasts were cultured on the surfaces of specially made silica prisms having ridge angles (θ) of 1°, 2°, 4°, 8°, 16°, and 32°. Cells were inhibited from crossing ridge angles of 8° or greater and cytoplasmic fiber networks were disrupted across angles of 16° or greater.

cell's thickness) presumably these cells should be less susceptible to contact guidance (and vice versa). On the other hand, various published observations suggest that cells can orient along fibrin and collagen of almost molecular dimensions and Rosenberg (68) found that cells can even orient along grooves cut in layers of fatty acid deposited with a Langmuir trough only a few tens of nanometers thick. Recently O'Hara and Buck (69) have described the alignment of fibroblasts along grooves cut as well as molded in plastic substrata. They observed (with scanning electron microscopy) that the fibroblasts bridge over the grooves rather than extending down into them. This seems to contradict both the Weiss and the Dunn hypotheses, unless perhaps the greater shrinkage of the cells relative to substratum during scanning electron microscopy preparation has pulled the cells flat and out of the grooves.

O'Hara and Buck propose yet another mechanical explanation based on the shape of the (usually elongate) adhesion plaques between fibroblasts and substrata. The idea is that these adhesions will have to fit in between the substratum adhesions. Thus, when the grooves are closer together than the length of those adhesion plaques (or when the fibers are narrower than this distance), the adhesions will be strongest along this axis. This simple geometric restriction on the elongate adhesions may indeed play a role in the phenomenon under some circumstances but is of no help in explaining why cells should elongate along glass fibers several tens or even hundreds of micrometers in diameter. More attention should also be given to contact inhibition in cells that lack organized linear stress cables of actomyosin or the elongate adhesion plaques in which these cables end. Macrophages, for example, are supposed to lack both, but are high subject to contact guidance. We should also not ignore the possibility that contact guidance may actually have several different complementary mechanisms, some acting in certain cell types or conditions and some in others.

5. PROSPECTS

There appears to be a convergence between studies of the mechanics of cell locomotion and studies of its directional guidance. We saw an example of this in the discussion of contact guidance, where some of the explanatory hypotheses involved direct effects of substratum curvature upon the cytoplasmic actomyosin network. In the case of chemotaxis, such an approach is still needed. There is little knowledge or even speculation concerning the connections that must exist between the propulsive actomyosin system of cells and the cellular machinery for comparing receptor occupancy at different points on the cell surface as cells detect attractant gradients. Although the machinery for gradient detection may be quite distinct from the actual machinery for propulsion, with the former merely providing directional instructions to the latter, we should not simply just assume so. New functions are often established during evolution by modifying the components of existing functions, and there are many possible ways in which the propulsive machinery of cells might have been modified to serve functions such as comparing attractant concentrations in order to respond to them. This type of connection between steering and propulsion should be a focus of much further work.

In fact, the aspect of propulsion itself which remains least well understood is directional coordination. No amount of cytoplasmic contractility can suffice to move a cell from place to place unless this contractility is adequately coordinated, so that it occurs sequentially at the right times and places in the cell.

As we approach an understanding of how a tissue cell coordinates its own internal contractility in order to achieve directional locomotion, it should become easier to comprehend how a cell can detect and respond to external guidance during processes such as chemotaxis. Conversely, an increased understanding of the mechanisms of external guidance should provide new approaches to analyzing the intracellular regulation of contractility and locomotion.

REFERENCES

1. R. D. Allen, Differences of a fundamental nature among several types of amoeboid movement, in *Aspects of Cell Motility*, Society of Experimental Biology Symposium No. XXII, 1968, p. 151.
2. D. L. Taylor, S. B. Hellewell, H. W. Virgin, and H. Heiple, The solation-contraction coupling hypothesis of cell movements, in S. Hatano, H. Ishikawa, and H. Sato, eds., *Cell Motility: Molecules and Organization*, University Park Press, Baltimore, 1979, pp. 363–377.

3. J. P. Trinkaus, On the mechanism of metazoan cell movements, in G. Poste and G. L. Nicolson, eds., *The Cell Surface in Animal Embryogenesis and Development*, Elsevier/North Holland, Amsterdam, 1976, pp. 225–329.
4. A. Cloney, Echinoderm Development (16 mm film or 8 mm film loops), Kalmia Company, 1971.
5. T. Gustavson and L. Wolpert, *Biol. Rev.* **42**, 442 (1967).
6. E. R. Clark, *Am. J. Anat.* **23**, 37 (1918).
7. E. R. Clark, H. T. Kirby-Smith, R. O. Rex, and R. G. Williams, *Anat. Rec.* **47**, 187 (1930).
8. W. S. Ramsey, *Exp. Cell Res.* **72**, 489 (1972).
9. M. A. Abercrombie, J. Heaysman, and S. Pegrum, *Exp. Cell Res.* **59**, 393 (1970).
10. M. H. Gail and C. W. Boone, *Biophys. J.* **10**, 980 (1970).
11. G. Albrecht-Buehler, *J. Cell Biol.* **72**, 595 (1977).
12. R. G. Harrison, *Anat. Rec.* **1**, 116 (1907).
13. T. Elsdale and J. Bard, *J. Cell Biol.* **54**, 626 (1972).
14. S. B. Carter, *Exp. Cell Res.* **48**, 189 (1967).
15. A. K. Harris, D. Stopak, and P. Wild, *Nature* **290**, 249 (1981).
16. A. S. G. Curtis, *J. Cell Biol.* **20**, 197 (1964).
17. C. S. Izzard and L. R. Lochner, *J. Cell Sci.* **42**, 81 (1980).
18. M. Abercrombie, J. E. M. Heaysman, and S. M. Pegrum, *Exp. Cell Res.* **60**, 437 (1970).
19. A. K. Harris, Cell surface movements related to cell locomotion, in *Locomotion of Tissue Cells*, CIBA Foundation Symposium 14, Elsevier, Amsterdam, 1973, pp. 3–26.
20. N. K. Wessells, B. Spooner, and M. A. Luduena, Surface movements, microfilaments and cell locomotion, in *Locomotion of Tissue Cells, CIBA Foundation Symposium 14* (new series), Elsevier/North Holland, Amsterdam, 1973, pp. 53–82.
21. Z. H. Price, *Exp. Cell Res.* **48**, 82 (1967).
22. M. Abercrombie, J. E. M. Heaysman, and S. M. Pegrum, *Exp. Cell Res.* **62**, 389 (1970).
23. M. Edidin and A. Weiss, *Proc. Natl. Acad. Sci. USA* **69**, 2456 (1972).
24. R. Taylor, P. Duffus, M. Raff, and S. dePetris, *Nature (New Biol.)* **233**, 225 (1971).
25. G. Mendoza and H. Metzger, *Nature* **264**, 548 (1976).
26. H. Harris, E. Sidebottom, D. M. Grace, and M. E. Bramwell, *J. Cell Sci.* **4**, 499 (1969).
27. P. I. Marcus, *Cold Spring Harbor Symp. Quant. Biol.* **27**, 351 (1962).
28. M. Clarke and J. A. Spudich, *Ann. Rev. Biochem.* **46**, 797 (1977).
29. R. D. Goldman, T. Pollard, and J. Rosenbaum, eds., *Cell Motility*, Vol. 3, Cold Spring Harbor Conference on Cell Proliferation, Cold Spring Harbor Laboratory, N.Y., 1976.
30. S. E. Hitchcock, *J. Cell Biol.* **74**, 1 (1977).
31. E. D. Korn, *Proc. Natl. Acad. Sci. USA* **75**, 588 (1978).
32. T. D. Pollard and R. Weihing, *CRC Crit. Rev. Biochem.* **1**, 1 (1974).
33. H. Ishikawa, Identification and distribution of intracellular filaments, in S. Hatano, H. Ishikawa, and H. Sato, eds., *Cell Motility: Molecules and Organization*, University Park Press, Baltimore, 1979, pp. 417–444.
34. D. C. Lin, K. D. Tobin, M. Grumet, and S. Lin, *J. Cell Biol.* **84**, 455 (1980).
35. J. Wehland, M. Osborn, and K. Weber, *Proc. Natl. Acad. Sci. USA* **74**, 5613 (1977).
36. J. W. Sanger, *Cell Tiss. Res.* **161**, 431 (1975).
37. E. Lazarides, *J. Cell Biol.* **68**, 202 (1976).

38. M. Verderame, D. Alcorta, M. Egnor, K. Smith, and R. Pollack, *Proc. Natl. Acad. Sci. USA* **77**, 6624 (1980).

39. L. G. Tilney, S. Hatano, H. Ishikawa, and M. Mooseker, *J. Cell Biol.* **59**, 109 (1973).

40. E. Gruenstein, A. Rich, and R. R. Weihing, *J. Cell Biol.* **64**, 223 (1975).

41. I. M. Herman and T. D. Pollard, *J. Cell Biol.* **88**, 346 (1981).

42. M. C. Willingham, S. S. Yamada, P. J. Bechtel, A. V. Rutherford, and I. H. Pastan, *J. Histochem. Cytochem.* **29**, 1289 (1981).

43. R. G. Painter, M. Sheetz, and S. J. Singer, *Proc. Natl. Acad. Sci. USA* **72**, 1359 (1975).

44. Y. Rikihisa and D. Mizuno, *Exp. Cell Res.* **110**, 87 (1977).

45. J. V. Small, G. Isenberg, and J. E. Celis, *Nature* **272**, 638 (1978).

46. B. M. Jockusch, Alpha actinin-like proteins, in S. Hatano, H. Ishikawa, and H. Sato, eds., *Cell Motility: Molecules and Organization*, University Park Press, Baltimore, 1979, pp. 121.

47. E. Lazarides and K. Burridge, *Cell* **6**, 289 (1975).

48. B. M. Jockusch and G. Isenberg, *Proc. Natl. Acad. Sci. USA* **78**, 3005 (1981).

49. B. Geiger, K. T. Tokuyasu, A. H. Dutton, and S. J. Singer, *Proc. Natl. Acad. Sci. USA* **77**, 4127 (1980).

50. J. A. Wilkins and S. Lin, *Cell* **28**, 83 (1982).

51. F. E. Oldfield, *Exp. Cell Res.* **30**, 125 (1963).

52. R. W. Gunderson and J. N. Barrett, *J. Cell Biol.* **87**, 546 (1980).

53. S. H. Zigmond and J. G. Hirsch, *J. Exp. Med.* **137**, 387 (1973).

54. B. M. Glaser, P. A. D'Amore, H. Seppa, S. Seppa, and E. Schiffmann, *Nature* **288**, 483 (1980).

55. S. H. Zigmond, *J. Cell Biol.* **88**, 644 (1981).

56. W. S. Ramsey, *Exp. Cell Res.* **70**, 129 (1972).

57. S. H. Zigmond, *Nature* **249**, 450 (1974).

58. E. Schiffmann, B. A. Corcoran, and S. M. Wahl, *Proc. Natl. Acad. Sci. USA* **72**, 1059 (1975).

59. S. H. Zigmond and S. J. Sullivan, *J. Cell Biol.* **82**, 517 (1979).

60. D. E. Koshland, *Science* **196**, 1055 (1977).

61. M. McCutcheon, *Physiol. Rev.* **26**, 319 (1946).

62. S. H. Zigmond, A new visual assay of leukocytes chemotaxis, in J. Gallin and P. Quies, eds., *Leukocyte Chemotaxis: Methods, Physiology and Clinical Implications*, Raven Press, New York, 1978, pp. 57–66.

63. P. Weiss, *Exp. Cell Res. Suppl.* **8**, 260 (1961).

64. R. D. Campbell and B. A. Marcum, *J. Cell Sci.* **41**, 33 (1980).

65. P. Weiss, Neurogenesis, in B. H. Willier, P. A. Weiss, and V. Hamburger, eds., *Analysis of Development*, Hafner Publishing, New York, 1955.

66. G. A. Dunn and J. P. Heath, *Exp. Cell Res.* **101**, 1 (1976).

67. G. A. Dunn and T. Ebendal, *Zoon* **6**, 65 (1978).

68. S. V. Boyden, *J. Exp. Med.* **115**, 453 (1962).

69. P. T. O'Hara and R. C. Buck, *Exp. Cell Res.* **121**, 235 (1979).

6

Regulation of Neural Crest Cell Migration and Differentiation

JAMES A. WESTON
Department of Biology
University of Oregon
Eugene, Oregon 97403

Recent successes in understanding the structure of individual genes and how they function (1) suggest that we may soon be able to answer the central question of development—how the use of genetic information in cells is regulated. Before we fully understand how higher eukaryotes develop, however, we must determine what regulates the progressive restriction of the developmental repertoire of embryonic cells so that cell lineages proceed toward the stable expression of a particular phenotype. We must ask, in other words, how genetic determinants are selected to be regulated in a specific differentiating cell population.

Another important question concerning the development of multicellular organisms is raised by the fact that differentiated cells do not exist as randomly distributed individuals, but reside within a complex, spatially organized tissue structure. We must inquire, therefore, how this pattern of spatial organization arises during development and how it is controlled.

Obviously, these two questions are related. The goal of developmental control mechanisms is to generate precisely localized, uniquely differentiated cell populations within a multicellular organism. During development, the history of a cell (i.e., its lineage and the experiences of its ancestors) is important. Likewise, a cell's own environment (which may be other cells or highly organized, noncellular components of the extracellular matrix), and its ability to detect developmental cues in its neighborhood, will affect its ability to use some genetic determinants and to ignore the existence of others (see also Chapter 4).

To analyze the mechanisms involved in such developmental processes, a system should be employed that provides an identified stem cell population giving rise to a variety of cellular phenotypes in defined spatial patterns. Vertebrate neural crest cells appear to satisfy these requirements. This chapter, therefore, will define several developmental events occurring at the cellular level during neural crest morphogenesis, and then will describe specific macromolecular changes in the microenvironments of these cells. Then, it might be possible to pose specific questions about the molecular mechanisms that underlie the complex developmental processes leading to the choice of cellular phenotypes and the positioning of differentiated cells in the organism.

1. THE ORIGIN AND FATE OF NEURAL CREST CELLS

As a vertebrate embryo develops, a region on its surface (the medullary plate) forms and folds in a complex fashion to produce the neural tube. The lateral folds of this medullary plate contain a population of cells with remarkable morphogenetic and differentiative abilities. This population of cells is segregated from the neural epithelium at, or slightly before, the time of fusion of the folds to form the brain and the neural tube. Thus segregated from the neural epithelium, these cells—called neural crest cells—embark on an extensive migration through the embryo (2–5). Ultimately, these neural crest cells will stop migrating in precisely defined locations in the developing embryo.

1.1. Neural Crest Derivatives

The embryonic neural crest provides stem cells that differentiate into a number of derivatives. As Table 1 illustrates, differentiated cells derived from the neural crest include skeletal and connective tissues of the head and face. The neural crest is the major source of cells that form the jaw, teeth and palate, and other facial structures. Crest cells also produce the pigment cells found in the iris of the eye and in the skin and its specializations (hair and feathers). Finally, the entire peripheral nervous system, including the sensory and autonomic neurons, the supportive cells of the peripheral nerves, and many

TABLE 1. Derivatives of the Neural Crest

A. Cranial neural crest only:
   1. Skeletal and connective tissue
      a. Bones and cartilages of the head and face
         Upper and lower jaw
         Dental papilla (odontoblasts)
         Palate
         Cranial vault floor
      b. Connective tissues
         Corneal stromal fibroblasts
         Contributions to dermis and subcutaneous adipose tissue of face, jaw, and upper neck
         Lining of forebrain (meninges)
      c. Muscles
         Ciliary muscles (striated)
         Smooth muscle of cranial blood vessels and dermis
   2. Endocrine and exocrine tissues
      Mesenchyme of thyroid, parathyroid thymus, and salivary glands
B. Cranial and trunk neural crest:
   1. Neural derivatives
      a. Sensory neurons of head and trunk (peptidergic and ?)
      b. Sympathetic neurons (noradrenergic)
      c. Parasympathetic neurons (cholinergic and ?)
      d. Neurosecretory cells
         Calcitonin-producing (C) cells of the thyroid
         Adrenal medullary cells (adrenergic)
      e. Supportive cells of the peripheral nervous system
         Glia
         Schwann sheath cells
         Satellite cells of ganglionic neurons
   2. Pigment cells of skin, hair (feathers), and iris

neurosecretory cells, originate from this identified embryonic cell population.

In order to study the differentiation of any cell, we require characteristic *phenotypic markers* by which the process can be recognized and followed. It is gratifying, therefore, that so many of the different cell types derived from the neural crest possess traits by which their differentiation can be characterized. Some of the phenotypic markers that have been utilized to date are summarized in Table 2.

1.2. Neural Crest Cell Migration

Before they express their definitive phenotype in characteristic locations in the embryo, crest cells normally migrate for considerable distances from their origin in the neural folds or at the top of the neural tube. Clearly, the ultimate distribution of crest cells in the embryo will be affected by the pathways they take.

1.2.1. Patterns of Migration. The pathways at neural crest migration appear to depend on their position along the anterior-posterior axis of the embryo. In the region of the brain, there are several different patterns of crest cell migration (6, 7) (Fig. 1). In the trunk region of the embryo, the pathways of migration partially resemble the pattern seen in the region of the hindbrain and midbrain. Here, the crest cells initially follow two migratory paths leading into cell-free spaces—one laterally toward overlying surface ectoderm and one down into the body of the embryo between the neural tube and the epithelial somite (Fig. 1). The cells that migrate in association with the ectoderm are uniformly distributed (8, 9), whereas the cells on the pathway into the embryo become segmentally distributed when they encounter the cells of the developing somite (8). These and other observations suggest that the distribution pattern is affected by environmental cues in the embryonic spaces where the crest cells migrate. This inference has been tested by a number of ingenious grafting experiments.

1.2.2. Environmental Determination of Pathways. The crest migratory pathways are determined, in part, by the embryonic environment. Thus, when donor neural crest cells from one level of the embryonic axis (e.g., the midbrain or trunk) are grafted in place of the host's crest at another axial level—the hindbrain, for example—the pattern of migration of the grafted crest cells is observed to correspond to the graft site, not its origin (7, 10, 11). Other experiments also support the idea that the local embryonic environment determines the pattern of crest cell distribution. For example, the discontinuous (segmental) distribution of crest cells that occurs when cells

TABLE 2. Diagnostic Cellular Markers for Some Neural Crest Phenotypes

Neural Crest Derivative	Morphological/Structural	Biochemical/Cytochemical	Physiological[a]
Pigment cells	Pigment granule (melanosomes)	Melanoprotein enzymes: DOPA, oxidase tyrosinase	*Tyr (DOPA) uptake
Neurons[b]	Cell shape: large cell bodies, long cell processes, synaptic structures (varicosities); neurofilaments; Nissl substance	Complex gangliosides; Neurofilament protein; NSE; Immunocytochemical staining (MCAb)	Cholera-, tetanus-toxin binding; NGF-binding and response; Electrical activity; Na-channels (tetrodotoxin binding; Na-flux stimulated by veratridine)
Aminergic	Dense-core vesicles Histochemical or immunocytochemical staining	FIF (catecholamines) Transmitters: NE, DA, 5HT Enzymes: TH, DBH, PNMT, MAO Receptors: AChR; NGF Antigens for MCAb	Transmitter release; High-affinity CA uptake; Toxin (e.g., bungarotoxin)-binding; NGF-binding and response
Cholinergic	Histochemical or immunocytochemical staining Synaptic vesicles	Transmitter: ACh Enzymes: CAT, AChE Receptors: AChR; NGF Antigens for MC Ab	Transmitter release; Toxin binding (see above)

Peptidergic	Histochemical or immunocytochemical staining	Neuroactive peptides/ hormones Somatostatin Substance P Calcitonin
Glia/support cells	Association with neurons; small, stellate morphology; immunocytochemical staining	Proteins S100 GFAP Galactocerebrosides
Ectomes-enchyme, cartilage	Fibroblastic cell shape	Cartilage matrix Type-II collagen; matrix proteoglycans

[a] Abbreviations: DOPA, dihydroxyphenylalanine; Tyr, tyrosine; FIF, formaldehyde-induced fluorescence; NE, norepinephrine; DA, dopamine; 5HT, 5-hydroxytryptamine (serotonin): CA, catecholamine; TH, tyrosine hydroxylase; DBH, dopamine-β-hydroxylase; MAO, monoamine oxidase; AChR, acetylcholine receptor; MC Ab, monoclonal antibody; NGF, nerve growth factor ACh, acetylcholine; CAT, choline-acetyl transferase; AChE, acetyl-choline esterase; GFAP, gliofibrillar acidic protein; PNMT, phenylethanolamine-N-methyltransferase; NSE, neuron specific enolase.

[b] Neurons classified by secretory product: Aminergic neurons produce biogenic amines; cholinergic neurons produce ACh; Peptidergic cells produce neuroactive peptides.

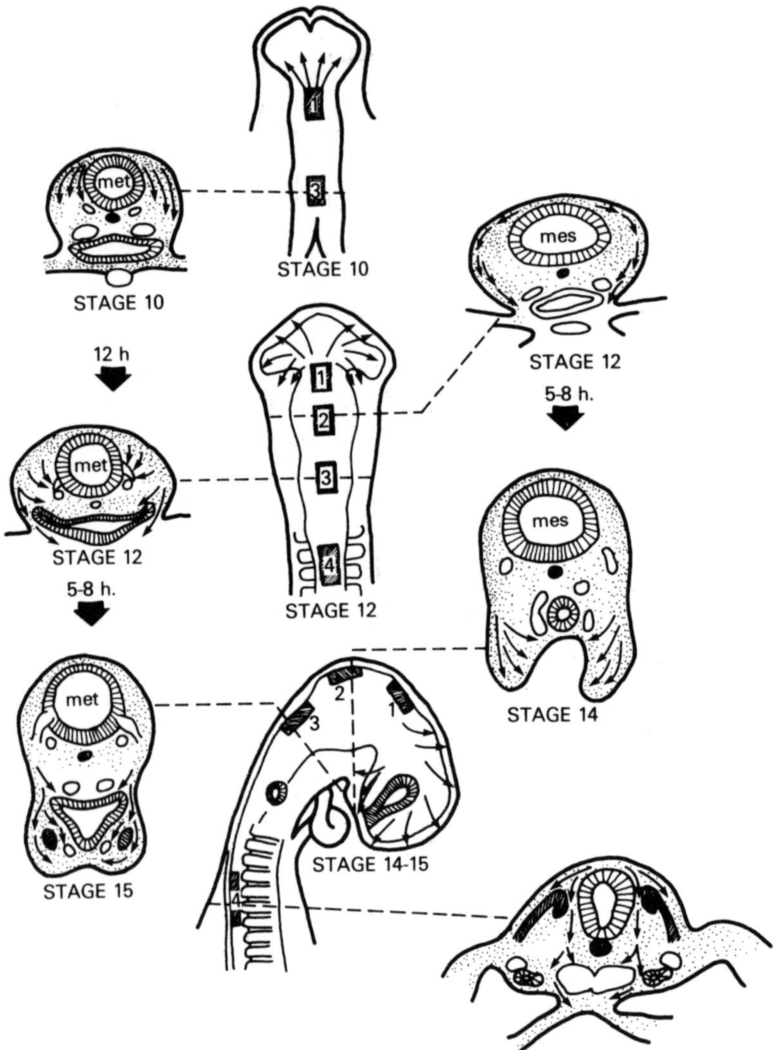

FIGURE 1. Pathways of neural crest cell migration at different axial levels of the avian embryo. Shaded, numbered rectangles designate source of crest cells whose normal migratory pathways are illustrated: (1) Anterior midbrain- and posterior forebrain-derived crest cells give rise mainly to skeletal and connective tissue cells in the head; (2) posterior midbrain crest contributes primarily to the upper jaw, a ganglion (ciliary) associated with the eye, and various glial derivatives; (3) the anterior hindbrain-derived crest is the major source of cells of the lower jaw and the fifth cranial (trigeminal) ganglion; (4) the trunk neural crest cells begin migrating from the neural tube relatively late, and initially follow two pathways in interstitial spaces—one laterally beneath the embryonic "skin," to form pigment cells, and one ventrally into the embryo, to form components of the peripheral nervous system. Crest cells from anterior levels of the neural tube associate with the newly forming gut and then move posteriorly along the gut to form intrinsic ganglia. *met*, metencephalon; *mes*, mesencephalon. Adapted from (6) (see also 2, 5).

on the ventral pathway encounter the somitic mesenchyme leads to the segmental distribution of crest-derived spinal ganglia (2, 8). This pattern can be perturbed simply by altering the embryonic environment. Thus, surgically adding or removing somites in embryos causes correspondingly more or fewer ganglia to form (see reference 2).

1.2.3. Initiation and Termination of Migration. Developmental changes in the embryonic environment affect the onset and cessation of crest cell migration. Thus, the migration of crest cells at the level of the midbrain in avian embryos is preceded by the appearance of relatively large cell-free spaces lateral to the neural folds (12). If these spaces fail to form, for whatever reason, crest migration fails to occur. Two other examples also indicate that extended crest cell migration requires new spaces in the embryonic periphery to "open" to receive the cells. Thus, crest-derived cells that eventually migrate into the developing eye to form a loose connective tissue *stroma* await the appearance of an expanded cell-free space between the developing lens and the corneal epithelium (see Chapter 8). Likewise, Derby (13, 14) has concluded that the lateral migration of pigment-forming cells is delayed in the mouse embryo until the close apposition of somite and ectodermal epithelia is disrupted to open a migratory pathway.

Finally, changes arise in the crest migratory pathways during development that appear to limit the direction and extent of migration. Thus, when premigratory crest is grafted into progressively older host embryos, the extent of migration appears to be progressively limited in older hosts (15). Further, when crest cells are grafted into unusual (*ectopic*) locations deep in the host embryo, their migration tends to continue away from the source of host crest cells rather than moving against the stream of migrating host crest. In contrast, if crest cells are grafted deep into a very young host embryo, before its own crest cells have begun to migrate, the grafted cells are able to migrate back toward the source of the host crest cells on their migratory pathway. These observations demonstrate that crest migratory spaces change in some way as normal migration proceeds, and it has been suggested that crest cells themselves may be partially responsible for this change (16).

These experiments imply that the pattern of crest cell migration (i.e., onset, pathway, and extent) may be determined largely by the environment (e.g., by providing a defined space and a suitable substratum) and that the crest cells themselves play only a minor role in determining their own morphogenetic behavior. If this be correct, we would predict that any motile cell population placed into the crest migratory spaces would disperse, just as they would on any appropriate tissue culture surface, and would become distributed in a pattern resembling that of crest cells on the same pathway. This experiment has recently been done (16–18) and the results refute the prediction: crest-derived cells disperse reasonably normally in host embryos, whereas other embryonic cells (e.g., somite cells, embryonic heart, or limb bud fibroblasts) remain at the graft site and do not disperse.

An exceptional result is obtained, however, if a neoplastic cell type (Sarcoma 180) is placed in the crest migratory space. These cells disperse and become distributed along the crest pathway. Such cells also disperse if implanted into other embryonic tissues (i.e., at the base of the limb bud of a chick embryo). It is particularly interesting that crest cells, but not fibroblasts, also disperse when grafted "ectopically" in host embryos (i.e., placed directly into the somite, or in the lateral plate and limb bud mesenchyme) (16, 17).

Taken together, these results suggest that crest cells possess remarkable abilities to migrate through embryonic tissues, in contrast to certain other embryonic cells, and that this property combines with developmental changes in the migratory environment to determine the pattern of crest cell migration. The similarity of the behavior of neoplastic cells and crest cells, moreover, suggests that their "invasive" abilities might result from common mechanisms. This idea will be considered below (Section 2.1.1).

2. THE STRUCTURE OF THE CREST MIGRATORY SPACES

If we want to analyze how crest cell morphogenetic behavior is affected by environmental cues, we must first determine the nature of the environments encountered by migrating crest cells. Next, we should examine the sources of the environmental components and determine how these components can be modified. We must then attempt to establish how observed patterns of morphogenetic behavior correlate with measurable changes in the composition of the extracellular matrix. This can be done either in vivo or in vitro. Although much still remains to be done, such an approach is beginning to establish a basis for understanding the regulation of morphogenesis of the neural crest.

2.1. The Extracellular Matrix in Crest Migratory Spaces

A number of workers have observed that neural crest, or crest-derived cells, migrate into cell-free spaces bounded by basement membranes and filled with extracellular matrix (3, 12, 14, 19). The major constituents of such matrices are described in Chapters 8 and 9. Basement membranes, for example, include characteristic collagen molecules, various glycosaminoglycans (GAGs), and a number of other structural macromolecules, such as laminin and fibronectin. These macromolecules interact with each other in specific ways to establish stable structures that associate with epithelia and that fill cell-free (interstitial) spaces. They are all discussed in detail in other chapters of this book.

2.1.1. Fibronectin in the Crest Migratory Spaces. The glycoprotein fibronectin has recently been shown to be a significant component of the extracellular matrix in the crest migratory spaces (20–23). Although the source of this material is not yet known, and local amounts of fibronectin have not yet been measured, this major component of the extracellular matrix seems to play an important role in the morphogenetic and differentiative behavior of neural crest cells (see Section 3.1.2.3). In view of the similarity between the behavior of crest cells and a neoplastic cell type when they were grafted ectopically into host embryos (see Section 1.2.3), it is interesting that both crest cells in culture and many transformed cells lack cellular fibronectin (24–26; see also reference 27). We do not yet know whether or how depletion of cellular fibronectin imparts "invasive" abilities to crest and other cell types. It may be that lack of fibronectin or the absence of fibronectin binding sites on the cell surface facilitates cellular dispersal. It is also possible that, just as transformed cells, which lack surface fibronectin, respond to the presence of this molecule when it is supplied from elsewhere (27) (Chapter 9), neural crest cells may respond to the presence of fibronectin in their migratory spaces precisely because they lack this macromolecule at their surface (3, 28).

2.1.2. GAGs in the Crest Migratory Spaces. GAGs represent the predominant class of macromolecules in the crest migratory spaces. Indeed, GAGs are thought to be responsible for the dramatic developmental changes in the extracellular spaces in embryos, and cell-associated and extracellular GAGs have been implicated as regulatory factors in a number of developmental processes. These include cell proliferation, cell migration, and cell differentiation (see Chapter 8). Before it can be established which, if any, of these molecular components of the extracellular matrix affect crest cell morphogenetic behavior, however, we must document the kinds and amounts of some of the various components that these cells encounter.

2.1.2.1. Regional Differences in the Amounts of GAG. The amount of GAG present in various extracellular spaces in the embryo can be measured using a microscope photometer in conjunction with histochemical staining of embryonic tissue. Such analyses show that there are precise regional differences in the amounts of GAG in the extracellular spaces in the embryo (13, 14) (Fig. 2). Such differences between closely adjacent regions of the embryo could not readily have been detected by biochemical analyses. Moreover, although measurements are difficult to make, the results do reveal that the concentration of GAG in the intertissue spaces is astoundingly high

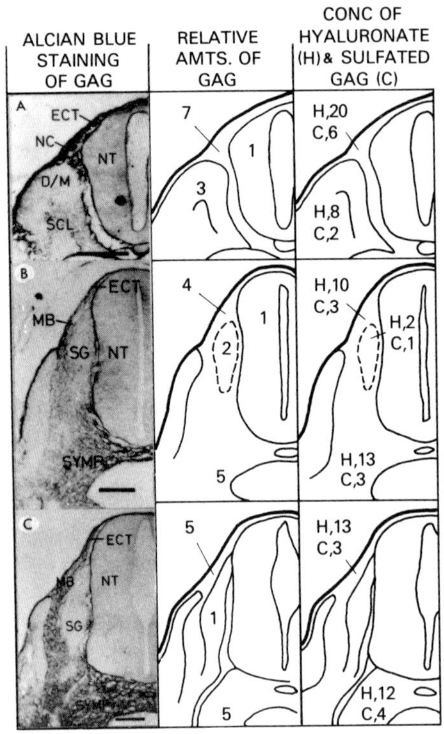

FIGURE 2. Three developmental stages in the trunk of the embryonic mouse. The first column shows transverse sections of embryos stained specifically for glycosaminoglycans (GAGs) with alcian blue. Microscopic photometry permits relative amounts of alcian blue-staining material to be measured in various adjacent locations in these sections. When such measurements are "normalized" to a part of the section (the neural tube) with little extracellular space, and therefore only background staining, characteristic regional differences, as well as developmental changes in GAG distribution can be documented (second column). If, prior to staining, sections are digested with enzymes that selectively remove one class of GAG (e.g., *Streptomyces* hyaluronidase, which under appropriate conditions removes only hyaluronic acid), the relative contribution of the different GAG classes to the total GAG in the ECM can be determined. Such proportions can be measured with the microscope photometer, and with suitable standards, these measurements can be used to estimate roughly the concentration of individual GAGs in the interstitial spaces. The numbers on the diagrams in the third column represent mg of hyaluronic acid (H) or sulfated GAG (C) per cc of interstitial space. Note the striking concentration differences within a section and local changes in GAG concentrations at different developmental stages. From (13, 14). ECT, ectoderm; NC, neural crest; NT, neural tube; D/M, dermatome/myotome; SCL, sclerotome; MB, presumptive melanoblasts; SG, sensory ganglion; SYMP, presumptive sympathetic ganglion cells.

(more than 20 mg/cm^3 of interstitial space). This discovery should emphasize why such measurements are useful: They permit us to construct more accurate experimental systems with which to analyze environmental control of cell morphogenetic and differentiative behavior.

2.1.2.2. GAG Composition. The proportions of various kinds of GAGs are also characteristic of different embryonic regions. In addition to semiquantitative measurements, the composition of GAG in the matrix can be estimated by selectively removing dye-binding macromolecules with enzymes of known specificity. Thus, *Streptomyces* hyaluronidase, which can be shown to digest only hyaluronic acid, removes about 55% of the dye-binding material in the region first encountered by migrating crest cells. After demonstrating that the enzyme does in fact degrade the designated molecule, that it does not affect the ability of remaining molecules to bind dye, and that other classes of GAG molecules are not removed nonspecifically, we can conclude that the proportion of hyaluronic acid in the spaces into which crest cells migrate is about 55% of the total GAG. In contrast, the proportion (and the absolute amount) of hyaluronic acid is significantly less in regions that crest cells have not yet entered (Fig. 2).

Likewise, Figure 2 shows that the proportions and the absolute amounts of sulfated GAGs (predominantly chondroitin sulfate and heparan sulfate) are different in closely adjacent embryonic spaces.

It must be emphasized that we do not yet know the significance to morphogenesis of these values. We should also emphasize, however, that one could not begin to pose questions about the role of such extracellular components in controlling morphogenesis until the facts about their amounts, kinds and distribution had been established.

2.1.3. Collagens in the Crest Pathway. Several types of collagen are found in embryonic extracellular spaces. Lining the spaces in association with embryonic epithelia, for example, type-IV collagen forms a major constituent of basal laminae (see reference 29). In cell-free spaces, other collagen types (I and II) are present as fibrillar components associated with other matrix macromolecules. Regional differences in the kinds, amounts, and structural associations clearly exist in the embryo, but the details of such differences and their relevance to crest morphogenetic and differentiative behavior have not yet been fully characterized. Some of these differences are undoubtedly due to tissue-specific patterns of collagen productions (see reference 30 and Section 2.2.2).

2.2. The Source of Extracellular Matrix Components

Having observed that there is a precise pattern in the amounts and kinds of extracellular material in the early embryo, it is natural to ask how these differences arise. Several obvious mechanisms are possible: First, different embryonic tissues may differ in the kinds or amounts of extracellular matrix (ECM) macromolecules (e.g., GAG, fibronectin, or collagen) that they produce. Then, since different embryonic tissues are present in different locations in the embryo, the contribution of each to adjacent spaces would clearly result in local differences.

Another way to account for local distribution patterns of extracellular matrix might be differential deposition or removal of the components of the matrix. Since these molecules are so large (see Chapters 8 and 9), and since they are known to interact specifically to form characteristic matrix fibrils and other structures (e.g., basement membranes), diffusion of these matrix components probably has little effect on the rate of accumulation. Instead, the rate of deposition, the composition and therefore the extent of molecular interaction, and finally the rate of degradation of the macromolecules will determine the rate of local accumulation of matrix.

Thus, at a minimum, it would be useful to know the net production of matrix components for each of the embryonic tissues associated with the crest migratory spaces. To obtain such data, two complementary approaches can be used. One approach, local incorporation of specific precursors into identifiable products by embryonic tissues in situ, can give an indication of qualitative differences in matrix production in embryonic regions (14, 31). To obtain quantitative data on the proportions of the various matrix components produced by each tissue, however, we must sacrifice tissue "integrity" and measure incorporation by homogeneous populations of embryonic cells in vitro. When this was done (31), striking differences in GAG production were revealed between neural crest and other embryonic cells.

2.2.1. Hyaluronic Acid Production. Since the region into which crest cells migrate is rich in hyaluronic acid, it is tempting to speculate that crest cells themselves might play a role in altering the environment in which they move. Neural crest cells are not, however, the sole source of hyaluronic acid. Nor, as Toole (32) has pointed out, is it likely that the production of large amounts of hyaluronic acid is sufficient to explain the sudden appearance of the large cell-free spaces into which the crest cells move. A close look at the kinds of attachments between adjacent epithelial layers in the embryo and the events in development leading to a separation of these tissues to form the crest migratory spaces would provide valuable information.

2.2.2. Sources of ECM Glycoproteins. The matrix glycoproteins found in the crest migratory spaces are probably produced by cells of the somite and of the neural and epidermal epithelia. A number of authors have now reported that crest cells lack and probably do not make fibronectin (20, 24, 25, 33). The lack of fibronectin at crest cell surfaces might also be due, however, to proteolytic activity by specialized enzymes at the cells' surface (34; Valinsky et al., in preparation; Reich, personal communication). Likewise, it is possible that crest cells produce fibronectin, but during their migratory phase, are unable to bind it at their surfaces.

In contrast, incorporation studies in vitro indicate that somite fibroblasts, like other normal fibroblastic cells, do make this macromolecule and accumulate it at their periphery to form part of a fibrillar extracellular matrix in regions where crest cells migrate (20, 23).

Finally, the neural and somitic epithelia that line the crest migratory space are known to be able to synthesize collagens and to organize these macromolecules at their periphery (30). Crest cells can also produce collagen (35) and some crest-derived cells contribute this material to their pericellular matrix (36).

2.3. Developmental Changes in Crest Migratory Spaces

Although we do not yet know the mechanism by which they occur, the changes in the intertissue spaces in which crest cells migrate are likely to be related to the role these spaces appear to play in regulating the pattern of crest cell migration (see Section 1.2.2). It might, for example, be valuable to examine changes that occur prior to the onset of crest cell migration, or that precede the extension of migrating crest cells into newly formed spaces. It would also be of interest to discover physical changes in the embryonic spaces that correlate with the diminished ability to support crest cell migration in older embryos (see Section 1.2.3). Finally, when crest-derived cells localize and aggregate to form a ganglion, a close look at changes in extracellular matrix molecules might suggest how the two events are causally linked.

2.3.1. Extracellular Matrix GAG and Crest Migration. The onset and extension of crest cell migration is preceded by the appearance of GAG-filled cell-free spaces rich in hyaluronic acid. As illustrated in Figure 2, crest and crest-derived cells migrate into spaces filled with hyaluronic acid. If the formation of these spaces were caused by the presence of GAG, and if this, in turn, permitted crest cells to enter, we would predict that if GAG were removed, then the spaces would also be altered and crest cell migration would be abnormal.

Several observations are consistent with these predictions:

1. When chick embryos are treated with hyaluronidase just prior to the appearance of the cell-free space lateral to the midbrain, the space fails to form (37).
2. Moreover, and of great interest, crest cells fail to migrate.
3. Likewise, as Toole has discussed (see Chapter 8), hyaluronidase activity causes a reduction in the proportion of hyaluronic acid and shrinkage of cell-free spaces in embryonic tissue.
4. Finally, there are mutants in the mouse that result in abnormal development of crest-derived structures (see Section 4; Table 4). In one such mutant, Patch (*Ph*), homozygous embryos die in early development. Before doing so, however, the embryos exhibit characteristic abnormalities of the head and face structures derived from cranial crest cells. They also form abnormal spaces containing extracellular matrix, into which crest cells migrate and seem to become "trapped." The details of the migration and fate of crest cells in these mutants have not yet been analyzed. It is possible, however, that subtle temporal and regional alterations in the distribution of extracellular materials are responsible for the altered morphogenetic events that occur in early development (3, 26).

2.3.2. Changes in Extracellular Matrix GAG Distribution. Cessation of crest cell migration correlates with the removal of extracellular GAG. The apparent loss of the ability of older embryonic environments to support the migration of crest cells occurs at about the same time in embryonic development that hyaluronidase activity in the somite increases and the proportion of hyaluronic acid decreases. This decrease occurs just where crest cells are migrating on their ventral path, and where some will begin to localize to form spinal ganglia. Figure 2 confirms that there are progressive local changes in the amounts and kinds of GAG where crest cells are beginning to coalesce to form spinal ganglia. Such changes in GAG could be due, in part, to developmental changes in the ability of the cells involved to produce or degrade GAG. In fact, a number of workers have demonstrated differences in the pattern of GAG synthesis by undifferentiated and differentiated crest cells in vitro (38, 39). Likewise, local changes in the pattern of GAG degradation have also been noted (3, 40) that might lead to local reductions in extracellular matrix GAG.

3. CONTROL OF NEURAL CREST CELL DIFFERENTIATION

Tables 1 and 2 dramatically illustrate that during normal development the neural crest produces a variety of cellular phenotypes—many with useful markers by which the process of differentiation can be characterized. Since the time and location of overt cell differentiation is largely known for many crest derivatives, and the morphogenetic history of these cells has also been well characterized, we can experimentally intervene in these processes to analyze the causal mechanisms involved. In general, such analyses have indicated that many embryonic crest cells are developmentally labile, that they differentiate in response to environmental stimuli, and that their differentiative abilities become progressively restricted as development proceeds.

3.1. Developmental Lability of Crest Cells

There are two basic kinds of experiments that demonstrate the extent to which neural crest cell differentiation can be altered by environmental stimuli. One approach is to transplant crest cells from one embryonic location to another (called *heterotopic grafting*). If differentiation were unaffected by such a procedure, one might conclude that the differentiative program was fixed for that cell. If, on the other hand, cell differentiation were different from what would have been expected in their original location, then we could conclude that the crest cells retained the ability to respond to cues in their new environment.

The second approach is to remove crest or crest-derived cells from the embryo and culture them in an artificial environment (i.e., in nutrient media on artificially contrived substrata). In these circumstances, some experimental control can be exerted over the environment in order to assess the effects of such environmental stimuli on cell differentiation.

3.1.1. Experimental Control of Differentiation In Vivo. After heterotopic grafting, some crest cells differentiate according to their new location in the host embryo rather than their original source. Some neural crest cells from the neck region of an avian embryo, for example, normally migrate into the developing gut to form parasympathetic (cholinergic) neurons that produce and use acetylcholine as their neurotransmitter (see Table 2). If crest cells from this region of a donor embryo are placed in the trunk level of a host [where crest cells would normally produce only sympathetic (adrenergic) neurons that make and use other neurotransmitters, e.g., adrenalin or noradrenalin], they differentiate according to their new location rather than their source. This experiment, therefore, clearly demonstrates that crest cells can differentiate in response to local environmental stimuli (11, 41). Similar heterotopic grafting experiments have also demonstrated that some cells of already formed parasympathetic ganglia can remigrate when placed back into crest migratory spaces of younger host embryos, and localize to form sympathetic ganglia. This result again indicates that the differentiation of at least some cells in a cholinergic ganglion can be redirected to become adrenergic neurons (42) (see also reference 5).

3.1.2. Environmental Control of Crest Cell Differentiation In Vitro. There are a number of experiments that clearly indicate that the differentiation of cultured crest cells is also altered by the culture environment encountered by the cells. Some of these experiments have been summarized in Table 3.

3.1.2.1. Melanogenesis. For example, although crest cells will produce a variety of cellular phenotypes in vivo, they often express only one phenotype—melanin pigment—when they are cultured. This common result sug-

TABLE 3. Responses of Some Neural Crest-Derived Cells to Environmental Cues in Culture

Cell Type	Environmental Stimulus[a]	Response[a]
Neural folds (crest)	Ectodermal epithelia ?	Melanogenesis
	Somite mesenchyme, notochord, ECM (collagen + Fn) } +NGF	FIF (adrenergic differentiation)
Cranial crest cells	Optic vesicle	Cartilage matrix
	Mandibular epithelium	Bone
	?	Melanogenesis
Nonneural cells of embryonic spinal ganglia	Loss of contact with neurons + ?	Melanogenesis
	Neuron cell surfaces	S100 protein (glial cell differentiation)
Melanocytes (–blasts)	ECM (collagen/Fn) + NGF	FIF (adrenergic differentiation)
Adrenergic neurons	Various conditioned media	ACh (cholinergic differentiation)

[a]Abbreviations: ECM, extracellular matrix; Fn, fibronectin; FIF, formaldehyde-induced-fluorescence; NGF, nerve growth factor; ?, source/nature of stimulus unknown.

gests that the culture environment can alter the pattern of differentiation and promote the expression of a single phenotype in cells that would normally have differentiated into other cell types if allowed to remain in the embryo. Here, unknown molecular components or physical conditions in the culture promote melanogenesis at the expense of other phenotypes.

Likewise, if crest-derived spinal ganglia are explanted in vitro, some cells, which would not normally do so, differentiate into pigment cells. During normal development, such crest-derived cells would probably have associated with precociously differentiating neurons and differented as supportive (glial) cells (43, 44). On the artificial culture substratum, however, the supportive cell precursors desert the neurons and are influenced by the culture environment to become melanocytes. On the other hand, when culture substrata were devised that discouraged the disruption of cell association in the ganglion, the differentiation of adventitious pigment cells was thwarted.

3.1.2.2. Gliagenesis. What is the fate of such cells when association with neurons is maintained? Recently, Holton has used sensitive immunological assays to show that interaction with nerves seems to promote the production of a characteristic glial cell protein (S100; see Table 2) by these crest-derived, nonneuronal cells. This clearly suggests that cells in the developing spinal ganglion make a choice between at least two possible phenotypes (pigment and supportive cell), and that this choice is influenced by what is probably a specific interaction with another cell population (45).

Neurons that had been killed with histological fixatives can also provide the stimulus needed by nonneuronal cells to produce S100. This suggests that these specific developmental cues are mediated by interactions at the cell surface or in the immediate environment surrounding the cells. The molecular basis for these interactions remains to be elucidated (see below, Section 3.2.3).

3.1.2.3. Adrenergic Neurogenesis. Neural crest-derived neurons also exhibit altered differentiative patterns in vitro. When crest cells are cultured in close proximity to developing somitic mesodermal cells, some of them acquire the ability to develop characteristic fluorescence when treated with formaldehyde vapor. This property, called formaldehyde-induced fluorescence, reflects the cells' ability to produce and/or store neurotransmitter molecules (catecholamines) characteristic of the adrenergic nervous system (46–48). Formaldehyde-induced fluorescence is expressed in these cells at the expense of the appearance of pigment granules.

For a variety of reasons, it seems likely that the crest cells in these experiments are being subjected to the influence of extracellular matrix produced by the cultured somite and perhaps notochordal tissue (5, 49) (Fig. 3). Re-

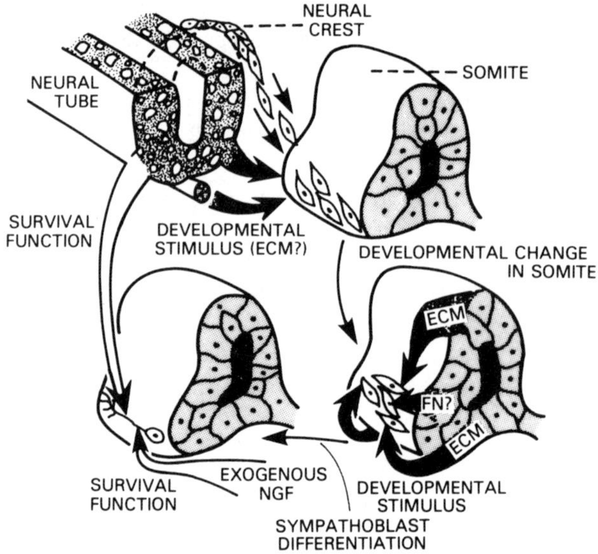

FIGURE 3. Diagrammatic representation of the progressive environmental stimuli encountered by crest cells migrating in the embryonic somitic environment. Such cues are postulated to influence adrenergic differentiation by these cells. Adapted from (48). ECM, extracellular matrix; FN, fibronectin; NGF, nerve growth factor.

cently, this inference has been tested in potentially interesting ways: If the effective components in this system are, in fact, one or more extracellular matrix constituents, and if, as seems likely, these macromolecular constituents interact to form an insoluble matrix around the cells, then it should be possible to allow cultured somite cells to establish their extracellular matrix, and then remove the cells without removing the matrix from the culture surface. This has been done (50). When the differentiative behavior of crest cells grown on such conditioned substrata was compared with cells grown in identical nutrient medium directly on control (plastic) substrata, the differences were striking (see Fig. 4). Some of the population that differentiates exclusively into pigment cells on plastic culture substrata will produce the histochemical fluorescence characteristic of catecholamine-containing cells when grown on surfaces composed of somite fibroblast-derived extracellular matrix.

Since we know what many of the extracellular components are in embryonic somite (Section 2.1), we can extend this analysis further by constructing artificial substrata with individual, isolated components of this extracellular matrix. Thus, cells grown on substrata of collagen, fibronectin, collagen + fibronectin, hyaluronic acid, or chondroitin sulfate exhibit characteristic patterns of differentiative behavior. Under otherwise identical nutrient conditions, collagen reduced the proportion of cells undergoing differentiation as pigment cells, and fibronectin promoted the appearance of catecholamine fluorescence when present at optimal concentrations (Fig. 4) (33, 50, 51). Although such analyses are clearly only in their initial stages, it seems reasonable to conclude that crest cells receive developmental cues from the substrata on which they reside, and that such cues may direct the selection of the phenotype of some crest cells. The exact identity of the developmental cues

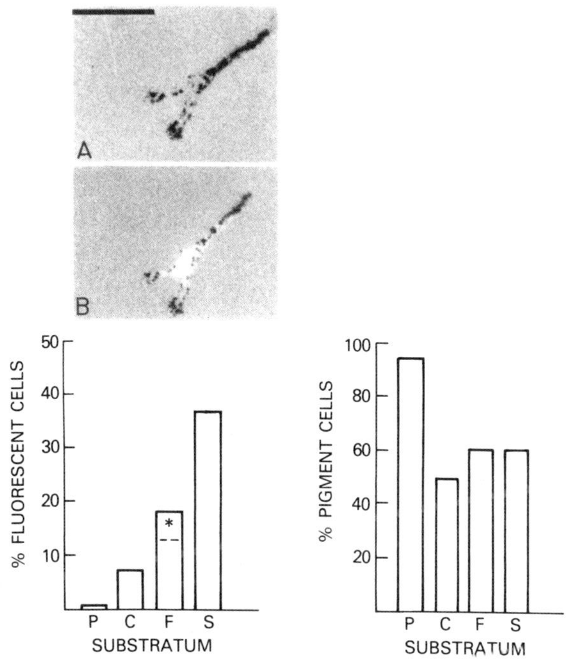

FIGURE 4. The proportion of crest cells exhibiting formaldehyde-induced fluorescence (FIF) or pigment granules is altered when crest cells are cultured on different substrata (P, tissue culture plastic; C, collagen-coated plastic; F, fibronectin added to culture dishes; S, substratum of "microexudate" produced by previously cultured embryonic somite cells). Under the conditions of culture, changes in differentiative behavior may probably be attributed to qualitative differences in substrata, including differences in the amounts of fibronectin. In general, FIF occurs in these cultured cells at the expense of pigment formation. It is interesting to note, however, that under some culture conditions (e.g., 5 μg exogenous Fn; see * on histogram "F"), some cultured crest cells are "confused" and express two phenotypes (FIF and pigment) simultaneously (see photo inset). *Inset*: A cell cultured in the presence of Fn that exhibits both pigment granules (A), and when fixed with formaldehyde vapor and stimulated with UV light, also shows the characteristic fluorescence of catecholamine-containing cells (B). [Adapted from Loring et al. (50).]

involved, their mode of action, and the mechanism of detection and response by crest cells all remain to be investigated.

3.1.2.4. Cholinergic Neurogenesis. It was mentioned above that transplantation experiments provided evidence that some autonomic ganglion cells could alter their phenotype when placed in different embryonic locations. An unequivocal demonstration that individual cells expressing the transmitter of one (adrenergic) neuron type could alter their phenotype and acquire the ability to make, store, release, and take up another neurotransmitter (acetylcholine) has been provided by the culture experiments of Patterson and his coworkers (52, 53). Such an alteration in cellular differentiation apparently occurs in response to diffusible components produced by various nonneuronal embryonic cells in culture. Some of these nonneuronal cells are likely to be the normal target organs of the "cholinergic" neurons.

3.2. Restrictions on the Ability of Some Crest Cells to Express Certain Crest Phenotypes, or to Change Their Developmental Pathway

Although it is clear that some crest-derived cells can continue to respond to new differentiative cues during development, there are also suggestive data to indicate that some stable developmental restrictions are imposed on the neural crest. Such observations should be carefully considered for what they will tell us about the process of progressive restrictions that occur in all developmental systems, and that appear to operate particularly in establishing the various neural crest lineages.

3.2.1. Restriction of Trunk Crest Cells. The ability of crest cells to make parts of the skull and connective tissue of the head is restricted in the embryonic trunk. Various "heterotopic" grafting experiments, in which neural crest from one axial level (e.g., the head) was transplanted to other locations (e.g., the embryonic trunk), have revealed that crest cells from the head can, under suitable conditions, give rise to every known crest derivative (2–5). In contrast, trunk neural crest cells are unable to contribute differentiated cells to the skull and connective tissue of the head. It appears, therefore, that developmental restrictions are imposed on some crest cells very early in their development so that they do not, and presumably cannot, respond to environmental stimuli that promote the development of components of the cranial skeleton.

Other experimental results (54) have indicated that cranial neural crest cells may have acquired the ability to make cranial skeleton and connective tissue even before they begin to migrate. Suitable (*permissive*) cues encountered during normal migration can apparently cause crest-derived cells to express these capabilities. However, these cells are not able to change this basic developmental program and embark on a different pathway toward such derivatives as pigment cells or neurons (see 36, 55, 56).

3.2.2. Restriction of Neuronal Precursors. Another inference can be made about the time of "specification" of other neural crest derivatives—namely, some of the neurons of the peripheral nervous system. We know that crest cells from all axial levels can produce sensory and autonomic ganglia (see Table 1), and that, after heterotopic transplantations, the type of ganglion formed corresponds to what normally develops at the site of the graft rather than the source of the grafted cells (41, 42, 57). Results from ectopic grafting experiments, however, indicate that cultured neural crest cells (as opposed to intact crest) and cells from crest-derived ganglia, other than spinal ganglia, are apparently unable to differentiate into sensory neurons even when given appropriate permissive environmental cues (16, 18). This may mean that another population of cells in the neural folds are also specified very early as neurons, and that other crest-derived cells have lost the ability to express this trait. Such a conclusion is certainly consistent with the observations that some neurons differentiate very precociously and that some are apparently already present in the newly formed ganglion, which can then provide the appropriate stimulus for nonneuronal crest-derived cells to differentiate into supportive (glial) cells of the ganglion (see Section 3.1.2.2).

3.2.3. Immunological Markers for Neuronal Differentiation. Recently, it has been possible to develop immunological reagents (so-called *monoclonal antibodies*) with exquisite sensitivity and specificity that recognize traits characteristic of peripheral neurons (58, 59). Using these antibodies, one can identify crest-derived cells expressing such traits even when other criteria do not reveal their latent developmental abilities. Such antibodies have been applied in a way that permits crest cells containing the corresponding antigenic determinant to be stained histochemically. The results reveal a small population of crest cells that have expressed this determinant, which is specific to neurons but is not yet characterized (60, 61) (Fig. 5). Such results provide further support, therefore, for the conclusion that some neurons are precociously differentiated. It is likely that monoclonal antibodies against more neuronal and many other cell types will provide additional valuable tools to analyze the changes in cell surfaces and cell phenotype during development.

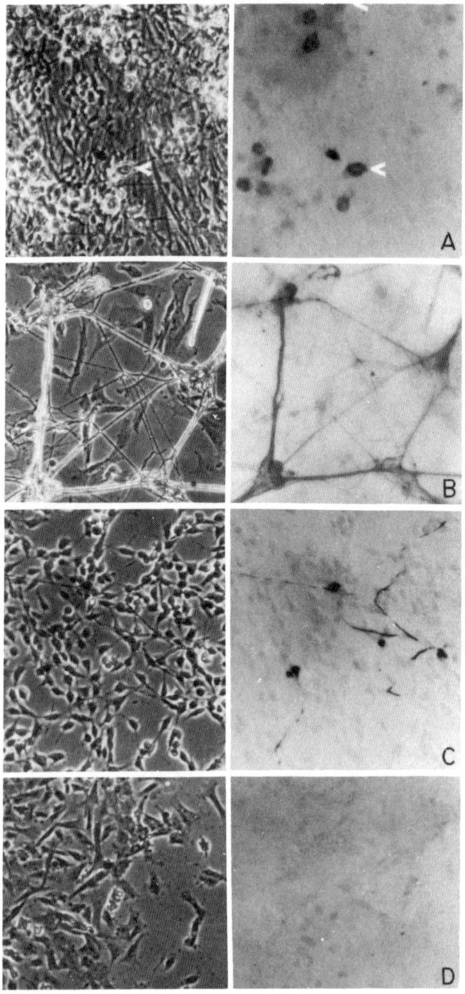

FIGURE 5. Monoclonal antibodies against different (as yet unknown) determinants in embryonic spinal (sensory) ganglion neurons were used as the basis for staining cultured embryonic cells. Antibody bound to cells was revealed by sensitive histochemical procedures that produce a dense precipitate on the cells. In all cases, a phase-contrast photo (*left* column) shows all cultured cells; bright-field photograph (*right* column) reveals antibody-mediated staining. (*A*) Antibody-mediated staining of some spinal ganglion nerve cell bodies, revealing cellular heterogeneity within the cultured ganglion. (*B*) A different antibody mediates staining of both cell bodies and fibers (and presumably, therefore, a different antigenic determinant) in similar cultures of ganglion neurons. Note absence of staining on fibroblastic cells in culture. (*C*) A third antibody, which can stain neuron cell bodies and fibers like the antibody-mediated staining in *B*, reveals the presence of its antigenic determinant in a subpopulation of cultured neural crest cells. This suggests that precocious neuronal differentiation occurs in a small proportion of these cultured crest cells. (*D*) The same antibody used in *C* was applied to cultured embryonic somite cells. The absence of staining is one of the normal controls for specificity of the antibody and the staining it mediates. [Adapted from Ciment and Weston (61).]

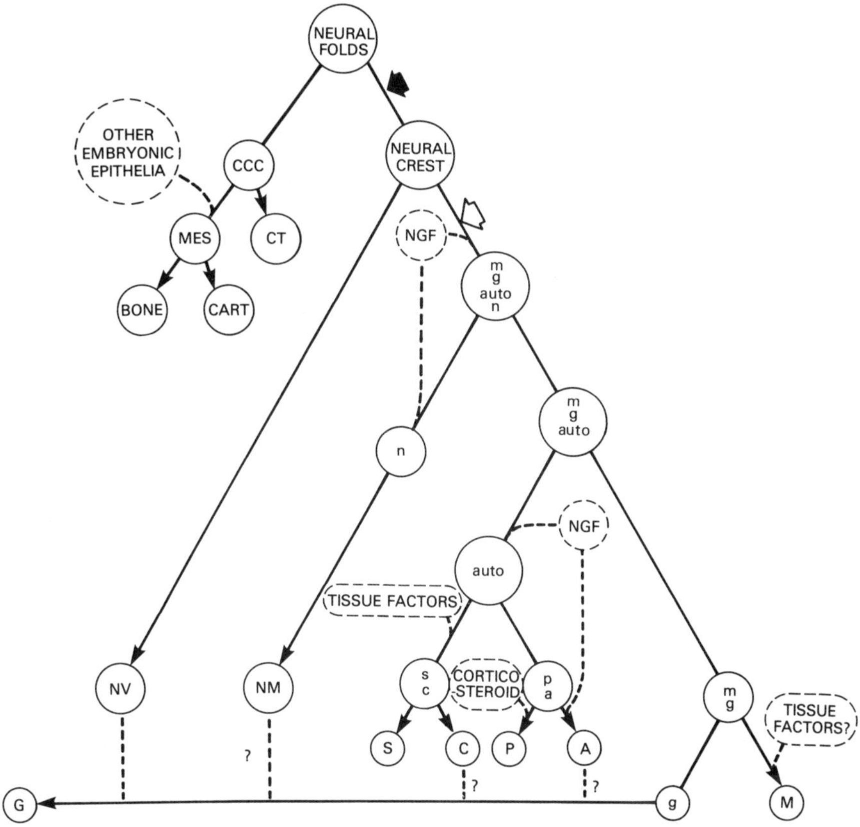

FIGURE 6. Diagrammatic scheme suggesting a possible sequence of binary "choices" leading to the segregation of various crest cell lineages (see Table 1). The circles represent cell populations that may be found in various embryonic locations. Lowercase letters within the circles represent the developmental capability remaining in each population. The time of some of the restrictions is postulated to occur prior to the onset of crest cell migration at a given axial level. Cranial crest cell migration, for example, is suggested to begin (solid arrow) after some ectomesenchymal cells have been specified. Likewise, the open arrow indicates the time of onset of migration of trunk crest cells, which is suggested to occur after some spinal ganglion neurons have been specified. Abbreviations: ccc, cranial crest cells; mes, mesenchymal cells; CT, connective tissue cells; Cart, cartilage; Neural Crest, trunk neural crest; NGF, nerve growth factor; m, melanoblasts; g, supportive cell precursors; auto, stem cells of the autonomic nervous system, probably positive for formaldehyde-induced fluorescence; n, sensory neuroblasts; NV, ventrolateral neurons of spinal ganglia; NM, mediodorsal neurons of the spinal ganglia; s, serotonergic neuroblasts; S, serotonergic neurons of the gut; c, cholinergic neuroblasts; C, cholinergic neurons; p, precursor cells of adrenal medulla; P, adrenal medullary cells; a, adrenergic neuroblasts, which are positive for formaldehyde-induced fluorescence; A, adrenergic neurons; G, differentiated glial or satellite cells; M, melanocytes. The dotted lines denote interactions between neurons and supportive cell precursors that promote gliagenesis (see text). Arrow heads indicate terminal differentiations. As indicated in the text, the environmental cues leading to the proposed sequence of restrictions in the crest cell population are largely unknown. [Adapted from Weston (3).]

3.3. The Progressive Restriction of Crest Cell Differentiative Ability During Normal Development

We have seen that some crest cells can be shown to be developmentally labile in vivo and in vitro and to remain so for a considerable part of early embryogenesis. In contrast, other crest-derived populations appear to become restricted early in development. It is not particularly productive, however, to debate whether or not the crest cell population, or any other embryonic cell population, is pluripotent—that is, whether it is able to differentiate into any of a variety of derivatives characteristic of its lineage. A more suitable, and developmentally more meaningful concern, is the sequence, or temporal order, of developmental restrictions imposed on cells of a particular lineage during embryogenesis.

In the case of crest cells, the results that have been discussed above suggest that, in fact, there may be a characteristic order by which the developmental repertoire of the crest lineage is defined. In the head of the embryo, a population of cells present in the neural folds is able to give rise to skeletal and connective tissue structures of the head and face. If other crest cells ever have this capability, they apparently lose it early in development (3). Likewise, early in embryogenesis, cells that appear to acquire the ability to become neurons are segregated from the remainder of the crest cell population. Cells also remain in the crest, however, that retain the ability to respond to specific environmental cues to become neurons of the sensory and the autonomic nervous system. Finally, some remaining crest-derived cells appear to respond to specific cues by deciding whether to differentiate as supportive (glial) cells of the peripheral nervous system or as pigment cells (see above and reference 3). Figure 6 summarizes these inferences and suggests one possible sequence of restrictions in the neural crest lineage.

Attempts to verify this postulated sequence of restrictions, and the order of segregation of the various members of the crest cell lineage, may now be possible. Likewise, it may be possible to establish what causal mechanisms are operating that impose these restrictions. To do so, we may exploit our ability to isolate crest cell clones (51) or developmentally homogeneous crest cell populations (25, 62) and to present them with biologically meaningful environments in vitro that simulate the known composition of the extracellular matrix in the crest migratory spaces (33, 50, 51) (see Section 2.1.2.1). Under these conditions, crest cell adhesion to their substrata and to each other can be measured (see Chapter 3), crest cell morphogenetic behavior can be observed and characterized, and crest cell differentiation can be analyzed. For the latter task, it seems abundantly clear that the "library" of crest cell phenotypes with their specific, sensitive, and reasonably convenient assays will provide unusually fruitful opportunities.

4. PROSPECTS: ANALYSIS OF THE "CONVERSATION" BETWEEN CREST CELLS AND THEIR ENVIRONMENT AT THE CELL SURFACE

The neural crest provides a microcosm in which many of the events and processes that command the attention of developmental biologists are manifest. These include the regulation of morphogenetic cell movement, cell-surface properties, cell contact and adhesive interactions, the structure and function of the extracellular matrix and its association with cell surfaces, the progressive restriction of developmental capabilities in a cell lineage, and the regulation of expression of specialized cellular phenotypes. This is especially true for the development of neurons of the peripheral nervous system, where neurotransmitter and neuroactive peptide synthesis, transmitter release and uptake mechanisms, and cell-surface receptors can be studied.

It is particularly significant that the developing neural crest can provide a context in which to study the way embryonic cells perceive and respond to their environment. Thus, the properties of the crest cell surface—the identity and role of receptors of developmental cues in the cell surface—would seem to be the next "frontier." The availability of monoclonal antibodies (see Section 3.2.3) seems likely to provide an extremely powerful tool for analyzing surface components, their developmental regulation, and their role in receiving and responding to the differentiative cues provided by extracellular matrix macromolecules.

There are also mutants available that affect the morphogenetic and differentiative behavior of crest cells (see Table 4). Such mutants are recognized by their pleiotropic effects on neural crest derivatives. For example, there are mutants that effect the pattern of pigment cell distribution or expression that also influence the differentiation of the peripheral nervous system, or the morphogenesis of the cranial skeleton. Some of these mutants are lethal in homozygotes (2, 63, 64) (Table 4). Developmental abnormalities may be the result of a lesion in the crest cells themselves, or they may be caused indirectly by changes in environmental cues encountered by the migrating crest cells. In any case, such mutants may eventually provide additional opportunities for analyzing developmental processes. In the mouse, such mutants as Steel (*Sl*), and Dominant spotting (*W*), have very similar phenotypes (see Table 4), caused, in one case, by abnormal embryonic environments (*Sl*) (63, 64), and in the other case, by the inability of certain cells undergoing migration, proliferation, and differentiation to respond to such environmental cues (63, 64).

Perhaps these mutants, combined with cell culture and biochemical and immunochemical methods, will eventually enable us to identify and isolate specific cellular or extracellular macromolecules involved in regulating normal morphogenesis and differentiation. In any case, we should clearly be

TABLE 4. Some Mutants in the Mouse That Affect Neural Crest Development

Name	Genotype	Phenotype	Possible Developmental Defect or Alteration
Patch	Ph/+	Large belly spot (white belt); wider skull	?
	Ph/Ph	Cleft face; abnormal heart morphogenesis; open neural tube; large interstitial "blebs"; lethal (? "rescued" by +/+ environment)	Altered composition of interstitial matrix (ECM)
Rump–white	Rw/+	White hind legs, rump and tail	?
	Rw/Rw	Embryonic lethal	
Steel	Sl/+	White belly spot	Altered environmental stimuli to migrating and dividing stem cells
	Sl/Sl	White, anemic, sterile ("rescued" in +/+ environment)	
Dominant spotting	W/+	White belly spot	Abnormal migration and mitosis of stem cell populations
	W/W	Acoustic ganglion and inner ear defects; white, anemic, sterile (not "rescued" in +/+ environment)	

Splotch	Sp/+	White spots	
	Sp/Sp	Neurological defects; embryonic lethal (not "rescued" in +/+ environment)	Abnormal crest cell segregation and migration
Microphthalmia	mi/+ Mi^{wt}/+	White spots on head, belly, tail	
	Mi^{wh}/Mi^{wh}	White; small eyes; Failure of bone resorption; inner ear abnormalities	Abnormal cranial crest migration; defect in C-cell differentiation (?)
	mi/mi	More severe defects than Mi^{wh}/Mi^{wh}; usually embryonic lethal	
Grey-lethal	gl/gl	Abnormal deposition of melanin granules; failure of bone resorption; teeth fail to erupt; lethal after weaning	? Abnormal differentiation of crest-derived neurosecretory cells
Dilute-lethal	d^l/d^l	Reduced pigment; abnormal behavior; myelin degeneration siezures (lethal)	Defect in phenylalanine metabolism
Piebald (piebald-lethal)	s/s s^l/s^l	White, with occasional pigment patches, megacolon (reduced cholinergic neurons in gut); inner ear abnormal	Regional defects in crest cell migratory ability; reduced viability of crest cells (neural crest cell defect)
Lethal-spotting	ls/ls	Extensive white spotting; megacolon (see s^l/s^l)	Altered ability of some crest cells to migrate

able to exploit neural crest development to answer some of the important questions posed at the beginning of this chapter and in Chapter 4.

REFERENCES

1. D. Brown, *Science* **211**, 667 (1981).
2. J. Weston, *Adv. Morphogen.* **8**, 41 (1970).
3. J. Weston, Motile and social behavior of neural crest cells, in R. Bellairs, A. Curtis, and G. Dunn, eds., *Cell Behaviour*, Cambridge University Press, London, 1981, pp. 429–470.
4. D. Noden, Interactions directing the migration and differentiation of avian neural crest cells, in D. Garrod, ed., *The Specificity of Embryological Interactions*, Chapman and Hall, London, 1978, p. 4.
5. N. LeDouarin, *Curr. Top. Dev. Biol.* **16**, 31 (1980).
6. M. Johnston, *Anat. Rec.* **156**, 143 (1966).
7. D. Noden, *Dev. Biol.* **42**, 106 (1975).
8. J. Weston, *Dev. Biol.* **6**, 279 (1963).
9. M. Teillet and N. LeDouarin, *C. R. Acad. Sci. Paris* **270**, 3095 (1970).
10. N. LeDouarin and M. Teillet, *Dev. Biol.* **41**, 162 (1974).
11. N. LeDouarin, D. Renaud, M. Teillet, and G. LeDouarin, *Proc. Natl. Acad. Sci. USA* **75**, 728 (1975).
12. R. Pratt, M. Larson, and M. Johnston, *Dev. Biol.* **44**, 298 (1975).
13. M. Derby, *Dev. Biol.* **66**, 321 (1978).
14. J. Weston, M. Derby, and J. Pintar, *Zoon* **6**, 103 (1978).
15. J. Weston and S. Butler, *Dev. Biol.* **14**, 246 (1966).
16. C. Erickson, K. Tosney, and J. Weston, *Dev. Biol.* **77**, 142 (1980).
17. M. Bronner and A. Cohen, *Dev. Biol.* **77**, 130 (1980).
18. C. LeLievre, G. Schweizer, C. Ziller, and N. LeDouarin, *Dev. Biol.* **77**, 362 (1980).
19. B. Toole and R. Trelstad, *Dev. Biol.* **26**, 28 (1971).
20. D. Newgreen and J-P. Thiery, *Cell Tissue Res.* **211**, 269 (1980).
21. J. Wartiovaara, I. Leivo, and A. Baheri, *Dev. Biol.* **69**, 247 (1979).
22. J. Wartiovaara, I. Leivo, and A. Vaheri, in S. Subtelny and N. Wessells, eds., *The Cell Surface: Mediator of Developmental Processes*, Academic Press, New York, 1980, p. 305.
23. B. Mayer, E. Hay, and R. Hynes, *Dev. Biol.* **82**, 267 (1981).
24. J. Loring, C. Erickson, and J. Weston, *J. Cell Biol.* **75**, 71a (1977).
25. K. Ranney, J. Loring, and J. Weston (in preparation).
26. J. Weston, Role of the embryonic environment in neural crest morphogenesis, in R. Pratt and R. Christiansen, eds., *Current Research Trends in Prenatal Craniofacial Development*, Elsevier/North-Holland, New York, 1980, p. 27.
27. K. M. Yamada, K. Olden, and L. E. Hahn, Cell surface protein and cell interactions, in S. Subtelny and N. Wessells, eds., *The Cell Surface: Mediator of Developmental Processes*, Academic Press, New York, 1980, p. 43.
28. J. Weston, K. Yamada, and K. Hendricks, *J. Cell Physiol.* **100**, 445 (1979).

29. M. Bernfield, Organization and remodeling of the extracellular matrix in morphogenesis, in L. Brinkley, B. Carlson, and T. Connelly, eds., *Morphogenesis and Pattern Formation*, Raven Press, New York, 1981, pp. 139–162.
30. A. Cohen and E. Hay, *Dev. Biol.* **26**, 578 (1971).
31. J. Pintar, *Dev. Biol.* **67**, 444 (1978).
32. B. Toole, C. Underhill, Y. Mikuni-Takagaki, and R. Orkin, Hyaluronate in morphogenesis, in R. Pratt and R. Christiansen, eds., *Current Research Trends in Prenatal Craniofacial Development*, Elsevier/North-Holland, New York, 1980, pp. 263–276.
33. M. Sieber-Blum, F. Sieber, and K. M. Yamada, *Exp. Cell Res.* **133**, 285 (1981).
34. L. Ossowski, J. Quigley, G. Kellerman, and E. Reich, *J. Exp. Med.* **138**, 1056 (1973).
35. J. Greenberg, J. Foidart, and R. Greene, *Cell Differ.* **9**, 153 (1980).
36. D. Newsome, *Dev. Biol.* **49**, 496 (1976).
37. R. Pratt, G. Morriss, and M. Johnston, *J. Gen. Physiol.* **68**, 15–16a (1976).
38. J. Greenberg and R. Pratt, *Cell Differ.* **6**, 119 (1977).
39. B. Glimelius and J. Pintar, *Cell Differ.* **10**, 173 (1981).
40. M. Kinsella and J. Weston, *Am. Zool.* **18**, 643a (1979).
41. N. LeDouarin, M. Teillet, C. Ziller, and J. Smith, *Proc. Natl. Acad. Sci. USA* **75**, 2030 (1978).
42. J. Smith, P. Cochard, and N. LeDouarin, *Cell Differ.* **6**, 199 (1977).
43. D. Nichols and J. Weston, *Dev. Biol.* **60**, 217 (1977).
44. D. Nichols, R. Kaplan, and J. Weston, *Dev. Biol.* **60**, 226 (1977).
45. B. Holton and J. Weston, *Dev. Biol.* **89**, 64 (1982).
46. B. Falck and C. Owman, *Acta Univ. Lund 11*, p. 1 (1965).
47. A. Cohen, *J. Exp. Zool.* **179**, 167 (1972).
48. S. Norr, *Dev. Biol.* **34**, 16 (1973).
49. M. Teillet, P. Cochard, and N. LeDouarin, *Zoon* **6**, 115 (1978).
50. J. Loring, B. Glimelius, and J. Weston, *Dev. Biol.* **90**, 165 (1982).
51. M. Sieber-Blum and A. Cohen, *Dev. Biol.* **80**, 96 (1980).
52. P. Patterson and L. Chun, *Dev. Biol.* **60**, 473 (1977).
53. P. Patterson, *Ann. Rev. Neurosci.* **1**, 1 (1978).
54. B. Hall and R. Tremaine, *Anat. Rec.* **194**, 469 (1979).
55. M. Tyler and B. Hall, *Anat. Rec.* **188**, 229 (1977).
56. J. Bee and P. Thorogood, *Dev. Biol.* **78**, 47 (1980).
57. D. Noden, *Dev. Biol.* **67**, 296 (1978).
58. L. Chun, P. Patterson, and H. Cantor, *J. Exp. Biol.* **89**, 73 (1980).
59. R. McKay, M. Raff, and L. Reichardt, eds. *Monoclonal Antibodies against Neural Antigens*. Cold Spring Harbor Reports in the Neurosciences, Vol. 2, Cold Spring Harbor Laboratory, New York (1981).
60. K. Barald, Monoclonal antibodies to embryonic neurons: cell specific markers for chick ciliary ganglion, in N. Spitzer, ed., *Neuronal Development*, Plenum Press, New York, 1982, pp. 101–119.
61. G. Ciment and J. Weston, in R. McKay et al., eds., *Monoclonal Antibodies against Neural Antigens*, Cold Spring Harbor Reports in the Neurosciences, Vol. 2, Cold Spring Harbor Laboratory, New York (1981).

62. B. Glimelius and J. Weston, *Cell Differ.* **10**, 57 (1981).

63. M. Green, Mutant genes and linkages, in E. Green, ed., *Biology of the Laboratory Mouse*, 2nd ed., McGraw-Hill, New York, 1966, pp. 87–150.

64. W. Silvers, *The Coat Colors of Mice*, Springer-Verlag, New York, 1979.

7

Developmentally Regulated Lectins

SAMUEL H. BARONDES

Department of Psychiatry
School of Medicine
University of California—San Diego
La Jolla, California 92093
and
Veterans Administration Medical Center
San Diego, California 92112

Development of organisms results from a series of molecular interactions within and between cells. These follow from interactions of molecules synthesized at specific stages of differentiation. One type, the subject of this chapter, is a protein-carbohydrate interaction mediated by a class of proteins called lectins (1–3). These proteins have two or more carbohydrate binding sites that are complementary to specific saccharide receptors. Because of their valences, they can couple complementary saccharide-containing compounds within cells or on cell surfaces, thereby influencing cellular and tissue organization.

Lectins are of interest to developmental biologists not only because their properties can be exploited in shaping the organism, but also because there is considerable evidence that they are developmentally regulated in a variety of living systems. These proteins are synthesized abundantly in certain cells (4) or tissues (5–7) at specific stages in development and then may either persist at high levels or disappear with further differentiation. What they do is not yet clearly known, but is receiving increasing attention.

The purpose of this chapter is to provide an introduction to the study of the role of endogenous lectins in development. It should be emphasized that what we are concerned with here is the role of lectins in the tissues in which they are made, rather than the application of exogenous lectins to cells or tissues to map surface carbohydrates or induce cell division. As background, some general properties of lectins will be summarized that provide clues to their endogenous roles. Then the developmental regulation of lectins in several systems will be described. Finally, approaches to the analysis of the function of developmentally regulated lectins will be considered.

1. SOME GENERAL CONSIDERATIONS ABOUT LECTINS

1.1. Assays

Lectins, first discovered in the 19th century as agglutinins of erythrocytes, are often still assayed by this convenient property. An example of a hemagglutination assay for a lectin is shown in Figure 1. Here a purified lectin is serially diluted in a row of microtiter wells such that a ½ dilution is in the first well, a ¼ dilution in the second, a ⅛ in the third, etc. All the wells contain erythrocytes that have surface saccharides that bind the lectin and are agglutinated by it. Agglutination is detected by the fuzzy mat of erythrocytes on the bottom of the well that contrasts with the sharp dot in the absence of lectin. Addition of a relatively high concentration of a sugar that combines with the lectin's active site, in this case lactose, completely blocks the hemagglutination activity of the lectin (Fig. 1). In contrast, cellobiose, a sugar that does not react with the lectin, has no effect (Fig. 1) ruling out a nonspecific action of sugar on the hemagglutination.

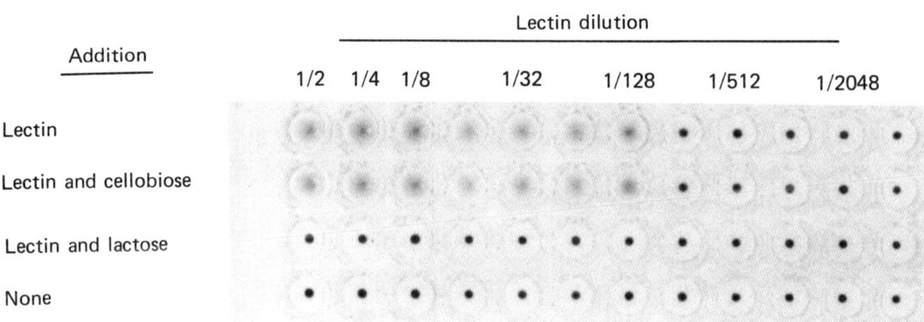

FIGURE 1. Assay of hemagglutination activity of a purified lectin. The starting solution for this assay contained 0.5 μg/ml of chicken-lactose lectin I, purified from embryonic chicken muscle. An aliquot of this solution was diluted with an equal quantity of buffer and 25 μl was added to each of the lectin-containing wells marked ½. Therefore, each of these wells contain 6.25 ng of lectin. Increasing dilutions of the starting solution were made for the subsequent series of wells as indicated. The disaccharides cellobiose or lactose were added to each well in the indicated row at a final concentration of 20 m*M*. Then glutaraldehyde-fixed trypsinized rabbit erythrocytes were added, and the hemagglutination patterns were determined after 1 hr of incubation at room temperature. Hemagglutination by the purified lectin is indicated by the fuzzy mat of cells at the bottom of the indicated wells. The final dilution that produced hemagglutination, 1/128, is the titer of the lectin solution. A specific activity of the lectin solution might be expressed as 128 divided by the initial lectin protein concentration (5×10^{-4} mg/ml), giving a specific activity of 256,000. Since lactose completely inhibits the hemagglutination activity of the lectin, producing the same type of dot as is found in the control wells which contain no lectin, it is taken to be a potent inhibitor of lectin activity. In contrast, cellobiose does not appear to interact with the active site of the lectin.

The maximal dilution of lectin that agglutinates the erythrocytes, called the titer, is determined by inspection of the microtiter plate (Fig. 1). In the case shown, the titer is 1/128. Activity is frequently expressed as the reciprocal of titer (titer^{-1}), which would be 128 in this case. The specific activity of the lectin might be expressed as 128 divided by the number of milligrams of protein per milliliter of the solution being assayed.

1.2. Sites of Action

The reaction measured by the hemagglutination assay is schematized in Figure 2. Lectin molecules, which contain more than one carbohydrate binding site, bind to complementary saccharides on the cell surface of adjacent cells leading to their association. In conceptualizing this reaction, it is important to realize that the affinity of the carbohydrate-binding site of the lectin for a complementary saccharide need not be very high to produce this agglutination reaction. This is due to the fact that many lectins are polyvalent, so that more than one binding site of the lectin molecule can become associated with complementary saccharides on the cell surface. This polyvalency can markedly increase the functional affinity of the reaction. Because cell-surface glycoproteins and glycolipids, with which lectins can interact, move freely in the plane of the membrane, close approximation of more than one complementary carbohydrate-containing substance in the membrane can result as the lectin successfully associates with them when they come within its range. This clustering of receptors by polyvalent lectins can lead to fairly strong associations between the lectin and the cell surface, even though the specific carbohydrates do not have a very strong affinity for the lectin. Indeed, these polyvalent interactions may make lectin-cell surface associations relatively nonspecific in that many monosaccharides or oligosaccha-

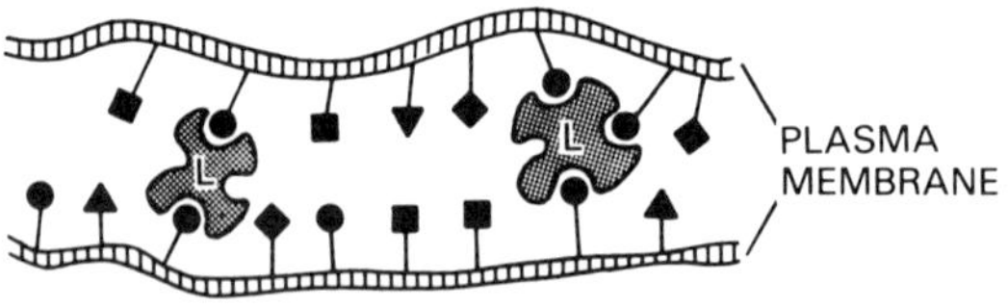

FIGURE 2. Conceptualization of agglutination of two cells by a tetravalent lectin. Two tetravalent lectin molecules (L) with binding sites complementary to saccharide residues, depicted as dark circles, are shown simultaneously binding to complementary saccharides on glycoprotein or glycolipid molecules protruding from the plasma membranes of two cells. Note that one lectin molecule is shown binding two different complementary molecules clustered in the membrane of one of the cells. This divalent binding would markedly increase the association of the lectin with the surface of that cell. Likewise the participation of many lectin molecules in binding two cells together could lead to a very strong association.

rides may bind well enough to the active site of the lectin to give significant polyvalent binding. This is not to dismiss the specific affinity of the active sites of lectins for complementary saccharides. However, there is a hierarchy of relative affinities; and polyvalent association with some of the less specific saccharides may be sufficient to lead to agglutination.

1.3. Possible Functions of Lectins

From studies of hemagglutination, some general properties of lectins and their possible implications for developmental functions can be inferred. These are listed in Table 1. The fact that lectins have specific carbohydrate-binding sites suggests that they may function to recognize complementary saccharide receptors either within cells or on the surface of like or different cells. The function of lectins must obviously include some type of specific carbohydrate recognition. In this regard, it is notable that some lectins contain more than one type of subunit with differences in their carbohydrate-binding sites. Such lectins could couple intracellular materials or cells containing different complementary saccharides.

The polyvalency of lectins and its implications for crosslinking receptors and for high-affinity binding have already been considered. Such properties would certainly be expected with the many lectins that are multimeric. However, some lectins are dimers, which would limit somewhat their potential biological functions. The fact that lectins are readily assayed as agglutinins indicates that they can not only bind to cell-surface saccharides but actually bridge cells together.

Two other properties of lectins have also been tabulated because they provide clues to function. Thus, lectins tend to be relatively abundant proteins, frequently comprising several percent of the total cellular protein. This suggests that lectins have a structural function, in contrast with the catalytic

TABLE 1. Lectin Properties That Might Mediate Developmental Functions

Property	Possible Functions
Specific carbohydrate-binding sites	Recognition of complementary glycoproteins or glycolipids within or between cells
Multiple carbohydrate-binding sites	Crosslinking glycoproteins or glycolipids; high affinity (multisite) binding
Agglutinin	Binding like or unlike cells together
Abundant	Structural organization
Soluble protein	Might involve movement between cellular compartments (e.g., interior to cell surface)

function characteristic of enzymes, which tend to be far scarcer. In addition, most lectins are isolated as soluble proteins upon homogenization in aqueous media. Therefore, these lectins, although frequently associated with the glycoproteins and glycolipids of membranes, are not constrained by being themselves integrated into membranes. This would allow for movement within or between cellular compartments that could be critical to their function.

Given these properties, it is easy to see how lectin-carbohydrate interactions could play important roles in a variety of developmental processes. Studying this is facilitated because lectins are easy to assay, abundant, and easy to purify to homogeneity by affinity chromatography. The pure lectins can be used to raise specific antibodies that are valuable in investigating lectin function. Aside from these theoretical and practical considerations, a developmental role for lectins has attracted attention because they often are extensively synthesized during development, as will now be considered.

2. LECTIN SYNTHESIS IN DEVELOPMENT

2.1. Slime Mold Lectins

The first study correlating lectin synthesis with differentiation was made with cellular slime molds (4). These organisms are useful for developmental research because they can be induced to differentiate rapidly from a unicellular vegetative form to a cohesive multicellular aggregate by depleting them of food (8). As they change to the aggregating state they synthesize lectins. For example *Dictyostelium discoideum*, a species of slime mold that is commonly studied, contains no detectable lectin in the vegetative state but abundant lectin within 9 hr of food deprivation (Fig. 3). At this time, cells are streaming into large aggregates and can be seen to adhere to each other. Cell–cell adhesion can be measured experimentally as described in Figure 3.

The lectins synthesized during development are very abundant proteins. In a number of species of cellular slime molds, lectins comprise at least 1% of the total cellular protein of aggregating cells (9–11). Their synthesis is directed by specific messenger RNA, which is made as the cells differentiate (12).

Slime molds contain more than one lectin. For example, *Dictyostelium discoideum* contains at least two, called discoidin I and discoidin II (13). The former, which is more abundant, is not synthesized at precisely the same stage as the latter (13). Another slime mold, *Dictyostelium purpureum*, contains seven distinct lectins, all functionally discriminable (14). Two are homotetramers of unique subunits and the other five are mixed tetramers of two other subunits. Although similar, they can be distinguished on the basis

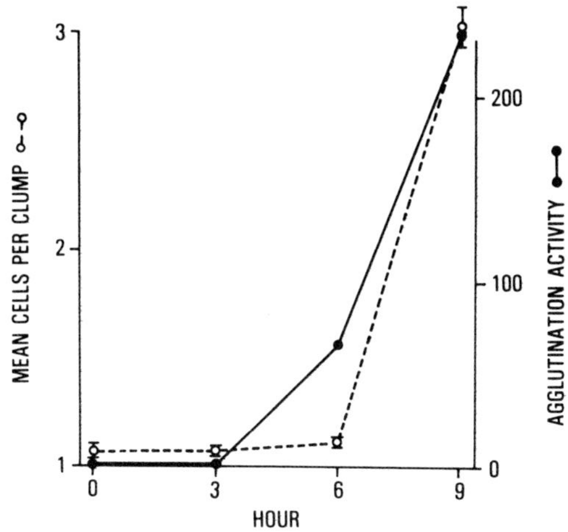

FIGURE 3. Lectin synthesis and development of cell–cell adhesion in differentiating *Dictyostelium discoideum*. *Dictyostelium* cells were deprived of food at time 0 and permitted to differentiate on a surface. At the indicated times aliquots of cells were removed, extracted and assayed for lectin activity and other aliquots were assayed for cell–cell adhesion, here measured as "mean cells per clump," as determined with an electronic particle counter. Note that agglutination activity is detectable before changes observed in cell–cell adhesion. (For details, see 4.)

of their relative affinity for column materials derivatized with specific saccharides, or by the relative potency of a series of different saccharides as inhibitors of their agglutination activity. In this species too, there is relatively greater synthesis of some lectins than others at different times in development.

2.2. Vertebrate Lectins

Developmentally regulated lectins are also found in higher organisms. For example, chicken embryos have three known developmentally regulated lectins, each named for a saccharide that markedly inhibits their hemagglutination activity. The first characterized, chicken-lactose lectin I, (CLL-I) has been purified by affinity chromatography (15, 16) and is a dimer with subunit molecular weight of approximately 15,000. As its name implies, its hemagglutination activity is markedly inhibited by lactose and other related β-galactosides. It shows striking developmental regulation in muscle, rising from low levels in extracts of 8-day embryos to high levels at 12–16 days with subsequent decline to much lower specific activities in extracts of adult muscle (Fig. 4). Another lactose-inhibitable lectin, chicken-lactose lectin II (CLL-II), is biochemically and immunologically distinct from CLL-I (17, 18). It is very prominent in developing kidney but declines to much lower

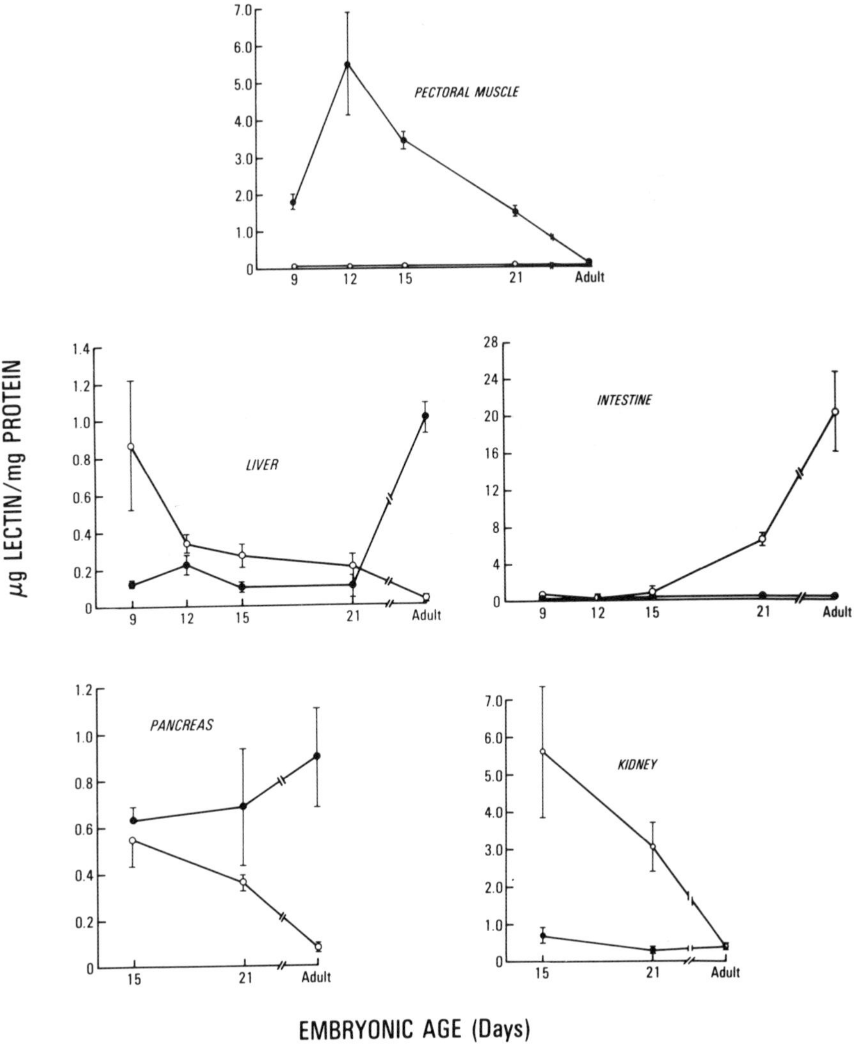

FIGURE 4. Chicken-lactose lectin I (CLL-I) and chicken-lactose lectin II (CLL-II) distribution in extracts of some embryonic and adult chicken tissues. CLL-I (●–●) and CLL-II (○–○) were quantified by radioimmunoassay using antibodies raised against highly purified lectin preparations. Since there is no detectable immunological crossreactivity between CLL-I and CLL-II, they can both be quantified in the same extract. Radioimmunoassay is more strictly quantitative than a hemagglutination assay, like that shown in Fig. 1, since it is not dependent on serial dilutions. It is also not interferred with by endogenous lectin inhibitors. (For details see reference 18.)

levels in the adult (Fig. 4). Kidney also contains significant amounts of CLL-I, whereas muscle contains little or no detectable CLL-II. Chicken-heparin lectin (CHL), which reacts with heparin as well as *N*-acetyl-D-galactosamine (19), is abundant in embryonic muscle but is present at lower levels later in development (20).

Identical lectins are also found in a number of other tissues at different stages of development. These lectins are not only immunologically cross-reactive but also have indistinguishable subunit molecular weights, isoelectric points, and peptide maps. CLL-I is very abundant in adult chicken liver (Fig. 4) where it is localized in cells lining the hepatic sinusoids (21) and is also prominent in adult pancreas (Fig. 4) where it is localized in the intercellular material surrounding the exocrine lobules (21). CLL-II, which is relatively rich in embryonic liver and pancreas, becomes especially abundant in adult intestine (Fig. 4). As will be shown later, the CLL-II in intestine is localized in goblet cells, which are continuously differentiating from stem cells in the intestinal mucosa. Therefore, its high concentration in adult tissue need not rule out a developmental function in differentiating populations of adult cells.

Developmentally regulated lectins have also been observed in other biological systems. For example, plant lectins appear with differentiation, as is discussed by Dazzo in Chapter 10. A lectin in rat lung that resembles CLL-I shows marked changes with development (22). Yet, despite their wide distribution, relatively little is known about the role these developmentally regulated lectins play in cellular and tissue differentiation. The evidence available suggests that in the case of the cellular slime molds, one function may be to mediate cell–cell adhesion. Evidence for this will now be considered.

3. LECTINS AS CELL-ADHESION MOLECULES IN SLIME MOLDS

3.1. Cell-Surface Lectins and Saccharides

The discovery that lectins are synthesized by slime mold cells before they aggregate and adhere (4) raised the possibility that the lectins mediate this cellular association (23). This seemed plausible since polyvalent carbohydrate-binding proteins could make cells adhere by binding to complementary saccharides on their cell surface, by analogy with their known action as agglutinins. Cellular slime molds do indeed contain cell-surface materials that bind slime mold lectins (24). The binding materials are presumed to be carbohydrates, since lectin binding to the cells can be blocked by saccharides that are specific for the carbohydrate-binding site. Furthermore, addition of either of the lectins from *Dictyostelium discoideum* to aggregating *Dictyostelium discoideum* cells that have been fixed with glutaraldehyde leads to their agglutination (24). In contrast, addition of these same lectins to glutaraldehyde-fixed vegetative *Dictyostelium discoideum* cells produces little or no agglutination (24). Presumably the surface changes of the developing *Dictyostelium discoideum* cells include the appearance of molecules that bind to the lectin.

Not only lectin-binding substances but lectins themselves appear on the surface of aggregating cells (4, 25–27). This was shown immunohistochemically by reacting antibody to purified slime mold lectins with intact slime mold cells and visualizing the bound antibody by either immunofluorescence or immunoferritin techniques (25, 26). It is also illustrated by the binding of living slime mold cells to fixed erythrocytes that contain complementary saccharides on their surface. An example of this interaction is shown in Figure 5. The erythrocytes, which are smaller than the slime mold cells, form rosettes around them. However, addition of a sugar that saturates the carbohydrate-binding sites of the cell-surface lectins blocks the association of the slime mold cells with the erythrocytes. In contrast, the rosettes are not affected by identical concentrations of a sugar that does not react with the endogenous cell-surface lectin. This study shows not only that lectin is present on the slime mold cell surface but also that carbohydrate-binding sites are available for interaction with another cell, in this case an erythrocyte. An interaction of this type provides a model for the possible role of these polyvalent carbohydrate-binding proteins in cell–cell adhesion.

Although endogenous lectins can be detected on the surface of aggregating slime mold cells by the techniques discussed above, studies using quantitative techniques show that only a small fraction of these very abundant proteins are normally found on the cell surface (28). For example, in both *Dictyostelium discoideum* and *Dictyostelium purpureum*, which contain about 5×10^6 lectin molecules per aggregating cell, only about 1 or 2% is detectable on the cell surface. This was determined by quantifying the binding of labeled univalent antibody fragments to living cells as well as by quantifying the amount of cell-surface lectin that was specifically eluted with a complementary sugar. In the latter experiments, living cells were shaken with a sugar that combines fairly well with the active site of the lectin (e.g., lactose in the case of *Dictyostelium purpureum*), and the lectin eluted into the medium was measured by radioimmunoassay and also identified by polyacrylamide gel electrophoresis (28). Virtually all the cell-surface lectin of aggregating cells, which had been labeled by the iodosulfanilic acid technique, was shown to be eluted in this manner. These experiments also show that the endogenous cell-surface lectin is held there by a carbohydrate-protein association rather than being integrated into the membrane bilayer.

FIGURE 5. Demonstration of active cell surface lectin in slime mold cells by rosette formation. Slime mold cells were reacted with formalinized sheep erythrocytes in the presence or absence of lactose, a saccharide that binds the carbohydrate binding site of their endogenous lectins. (*A*) In the absence of an active saccharide, spherical red cells have formed a rosette around the larger slime mold cells. (*B*) In the presence of lactose, the red cells are not associated with the slime mold cells. Scale, 1 cm = 5 μm. (For details see reference 25.)

3.2. Elicitation of Cell-Surface Lectins

If slime mold lectins function in cell-surface interactions, it was difficult to understand why only a very small percentage was normally found at this site. However, we have recently found that the amount of cell-surface lectin can be markedly augmented by a number of cell-surface reactions that either crosslink the available lectin or other cell-surface molecules. For example, divalent immunoglobulin directed against a slime mold lectin crosslinks the endogenous slime mold lectin and elicits the appearance of a large amount of additional lectin onto the cell surface (28). Although the precise amount elicited could not be computed with the techniques used, it is estimated to be at least five times as much as is normally present on the cell surface. In contrast, univalent antibody fragments, which do not crosslink cell-surface lectin, did not have this effect (28). Polyvalent oligosaccharide-containing compounds, such as bovine serum albumin derivatized with about 20 lactose residues per molecule, also increased cell-surface lectin, presumably as a result of crosslinking it (29). Purified slime mold lectins also had this effect, presumably by crosslinking available cell-surface receptors (29). In this instance, the elicitation reaction was studied with slime mold cells that had been extensively labeled with [^{14}C]acetate so that their endogenous lectin was radioactive. Addition of unlabeled exogenous lectin elicited the appearance of considerable additional labeled lectin on the cell surface.

Elicitation of additional cell-surface slime mold lectin also resulted from addition of plant lectins such as concanavalin A (29). This suggests that the elicitation reaction is not only produced by crosslinking of cell-surface slime mold lectin or its receptors. Here crosslinking of saccharides that react with concanavalin A apparently mediates the response. Polyvalent interaction of concanavalin A with cell-surface glycoproteins or glycolipids seems necessary since elicitation of surface slime mold lectin was not produced by binding succinyl-concanavalin A (29), a derivative of concanavalin A with a lower valence. However, crosslinking of surface-bound succinyl-concanavalin A, by addition of antibody that crosslinked this reagent, led to appearance of additional cell-surface slime mold lectin. Whether or not some of the cell-surface saccharides that bind concanavalin A are also receptors for slime mold lectins is not presently known.

The fact that various cross-linking reactions elicit the cell surface appearance of endogenous lectin from an intracellular pool could be relevant to its role in development. For example, interaction of available cell-surface receptors on one slime mold cell with the unoccupied carbohydrate-binding sites of cell-surface lectin on another slime mold cell could elicit the appearance of additional lectin on the surface of both cells. This additional lectin could both increase the cell–cell association through combining with available sites and also elicit additional cell-surface lectin. This positive feedback system could be important in augmenting cellular associations. The mechanism of this elicitation reaction and the process by which slime mold lectins cross the plasma membrane is not presently known.

3.3. Mutant With Defective Lectin

Participation of slime mold lectins in cell–cell adhesion is also supported by studies of the developmental effects of a mutant of *Dictyostelium discoideum*

(30) that synthesizes defective discoidin I, the major lectin of *Dictyostelium discoideum*. The defective lectin has an abnormal carbohydrate-binding site and does not agglutinate sheep erythrocytes or bind to the affinity column used to prepare the normal lectin. It is detected by radioimmunoassay because it cross-reacts with antibody raised against the normal lectin. Since defective lectin is synthesized as abundantly as the normal lectin during differentiation, the developmental program of the mutant is normal, up to that point. The mutant is markedly defective in cell–cell adhesion, forming loose aggregates but not progressing beyond that stage. This abnormality is presumed to be a direct consequence of the defect in the carbohydrate-binding site of discoidin I.

Evidence that there is a relationship between the abnormality in discoidin I and the abnormality in the mutant's cell–cell association also comes from studies with revertants (31). A number have been isolated in which there is partial or complete recovery of the carbohydrate-binding function of discoidin I along with recovery of cell–cell association. In the revertants with partial recovery of carbohydrate-binding function and lectin activity, there are abnormalities in colony morphology. The fact that the mutant appears to behave like a point mutant in the structural gene for discoidin I, rather than a mutation in some regulatory process that could secondarily effect cell–cell association, strongly supports a role for discoidin I in cell–cell adhesion. However, the results do not necessarily indicate that discoidin I actually functions to bind the cells together. Rather it remains possible that it indirectly influences cell–cell adhesion.

4. LECTINS IN DEVELOPING AND ADULT CHICKEN TISSUES

As with slime molds, levels of lectins change strikingly in many differentiating chicken tissues. However, little is presently known about their function in development. The only tissue studied in any detail is developing muscle. As shown in Figure 4, there are striking changes in the levels of CLL-I in developing muscle. The lectin, although predominantly intracellular, has been detected at the cell surface by immunohistochemical techniques (16, 32). This has raised the possibility that it, like slime mold lectins, plays a role in cell–cell interactions.

One characteristic cellular interaction in developing muscle is myoblast fusion. At a specific stage in development, the immature muscle cells fuse to form a polynucleated myotube. Since lectins are abundant during the period of myoblast fusion and much less prominent in mature muscle (Fig. 4), it seemed possible that lectins in some way influence membrane organization that could lead to myoblast fusion. The finding that thiodigalactoside, a

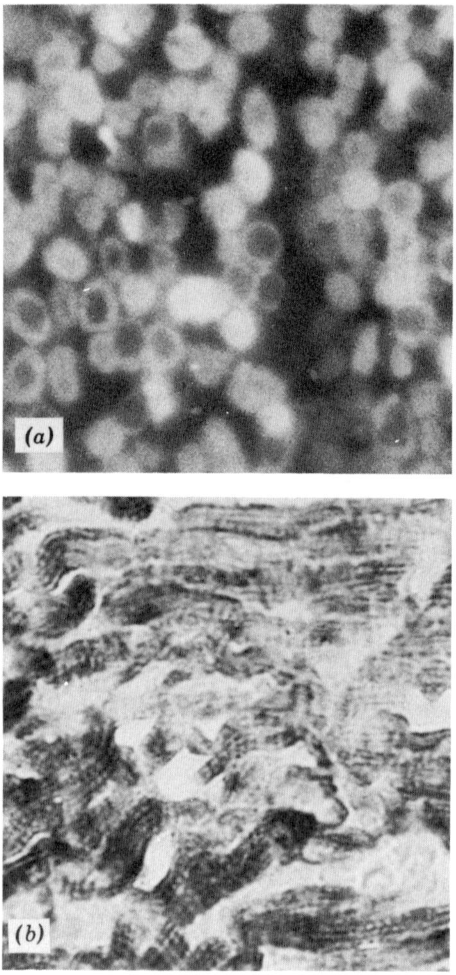

FIGURE 6. Distribution of CLL-I in pectoral muscle at different stages of development, as visualized by two immunohistochemical techniques. (*A*) Unfused myoblasts from 10-day embryonic muscle showing diffuse intracellular lectin as indicated by immunofluorescent staining. The dark areas within the cells are nuclei. (*B*) Localization by the immunoperoxidase technique in a region of 15-day embryonic muscle. Here the black reaction product is predominantly in discrete intracellular locations which have the same distribution as systems of membranes that may be continuous with the plasma membrane. (*C*) Cross-section of a region of pectoral muscle obtained 2 days after hatching showing immunofluorescence localization of CL1-I at the surface of and between more mature muscle fibers. Scale, 1 cm = 4 μm. (For details, see reference 36.)

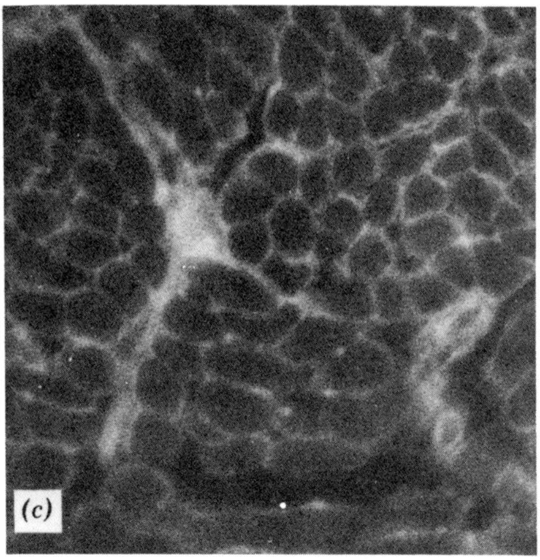

FIGURE 6. *(continued)*

potent hapten inhibitor of CLL-I, blocked the fusion of cultured rat myoblasts (33) seemed to implicate this lectin in some aspect of the fusion reaction. However, the fusion of myoblasts in primary chicken muscle cultures was not influenced by addition of thiodigalactoside (6). The discrepancy may be due to the time of addition of the hapten inhibitor to the cultures (34). Apparently the inhibitor must be added for a considerable period before fusion begins to block this process (34). The significance of this is not presently clear. Nor is the significance of the finding that repeated administration of low levels of CLL-I to primary chicken muscle cultures blocks myoblast fusion, whereas much higher levels of plant lectins are required to produce the same effect (35). Clearly the lectin and reagents that react with its active site can influence muscle development in culture. However, the relationship of these in vitro effects to the function of the lectin will require considerable additional investigation.

Immunohistochemical localization of CLL-I in muscle tissue from intact embryos or young tissues at various stages of development (36) (Fig. 6) does not suggest its participation in myoblast fusion, but cannot exclude it. One interpretation of the localization studies is that CLL-I plays a role in the development of the complex organization of muscle membranes (36). This is suggested by the finding that the lectin, which is diffusely distributed within myoblasts, becomes localized in a distribution like that of the developing intracellular membranes of myotubes (Fig. 6). Still later, the lectin is found predominantly at the surface of myotubes or between them, although some remains associated with the intracellular membranes. The relatively small

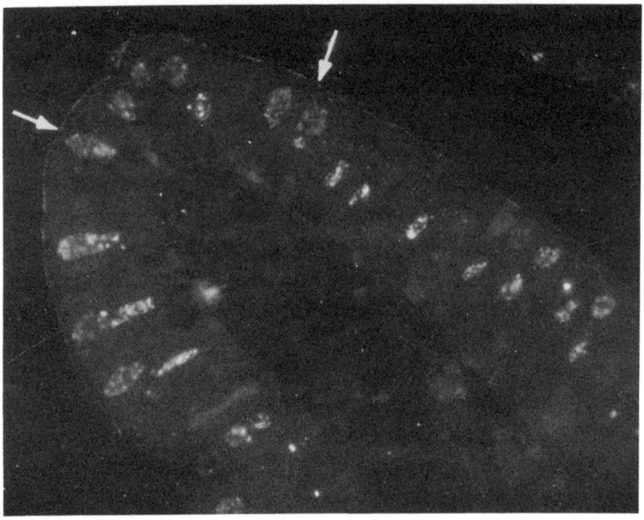

FIGURE 7. Localization of CLL-II in adult chicken intestine by immunofluorescence. Lectin is localized in clusters of highly fluorescent particles, which are the secretory granules of the goblet cells. Goblet cells are interspersed in the epithelium between other cell types which show no significant labeling. The line of immunofluorescent labeling at the intestinal mucosal surface (arrows) presumably derives from secretion of lectin along with the mucin of secretory granules. The secreted material apparently associates with the surface of the intestinal epithelium. Scale, 1 cm = 10 μm. (For details, see reference 37.)

amount of CLL-I in adult chicken muscle is also at these sites. It is not known if the function of CLL-I remains the same at different stages.

Nothing at all is yet known about the function of CLL-II or CHL in developing chicken tissues. However, work with CLL-II in adult chicken intestine has led to the discovery of its secretion, which could also be important for its role in development. CLL-II has been localized immunohistochemically to the secretory granules of the goblet cells of the intestine (37) (Fig. 7). The mucin within these secretory granules is a glycoprotein composed of about 85% carbohydrate and binds to CLL-II (37). Furthermore, CLL-II is found on the surface of the intestinal mucosa (Fig. 7), and secretion of mucin is accompanied by secretion of CLL-II (37). These studies show that this intracellular lectin can become extracellular under appropriate circumstances, which is reminiscent of the changes in CLL-I with muscle development and the elicitation of cell-surface slime mold lectin. They also show that CLL-II, which presumably plays a specifically developmental role in kidney and other embryonic tissues (Fig. 4), is also adapted for a role in the adult organism. Preliminary studies suggest that CHL is also secreted (38).

5. CONCLUSION

Because of their properties, it might be expected that the lectins found in tissues would interact with the carbohydrates on the surface of cells or between them. This could play an important role in tissue organization and development. Evidence for striking changes in lectin levels in developing tissues, ranging from slime molds to vertebrates, provides some circumstantial support for this idea. A growing body of evidence suggests that, in the case of cellular slime molds, lectins play a role in cell–cell adhesion. A similar role of some developmentally regulated lectins in the chicken has been suggested but has very limited support. One clear result of work with two developmentally regulated lactose-binding lectins from chicken, CLL-I and CLL-II, is that they appear to have different functions in different tissues, ranging from membrane organization to secretion. The same lectins are concentrated in some developing tissues and in other adult tissues of the same animal, but the relationship of their functions in development and adulthood are not yet known.

ACKNOWLEDGMENTS

This work was supported by grants from the USPHS, the McKnight Foundation, and by the Veterans Administration Medical Center. I am especially grateful to Dr. W. Springer, E. C. Beyer, P. Haywood-Reid, and D. Cooper whose recent work provides an important basis for this manuscript.

REFERENCES

1. N. Sharon, *Sci. Am.* **236**, 108 (1977).
2. I. J. Goldstein and C. E. Hayes, *Adv. Carbohydr. Chem. Biochem* **35**, 127 (1978).
3. S. H. Barondes, *Ann. Rev. Biochem.* **50**, 207 (1981).
4. S. D. Rosen, J. A. Kafka, D. L. Simpson, and S. H. Barondes, *Proc. Natl. Acad. Sci. USA* **70**, 2554 (1973).
5. T. P. Nowak, P. L. Haywood, and S. H. Barondes, *Biochem. Biophys. Res. Commun.* **68**, 650 (1976).
6. H. Den, D. A. Malinzak, and A. Rosenberg, *Biochem. Biophys. Res. Commun.* **69**, 621 (1976).
7. D. Kobiler and S. H. Barondes, *Dev. Biol.* **60**, 326 (1977).
8. W. F. Loomis, *Dictyostelium discoideum, A Developmental System*, Academic Press, New York, 1978.
9. D. L. Simpson, S. D. Rosen, and S. H. Barondes, *Biochemistry* **13**, 3487 (1974).
10. S. H. Barondes and P. L. Haywood, *Biochem. Biophys. Acta* **550**, 297 (1979).

11. S. D. Rosen, J. Kaur, D. L. Clark, B. T. Pardos, and W. A. Frazier, *J. Biol. Chem.* **254**, 9408 (1979).
12. G. C. L. Ma and R. A. Firtel, *J. Biol. Chem.* **253**, 3924 (1978).
13. W. A. Frazier, S. D. Rosen, R. W. Reitherman, and S. H. Barondes, *J. Biol. Chem.* **250**, 7714 (1975).
14. D. L. Cooper and S. H. Barondes, *J. Biol. Chem.* **256**, 5046 (1981).
15. H. Den and D. A. Malinzak, *J. Biol. Chem.* **252**, 5444 (1977).
16. T. P. Nowak, D. Kobiler, L. E. Roel, and S. H. Barondes, *J. Biol. Chem.* **252**, 6026 (1977).
17. E. C. Beyer, S. E. Zweig, and S. H. Barondes, *J. Biol. Chem.* **255**, 4236 (1980).
18. E. C. Beyer and S. H. Barondes, *J. Cell Biol.* **92**, 23 (1982).
19. H. Ceri, D. L. Kobiler, and S. H. Barondes, *J. Biol. Chem.* **256**, 390 (1981).
20. F. Mir-Lechaire and S. H. Barondes, *Nature* **272**, 256 (1978).
21. E. C. Beyer, K. Tokuyasu, and S. H. Barondes, *J. Cell Biol.* **82**, 565 (1979).
22. J. T. Powell and P. L. Whitney, *Biochem. J.* **188**, 1 (1980).
23. S. H. Barondes, in A. S. G. Curtis and J. Pitts, eds., *Cell Adhesion and Motility*, Cambridge University Press, Cambridge, United Kingdom, 1980, p. 309.
24. R. W. Reitherman, S. D. Rosen, W. A. Frazier, and S. H. Barondes, *Proc. Natl. Acad. Sci. USA* **72**, 3541 (1975).
25. C.-M. Chang, S. D. Rosen, and S. H. Barondes, *Exp. Cell Res.* **104**, 101 (1977).
26. C.-M. Chang, R. W. Reitherman, S. D. Rosen, and S. H. Barondes, *Exp. Cell Res.* **95**, 136 (1975).
27. C.-H. Siu, R. A. Lerner, G. C. L. Ma, R. A. Firtel, and W. F. Loomis, *J. Mol. Biol.* **100**, 157 (1976).
28. W. R. Springer, P. L. Haywood, and S. H. Barondes, *J. Cell Biol.* **87**, 682 (1980).
29. W. R. Springer and S. H. Barondes, *Exp. Cell Res.* **138**, 231 (1982).
30. J. Ray, T. Shinnick, and R. A. Lerner, *Nature* **279**, (1979).
31. T. Shinnick and R. A. Lerner, *Proc. Natl. Acad. Sci. USA* **77**, 4788 (1980).
32. T. R. Podleski and I. Greenberg, *Proc. Natl. Acad. Sci. USA* **77**, 1054 (1980).
33. T. K. Gartner and T. R. Podleski, *Biochem. Biophys. Res. Commun.* **67**, 972 (1975).
34. S. J. Kaufman and M. L. Lawless, *Differentiation* **16**, 41 (1980).
35. R. G. MacBride and R. J. Przybylski, *J. Cell Biol.* **85**, 617 (1980).
36. S. H. Barondes and P. L. Haywood-Reid, *J. Cell Biol.* **91**, 568 (1981).
37. E. C. Beyer and S. H. Barondes, *J. Cell Biol.* **92**, 28 (1982).
38. H. Ceri, P. J. Shadle, D. Kobiler, and S. H. Barondes, *J. Supramol. Struc.* **11**, 61 (1979).

8

Regulation of Morphogenesis by the Pericellular Matrix

BRYAN P. TOOLE and CHARLES B. UNDERHILL
Department of Anatomy and Cellular Biology
Tufts University School of Medicine
Boston, Massachusetts 02111

The acquisition of form and function during embryonic development involves several types of cell behavior, e.g., proliferation, movement, shape change, and adhesion, as well as expression of specific molecular products. These events take place within a pericellular matrix composed of characteristic macromolecules of three major classes—collagens, proteoglycans, and glycoproteins. These components are highly interactive, both with one another and with the surfaces of the cells which they surround or abut. Their composition varies greatly in relation to their source, their locale, and the stage of development of the particular organ or tissue within which they are contained. Over the past decade, it has become apparent that these extracellular macromolecules play an important role, not only in maintaining the physical integrity and structural properties of tissues, but also in regulating the behavior of their constituent cells.

This chapter will focus on the structure and function of pericellular matrices which are rich in hyaluronate. Two major aspects will be considered: (1) the occurrence and structure of these matrices, especially in regard to the macromolecular interactions that may be responsible for their integrity and biological properties; and (2) the possible roles of these matrices in regulating cell interactions in vitro and tissue morphogenesis in vivo.

1. EXTRACELLULAR MACROMOLECULES

The major components of extracellular matrices that have been well defined are the proteoglycans (1–3), collagens (4–6), and certain glycoproteins such as fibronectin and laminin (7–8) (Chapter 9). Since the structure and properties of these components have been adequately reviewed, only a very brief outline is given here. The reader is referred to a recent comprehensive volume (9), which details many of the biochemical and biological characteristics of these extracellular molecules.

The proteoglycans are composed of glycosaminoglycan side chains covalently linked to a core protein. For the purposes of this discussion, hyaluronate will be considered a proteoglycan, although the linkage to protein is not

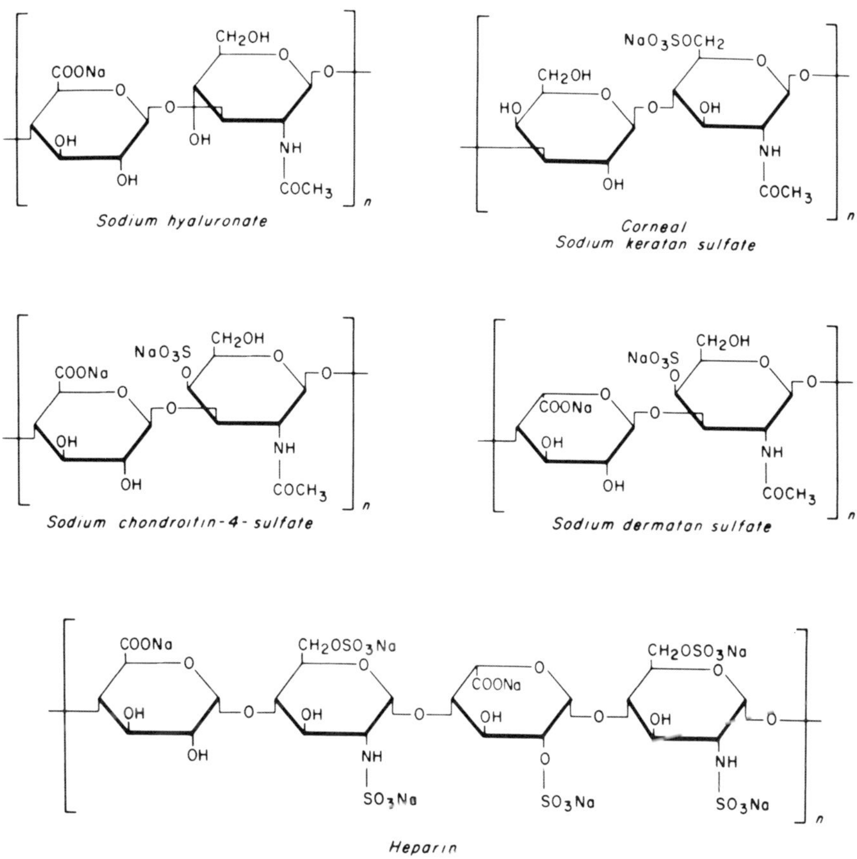

FIGURE 1. Repeating disaccharide components of some glycosaminoglycans. Hyaluronate is composed of glucuronic acid and *N*-acetylglucosamine; keratan sulfate of galactose and *N*-acetylglucosamine-6-sulfate; chondroitin sulfates of glucuronic acid and *N*-acetylgalactosamine-4 or -6-sulfate; dermatan sulfate of iduronic acid and *N*-acetylgalactosamine-4-sulfate (with small amounts of glucuronic acid); heparin of glucuronic acid or iduronic acid-2-sulfate and glucosamine with both *N*-sulfate and *O*-sulfate groups (iduronic acid comprises about 70% of the uronic acid moieties). In heparan sulfate, the uronic acid is chiefly glucuronic acid and some of the glucosamine moieties are *N*-sulfated while others are *N*-acetylated. Heparan sulfate contains about 1 mole of sulfate per disaccharide whereas heparin contains 2–3 moles. [Reprinted, with permission, from Toole (101).]

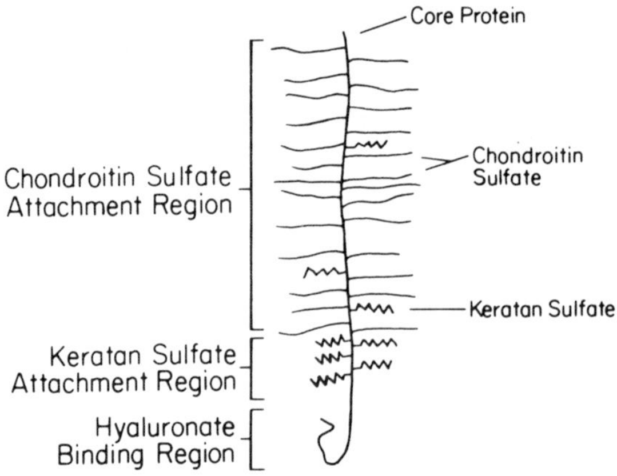

FIGURE 2. Model of the cartilage proteoglycan monomer, adapted from Hascall and Hascall (1). As indicated, the monomer is organized into three regions which are structurally and functionally different. The monomer is heterodisperse in structure and its detailed composition varies considerably in different cartilages. The much-studied proteoglycan from bovine nasal cartilage has an average molecular weight of $\sim 2.5 \times 10^6$ comprised of a core protein of molecular weight $\sim 2.0 \times 10^5$ to which are covalently attached ~ 100 chondroitin sulfate side chains of molecular weight $\sim 2 \times 10^4$, ~ 50 keratan sulfate side chains of molecular weight $\sim 5 \times 10^3$, as well as additional O-linked oligosaccharides in the keratan sulfate attachment region and *N*-linked oligosaccharides in the hyaluronate-binding region.

yet certain (10). The glycosaminoglycan side chains are polymers of negatively charged disaccharides, which, with the exception of keratan sulfate, are composed of hexuronate and hexosamine residues (Fig. 1). An example of a particularly well-studied proteoglycan is that which is found in cartilage (see Fig. 2). The core protein of cartilage proteoglycan is divided into specific regions, one of which is rich in chondroitin sulfate side chains, another which is rich in keratan sulfate chains, and still another which interacts noncovalently with hyaluronate. It is likely, however, that the proteoglycan components of different tissues, even when containing similar glycosamino-

FIGURE 3. Hierarchies of type-I collagen structure. The amino acids (*a*) are incorporated into the polypeptide pro-α chains (*b*), after which some of the prolines and lysines are hydroxylated. The pro-α chains interact to form the triple-chain procollagen molecule from which C- and N-terminal propeptides are subsequently cleaved. In the mature type-I collagen molecule, each α chain (*c*) is coiled into a helix, three of which (*d*, *e*) are wrapped around each other to form a superhelical rigid rod, 2800 Å in length and about 15 Å wide (*f*). Note the two short nonhelical regions, telopeptides, at each end of the molecule. These molecules pack in a "quarter stagger" array polarized in one direction to give rise to the striated collagen fibrils (*g*), which are found in geometrical patterns, parallel in tendon, orthogonally arrayed as in plywood in the cornea (*h*), and in more random bundles in mammalian skin. *c-e*, 9,625,000×; *f*, 181,500×; *g*, 66,000×; *h*, 27,500×. [Reprinted, with permission, from Gross (102).]

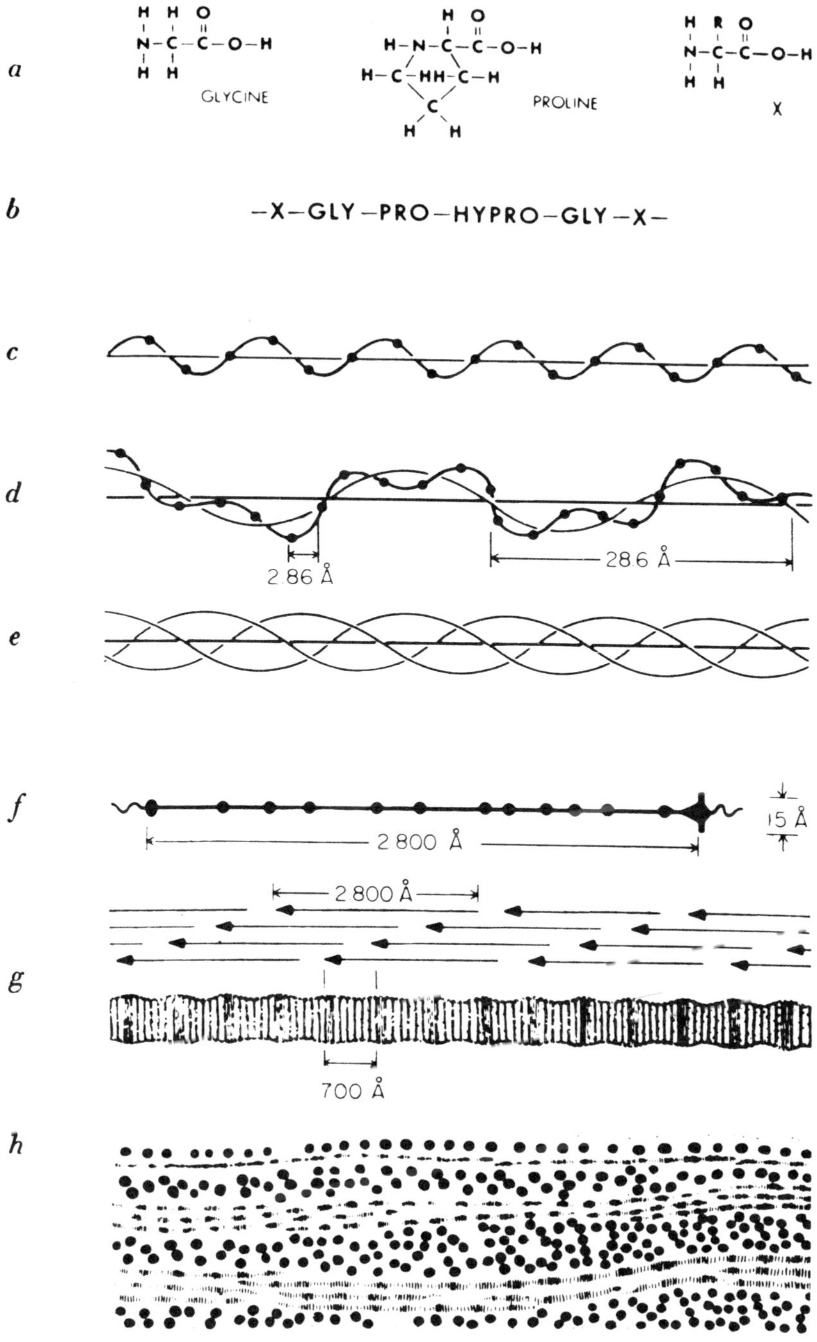
a
GLYCINE
PROLINE
X
b
—X—GLY—PRO—HYPRO—GLY—X—
c
d
2.86 Å
28.6 Å
e
f
2800 Å
15 Å
g
2800 Å
700 Å
h

glycan side chains, may differ considerably in their macromolecular structure and properties. It is apparent, for example, that the proteoglycan of cartilage is quite different from that found in the cornea (1, 11), even though the chondroitin sulfate and keratan sulfate side chains are alike.

At least five types of collagen have now been identified. The most widely distributed, type I, contains two α-1 type-I chains and one α-2 chain. Type-II and -III collagens are composed of three identical chains, α-1 type-II in the former and α-1 type-III in the latter. The individual chains of these three collagen types have a molecular weight of approximately 95,000 and in the native molecule are assembled in a triple helical configuration. Each of these types of collagen molecules has the ability to self-aggregate to form fibrils with the typical 64-nm cross-striations (see Fig. 3). Types-IV and -V collagens, which are components of basement membranes, are less well defined in terms of structure and aggregation properties.

Fibronectin and laminin are described in detail in Chapter 9 of this volume. Of special note, however, is the observation that these two glycoproteins have multiple reactive sites of different specificities. On fibronectin, separate sites recognize heparin, hyaluronate, and collagen (Chapter 9) (12). Laminin also interacts with heparin and collagen (8, 13). In addition, specific interactions between hyaluronate and proteoglycans (Section 3.2) and between proteoglycans and collagen (Section 3.3) have been demonstrated. All these classes of extracellular macromolecules have been shown to bind to specific sites on the surfaces of various cells (8, 14–16) (Chapter 9). These interactions are thought to be of considerable significance, not only in determination of the physical properties of connective tissues but also in the control of cell behavior during development.

2. VISUALIZATION OF THE PERICELLULAR MATRIX

Numerous types of experimental approaches have documented the presence of various classes of macromolecules and their organization in association with the external cell surface. Some of these approaches are discussed in other chapters of this volume (Chapters 1–3, 9). However, one particularly illustrative and simple method for demonstrating a pericellular zone or "coat" around cultured cells was developed by Clarris and Fraser (17). Particles such as formalin-fixed bacteria, red blood cells, or carbon black are added to the medium of the cultured cells and allowed to settle. Upon microscopic examination, clear areas, from which the particles are excluded, can be observed surrounding each cell. An example of this technique using rat fibrosarcoma cells, which produce a particularly large coat, is shown in Figure 4A. The extent of the pericellular region depends upon the cell type, the length of time in culture, and the physiological state of the cells. The

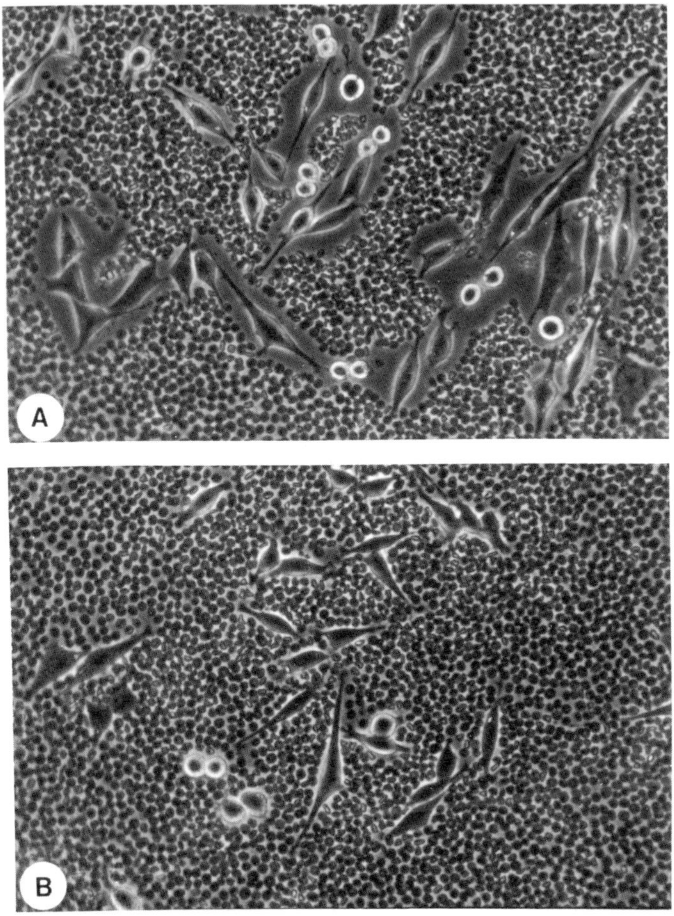

FIGURE 4. Pericellular coats of hyaluronate around rat fibrosarcoma cells; *A*, control; *B*, treated with streptomyces hyaluronidase. The coat of hyaluronate has been visualized indirectly due to its ability to exclude fixed red blood cells (70).

pericellular coats do not appear to be affected by treatment with ribonuclease, deoxyribonuclease, neuraminidase, EDTA, or dilute trypsin. However, as illustrated in Figure 4B, the coats are very quickly destroyed by hyaluronidase (17). This strongly suggests that hyaluronate is a major structural component of the cell coat.

Studies of these pericellular coats to date have mainly been directed towards their function as a physical barrier. For example, synovial cells, which have a thick coat (17), normally are not susceptible to infection by several types of viruses (Newcastle, vesicular stomatitis, and rubella). However, if the cells are first treated with hyaluronidase, then infection can take place (18, 19). The most likely explanation for this phenomenon is that the removal of the hyaluronate-containing coat allows the viruses to make contact with and penetrate the plasma membrane.

In addition to acting as a barrier to particles such as viruses, there is also some evidence that the cell coat can block the access of soluble macromolecules to the plasma membrane. Lippmann (20) has shown that the addition of hyaluronate to isolated tumor cells partially protects them from the cytotoxic effects of antiserum combined with complement. It is probable that the tumor cells bind hyaluronate to their surface, preventing access of the complement system to the antigenic sites. Darzynkiewicz and Balazs (21) found that the ability of phytohemagglutinin to induce cell division in lymphocytes was blocked by the addition of large amounts of hyaluronate and that the degree of suppression was a function of both the size and concentration of the hyaluronate. As in the previous example, the lymphocytes may bind hyaluronate to their surfaces, which then blocks the interaction of the lectin with its binding sites.

Later in this chapter we will consider the possible effects of the hyaluronate-containing pericellular zones on cell interactions, especially in relation to embryonic morphogenesis (Sections 4 and 5).

3. STRUCTURE OF THE PERICELLULAR MATRIX

The pericellular coats described in Section 2 are clearly associated with the cell surface, but it is difficult to define where within these coats the cell surface ends and the extracellular matrix begins. The two phases merge and share similar macromolecular components. In vivo, discrete pericellular zones are rarely distinguishable from the extracellular matrix, although in cartilage for example, the matrix distant from the cell clearly contains different proportions of components from that close to the cell (22). The most discrete form of peri- or juxtacellular zone is the basal lamina surrounding muscle cells and at the basal surface of epithelia. In all these cases, the three classes of extracellular macromolecules (i.e., collagens, proteoglycans, and glycoproteins) are found within the pericellular zones and extracellular matrix. However, the composition of each of these macromolecules and their proportions varies considerably for different cell types under various physiological conditions or at different stages of embryonic development of a given tissue. In this chapter, we will place particular emphasis on the hyaluronate component of the pericellular and extracellular matrix.

Figure 5 represents a highly speculative model demonstrating the possible modes of participation of hyaluronate in the structure of a pericellular coat. This model depicts the interactions of hyaluronate with both the cell surface and with other extracellular macromolecules, as well as other interactions

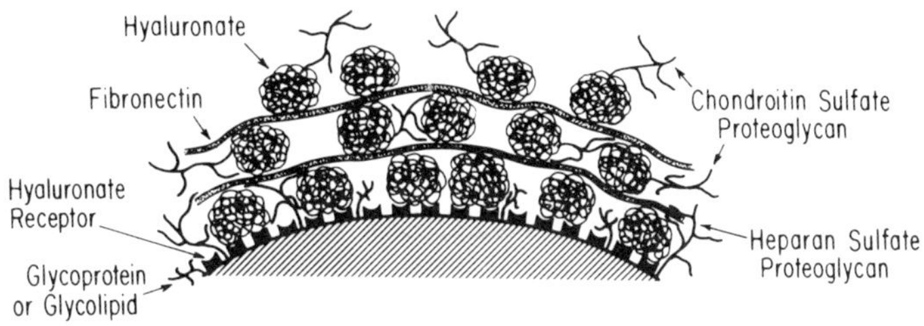

FIGURE 5. Hypothetical model for a hyaluronate-rich cell coat (see Fig. 4). A series of interactions between cell-surface and extracellular macromolecules leads to formation of an extensive coat, which is continuous with the extracellular matrix. This is but one of many possible arrangements of these and other substances such as collagen and laminin. Further discussion of the interactions depicted is provided in the text. [Reprinted, with permission, from Toole (85).]

not involving hyaluronate that are also thought to contribute to this structural network. The properties of such a coat will, of course, depend greatly on the proportions of each molecule and the exact nature of their interactions. Evidence for some of the various interactions depicted in the model are described in the following sections. Interactions of fibronectin with the cell surface and with other extracellular macromolecules are described in Chapter 9 and thus are not discussed here.

3.1. Glycosaminoglycan-Cell Surface Interactions

Hyaluronate is found on the surfaces of a wide variety of cell types (17, 23–26). The existence of binding sites for hyaluronate on the surface of cells was demonstrated by experiments using radiolabeled ligand (16, 27). [^{3}H]hyaluronate was added to suspensions or monolayers of different types of cells, and the amount of binding was then determined. The extent of binding was found to be a function of both the concentration of labeled hyaluronate and the cell number. Most of the binding took place within the first 2 min of the incubation; it was not influenced by the presence or absence of small amounts of divalent cations and could be prevented by the addition of a large excess of unlabeled hyaluronate but not other glycosaminoglycans (Fig. 6). High-molecular-weight preparations of hyaluronate bound more effectively than low-molecular-weight preparations, presumably because the repetitive structure of hyaluronate allows it to bind to more than one binding site at a time. As illustrated in Figure 7, the larger the molecule of hyaluronate the greater the number of binding sites with which it can interact and thus the stronger the binding. However, the binding of [^{3}H]hyaluronate to cells is inhibited by oligosaccharide fragments of hyaluronate consisting of six sugars or more, suggesting that this is the minimum size recognized. Labeled hyaluronate bound to the cells can be totally removed by incubating

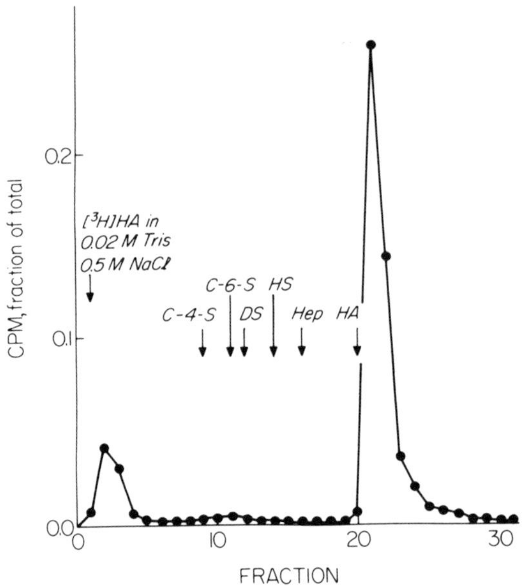

FIGURE 6. Specificity of binding of hyaluronate to the surface of SV-3T3 cells. [^{3}H]hyaluronate was applied to a small column of fixed SV-3T3 cells (0.5 ml containing 5×10^8 cells). The column was then treated sequentially with 0.5 mg each of the following unlabeled GAG in 0.5 ml buffer: chondroitin-4-sulfate (C-4-S), chondroitin-6-sulfate (C-6-S), dermatan sulfate (DS), heparan sulfate (HS), heparin (Hep), and hyaluronate (HA). The [^{3}H]hyaluronate was displaced only by the unlabeled hyaluronate. [Results derived from Goldberg et al. (103).]

the intact cells with streptomyces hyaluronidase or trypsin, indicating that the binding sites are located on the cell surface. Scatchard plot analysis of the binding data generated in these experiments demonstrates that for hyaluronate of molecular weight approximately 10^6, a single SV-3T3 cell binds at saturation approximately 3×10^3 molecules with a K_d of $1\text{–}2 \times 10^{-9}$ M. The affinity and extent of binding varies considerably for different cell types and is altered by viral transformation (see Table 1) (28).

The cell-surface binding sites for hyaluronate appear to be distinct from other proteins that are capable of binding hyaluronate, i.e., the proteoglycan monomer from cartilage (1) and fibronectin (Chapter 9). Chondroitin (desulfated chondroitin sulfate), as well as hexasaccharides and octasaccharides of hyaluronate, are effective binding competitors for the cell-surface binding sites. However none of these substances shows activity towards the hyaluronate binding sites on cartilage proteoglycan (1). In addition, the binding of labeled hyaluronate to the cell surface is not inhibited by antibodies directed against fibronectin and the binding is greater to virus-transformed cells than to their nontransformed (parent) counterparts (see Section 4.1), indicating that fibronectin is not the receptor for hyaluronate on the cell surface.

In addition to its role in maintaining a layer of hyaluronate on the cell surface, the binding sites may also have other physiological functions. The binding sites may mediate endocytosis of hyaluronate from the extracellular milieu, leading to degradation by lysosomal hyaluronidase found in these cells (see Table 1) (29, 30). As much as 50% of the [^{3}H]hyaluronate bound by SV-3T3 cells or chick embryo fibroblasts is degraded within 3 hr (31). On the other hand, 3T3 cells exhibit fewer binding sites and lower hyaluronidase levels than SV-3T3 cells (Table 1) (31). Another possible function of the binding sites is to induce changes within the cytoplasm of the cell. One possible example of this is that the addition of small amounts of hyaluronate to polymorphonuclear leukocytes dramatically stimulates their phagocytic function (32). Another is the inhibition of proteoglycan synthesis by chondrocytes in the presence of hyaluronate (33–36).

Kraemer (37–39) has shown that a heparan sulfate-containing proteoglycan is also tightly associated with the surfaces of many different cultured cell lines. Recent work has demonstrated specific sites that bind exogenously added heparan sulfate and heparin to the surface of hepatocytes (15, 40). These binding sites appear to be specific for the glycosaminoglycan portion

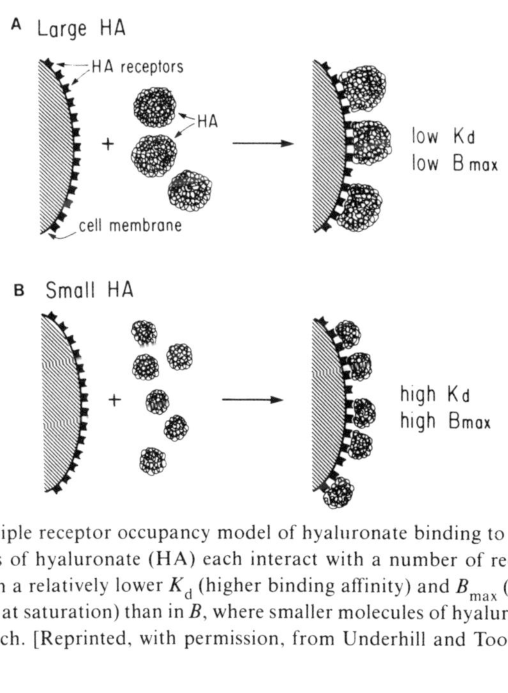

FIGURE 7. Multiple receptor occupancy model of hyaluronate binding to the cell surface. In *A*, large molecules of hyaluronate (HA) each interact with a number of receptors on the cell surface resulting in a relatively lower K_d (higher binding affinity) and B_{max} (smaller number of molecules binding at saturation) than in *B*, where smaller molecules of hyaluronate interact with fewer receptors each. [Reprinted, with permission, from Underhill and Toole (27).]

TABLE 1. Comparison of Hyaluronate Parameters for 3T3 and SV 3T3 Cells

Parameters	3T3	SV-3T3
Cell surface HA[a] (ng/10^6 cells)	330	10
Dissociation constant for HA binding[b] (μg HA/ml)	5.6	1.9
Maximum binding capacity for HA[b] (ng HA/10^6 cells)	7.6	46
Hyaluronidase activity[c] (units/mg cell protein)		
Cell layer	<1	1.8
Medium	<1	9.1

[a] From Underhill and Toole (70).

[b] From Underhill and Toole (28). K_d (dissociation constant) is inversely proportional to binding affinity.

[c] From Orkin et al. (31). A unit is defined as the release of 1 μg reducing terminal *N*-acetylglucosamine per 16 hr incubation in 0.15 *M* NaCl/0. 1 *M* Na formate, pH 3.7, at 37°C.

of the proteoglycan, as opposed to the protein core, since some of the endogenous heparan sulfate proteoglycan from hepatocytes (15) and Chinese hamster ovary cells (41) can be displaced from the cell surface by addition of large concentrations of exogenous heparin. In addition to this mode of attachment of the heparan sulfate polysaccharide chains to binding sites on the cell surface, however, it seems that the core protein of a separate population of endogenous cell-surface heparan sulfate-proteoglycan is attached directly to the plasma membrane. Kjellen et al. (42) have separated two distinct species of proteoglycan by treating hepatocyte plasma membranes first with heparin to remove the fraction bound to the glycosaminoglycan binding sites mentioned above and then with detergent. The detergent-extracted proteoglycan has a higher molecular weight than the heparin-displaced species and its protein core exhibits marked hydrophobic properties. These workers concluded that the larger proteoglycan is attached directly to the hepatocyte plasma membrane and that the smaller molecule may be derived from it by proteolytic cleavage. Thus, there appear to be two modes of interaction of heparan sulfate-proteoglycan with the cell surface; one mediated via binding of the carbohydrate moiety to a protein binding site, which may or may not be fibronectin (see Chapter 9), and the other by a direct interaction of the protein core of the proteoglycan with the lipid membrane.

Recent evidence suggests that a similar situation may also apply to hyaluronate. In addition to the hyaluronate binding sites described earlier, hyaluronate may also be attached to the cell surface via a hydrophobic protein that is covalently attached to the hyaluronate (10).

Several studies (24, 43–45) indicate that chondroitin sulfate proteoglycans are also intimately associated with the cell surface; however, the nature of this association has not yet been elucidated.

3.2. Proteoglycan-Hyaluronate Interactions

In addition to the interactions of proteoglycan and hyaluronate with the cell surface, there is widespread evidence for their interaction with one another. The most detailed evidence for such an interaction comes from the work on aggregation of cartilage proteoglycan monomers with hyaluronate to form enormous complexes of sizes equivalent to 100 million molecular weight (1, 46). This aggregation involves the binding of many monomers along the hyaluronate molecule at intervals of approximately 30 nm, which corresponds to about 50 disaccharide repeating units of hyaluronate (Fig. 8). The interaction takes place between a specific region of the proteoglycan core protein, a five-disaccharide length of the hyaluronate, and a third component termed *link protein*. The resulting complex is very stable and yields aggregates of several microns in length.

Recent evidence has indicated that similar interactions to the above take place between hyaluronate and proteoglycans of noncartilaginous tissues, e.g., chondroitin sulfate and heparan sulfate proteoglycans of aorta and lung (47–49). In addition, chondroitin sulfate proteoglycans derived from cul-

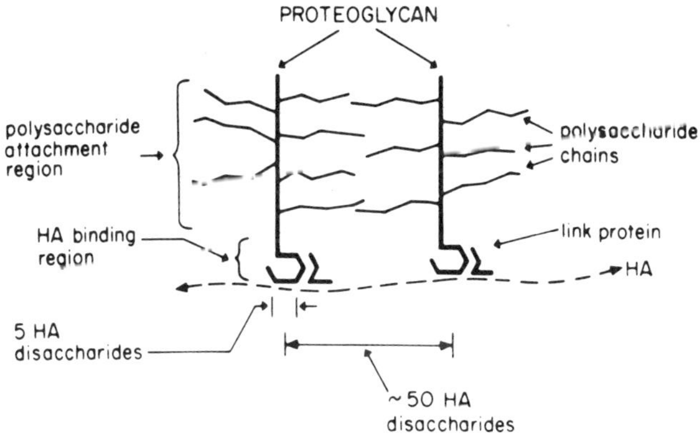

FIGURE 8. Interaction of cartilage proteoglycan monomers with hyaluronate (HA) and link protein to form a very large aggregate. The monomers are spaced approximately 50 hyaluronate disaccharides apart; however a decasaccharide is sufficient for binding of hyaluronate to the core protein. [Reprinted, with permission, from Heinegard and Hascall (104).]

tured glial cells and skin fibroblasts have been purified and shown to form large aggregates on addition of hyaluronate (50, 51). It should be pointed out, however, that treatment of the pericellular matrix associated with human embryo fibroblasts with leech hyaluronidase (which is specific for hyaluronate) does not result in loss of proteoglycans from the matrix (52), suggesting that other mechanisms must also be important in maintaining the chondroitin sulfate proteoglycan in the pericellular coat.

3.3. Proteoglycan-Collagen Interactions

Another form of interaction that may contribute to pericellular matrix organization is that between proteoglycans and collagens (53–56). Ruthenium red staining of several tissues has revealed an ordered association of proteoglycan with collagen fibers (57–59) and with basement membranes (59–61)

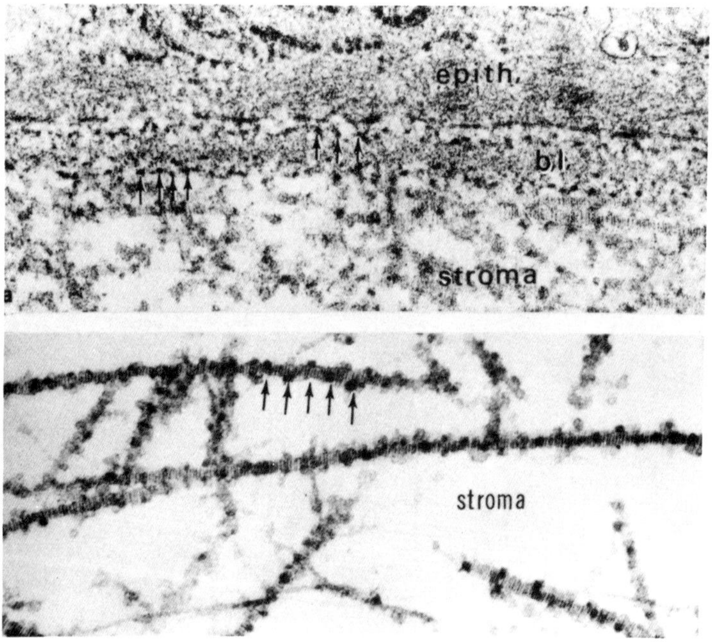

FIGURE 9. Relationship of proteoglycan to (*A*) basement membrane (85,000×) and (*B*) stromal (52,000×) collagens in the chick embryo cornea. In the basement membrane, ruthenium red-stained chondroitin sulfate-proteoglycan (arrows) is arranged in a regular pattern on either side of the electron-dense basal lamina (b.l.) (59). Proteoglycan is also associated with the stromal fibrils in a specific repeated location. Treatment of the cornea with chondroitinase ABC prior to processing eliminated the ruthenium red staining in both the basal lamina and stroma (59). [Reprinted, with permission, from Toole (101).]

(see Fig. 9). In vitro, chondroitin sulfate and dermatan sulfate proteoglycans from cartilage, heart valves, skin, and tendons have been shown to react with soluble collagen under physiological conditions to form insoluble aggregates that become organized into characteristic collagen fibrils with 64-nm striations (54, 56, 62). Aggregate formation takes place with types-I, -II, and -III collagen (63) and is dependent on both the protein and polysaccharide portions of the proteoglycans (56, 63, 64). The function of this interaction in fibril structure or formation has not been elucidated, but its possible contribution to pericellular matrix integrity should be assessed in future experiments.

4. REGULATION OF CELL INTERACTIONS BY THE PERICELLULAR MATRIX

4.1. Cell-Cell Interactions

Many types of cells display a hyaluronate-dependent type of aggregation in suspension that is independent of divalent cations. In the case of alveolar macrophages (65) and some types of lymphoma cell lines (66, 67), addition of exogenous hyaluronate induces aggregation. For other cell types (SV-3T3, PY-BHK), aggregation occurs spontaneously, being mediated by endogenous cell surface-bound hyaluronate (68). It is hypothesized that the hyaluronate crosslinks cells by attaching to free-binding sites on the surfaces of adjacent cells (16, 68) (see Fig. 10A). Consistent with this hypothesis is the fact that aggregation can be inhibited both by degradation of cell-surface hyaluronate (Fig. 10B) and by the addition of a large excess of exogenous hyaluronate (Fig. 10C). In the latter case, the endogenous hyaluronate would occupy all receptor sites, preventing crossbridging. Thus, the amount of hyaluronate present in the cellular environment can regulate adhesion of these cells to one another.

It is very likely that the receptors that bind labeled hyaluronate are identical to those that mediate the divalent-cation independent aggregation of cells. This is suggested by the fact that those conditions that promote aggregation also promote the binding of labeled hyaluronate (69). First, high concentrations of hyaluronate, fragments of hyaluronate consisting of six or more sugars, and chondroitin inhibit both aggregation and [^{3}H]hyaluronate binding, whereas the sulfated glycosaminoglycans have little or no effect on either phenomenon. Second, both aggregation and binding are optimal at pH 7 and decrease substantially at pH values on either side of neutrality. Third, ionic strength affects both phenomena in a very similar fashion. Neither aggregation nor binding occurs at NaCl concentrations of 0.05 *M* and below. As the concentration of NaCl is increased, aggregation and binding also

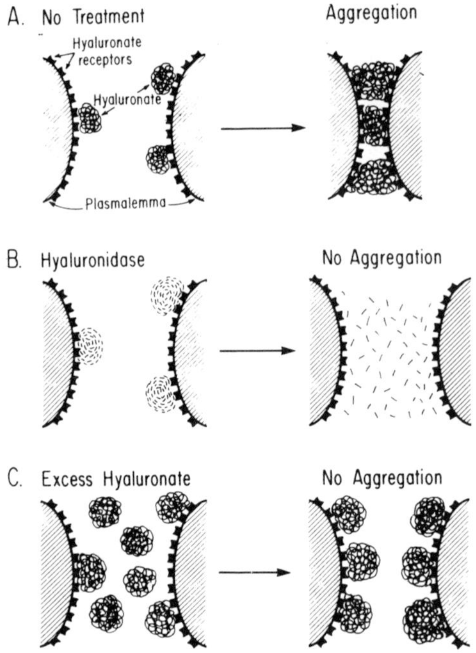

FIGURE 10. Hyaluronate-mediated, divalent cation-independent aggregation of cell lines in suspension (28, 68, 69). (*A*) Endogenous cell-surface-associated hyaluronate crossbridges cells by binding to receptors on adjacent cells. (*B*) Hyaluronidase treatment degrades the hyaluronate, inhibiting aggregation. (*C*) Addition of excess exogenous hyaluronate blocks all the receptors, inhibiting aggregation. [Reprinted, with permission, from Toole (85).]

increase until a plateau value is reached at 0.5 *M* NaCl. This effect appears to be due to the fact that higher salt concentrations neutralize the negative charge on the hyaluronate. Fourth, higher-molecular-weight preparations of hyaluronate are more potent inhibitors of aggregation and binding than are lower-molecular-weight preparations. This last effect can be accounted for by the multivalent nature of hyaluronate binding, which is diagrammatically represented in Figure 7. Finally, virus transformation of cells was found to potentiate both aggregation and binding (28). When compared in physiological saline, virus-transformed cells (SV-3T3 and PY-BHK) aggregate to a greater extent and bind more [^{3}H]hyaluronate than their nontransformed counterparts (3T3 and BHK). Furthermore, increasing the ionic strength to 0.5 *M* NaCl enhances both the aggregation and the binding of [^{3}H]hyaluronate in each cell line, but more so in the case of the transformed than the parent cells. These results suggest that each of these cell lines has binding sites on their surface and that virus transformation leads to an increase in the ability of these receptors for binding hyaluronate.

One striking finding is that there is an inverse relationship between those cell lines that bind the greatest amount of hyaluronate and those that have the largest cell coat (70). Figure 11 shows the cell coats of the parent and virus-transformed cell lines both before and after treatment with hyaluronidase. The coats of the parent cell types appear to be much larger than those of the virus-transformed cells. This conclusion has been verified by chemical

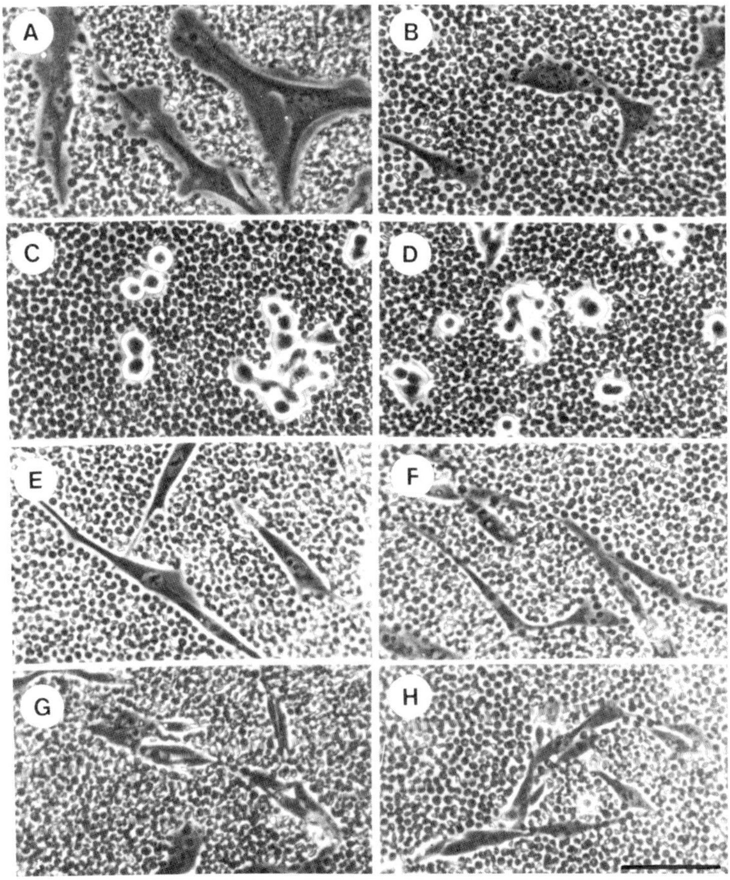

FIGURE 11. Effect of viral transformation on the size of pericellular, hyaluronate-rich coats surrounding 3T3 and BHK cells, visualized as in Figure 4. (*A*) 3T3; (*C*) SV-3T3; (*E*) BHK; (*G*) Py-BHK cells. Panels *B*, *D*, *F*, and *H* show the corresponding cells after treatment with streptomyces hyaluronidase. The size of the coat around the 3T3 cells is larger than that around the BHK cells. No discernible coat is present around the SV-3T3 and Py-BHK cells. Chemically measured amounts of hyaluronate associated with these cells correlate with the size of the coats (70).

measurements of the amount of hyaluronate associated with each cell type (see Table 1). Thus, while the parent cell types have the greatest amount of endogenous hyaluronate, they also have the least ability to bind labeled hyaluronate and to aggregate. Possibly the large amount of endogenous hyaluronate around the parent cells inhibits both aggregation and binding by occupying or hindering access to the cell-surface binding sites.

There are several other examples in which the cell coat appears to inhibit cell-to-cell interactions. Treatment of cultured fibrosarcoma cells (23) or synovial cells (71) with hyaluronidase results in an increase in cell lysis on exposure to lymphocytes from immunized animals. The pericellular coat normally prevents close contact between the lymphocytes and the fibrosarcoma cells, but when this coat is enzymatically removed the lymphocytes can interact with the surface of the tumor cell and exert their cytotoxic effect. Also, pretreatment of Rous sarcoma virus-transformed chick embryo fibroblasts with hyaluronidase significantly enhances the aggregation of these cells by concanavalin A and wheat germ agglutinin (72). In this case, the hyaluronate-containing cell coat may act by preventing the close contact between cells necessary for aggregation or by covering lectin-binding sites.

The inhibition of cell–cell interactions by large amounts of hyaluronate in the extracellular milieu may also apply to embryonic cells in vivo. High concentrations of hyaluronate are present in the pathways of migration of embryonic cells. This hyaluronate may act to prevent the cells from adhering to one another, which in turn would facilitate movement by preventing interactions that lead to precocious immobilization or differentiation (see Section 5).

4.2. Cell–Substratum Interactions

Based primarily on studies in vitro, it has been concluded that cell movement requires attachment and detachment of localized areas of the cell surface with respect to an extracellular substratum. The focal attachment points are specialized regions of the pericellular zone that in vitro are restricted to the underside of the cells. These sites of attachment of cells to artificial substrata, sometimes termed foot pads, are enriched in the glycoprotein fibronectin (Chapter 9) and in glycosaminoglycans (45). The composition of the glycosaminoglycan components has been shown to vary according to the metabolic age of the foot pads. Isotope incorporation experiments suggest that when 3T3 cells are in the process of attaching to an artificial substratum, the foot-pad material is highly enriched in labeled heparan sulfate; whereas more mature foot pads and foot prints, i.e., material left behind by migrating cells, contain more hyaluronate and undersulfated chondroitin-4-sulfate proteoglycan (73, 74). On the basis of these and other experiments, Culp and colleagues (45, 73, 74) have proposed that interaction of cell-surface heparan sulfate-proteoglycan and fibronectin mediates substratum attachment but that the weakening of this initial adhesion, which is necessary for cell movement, involves accumulation of hyaluronate and chondroitin sulfate-proteoglycan. The latter glycosaminoglycans may compete with heparan sulfate for binding to fibronectin.

Unfortunately, little information is available with respect to levels of heparan sulfate during morphogenetic cell movements in vivo, but hyaluronate concentrations are clearly elevated (see Section 5). Consequently, the suspected role of hyaluronate in cell motility in vitro is of clear relevance to morphogenesis. Other observations in addition to those of Culp's group also point to the possibility that hyaluronate can form a bridge which keeps cells and their substratum "at arm's length," an action that may facilitate focal detachment. Involvement of hyaluronate in cell-substratum interactions is indicated by the following observations: (1) SV-3T3 cells plated on hyaluronate-derivatized substrata glycosylate the hyaluronate (75); (2) Chinese hamster ovary cells cultured on a solid substratum produce larger amounts of hyaluronate than the same cells grown in suspension (25); (3) hyaluronate interacts in a specific manner with fibronectin (12), an important component of foot pad material, which in itself is stimulatory to cell migration (76); (4) Rous sarcoma virus-transformed chondrocytes are detached from their substratum by treatment with streptomyces hyaluronidase (77).

That this involvement of hyaluronate in cell–substratum interactions is related to weakened adhesion can be concluded from the following observations. First, loosely-attaching variants of Chinese hamster ovary cells produce more cell surface-associated hyaluronate than tightly attaching variants (78–79). Second, the spreading of mesenchymal cells on the surface of collagen gels is retarded by the presence of hyaluronate in the gels (80). And third, the incorporation of labeled precursors into hyaluronate located in the cell adhesion sites increases in association with cell detachment as compared to initial attachment (45, 73, 74).

In addition to the above evidence for involvement of hyaluronate in cell-substratum detachment, direct evidence for promotion of cell movement by hyaluronate comes from the experiments of Bernanke and Markwald (81), who showed that the addition of hyaluronate to collagen gels causes an increase in the rate of movement of chick embryo cardiac mesenchyme cells in these gels and in the number of filopodia associated with the cells.

5. HYALURONATE AND MESENCHYMAL CELL MOVEMENT

A common pattern in the early development of tissues and organs is accumulation of hyaluronate at the sites of cellular movement, followed by its removal and replacement by a structural matrix during subsequent differen-

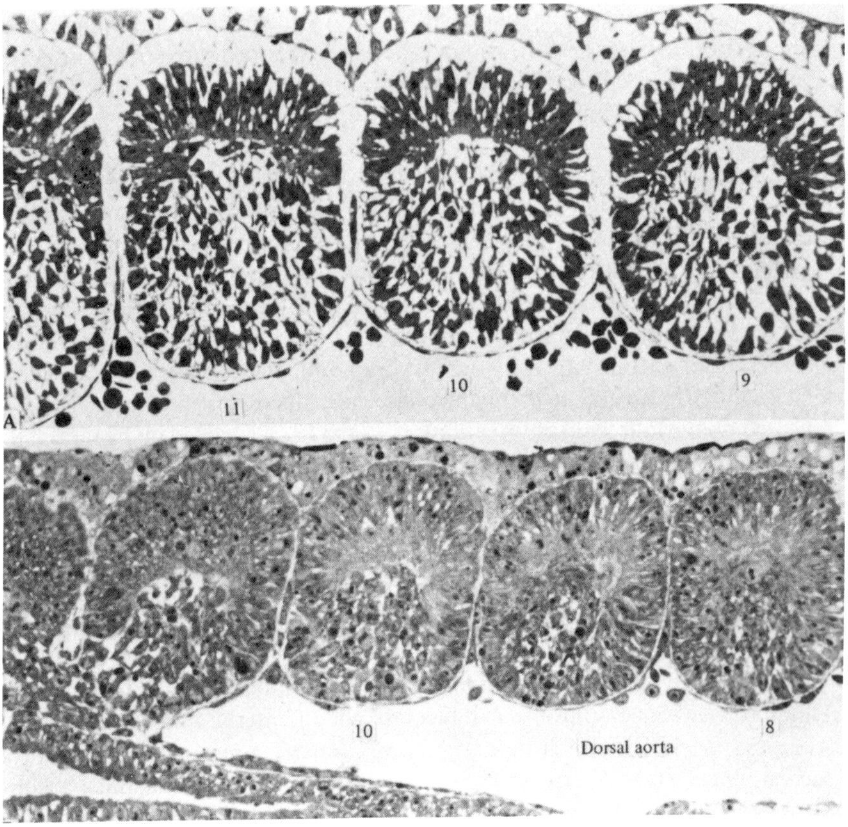

FIGURE 12. Effect of removal of hyaluronate on the structure of the dispersing sclerotome of the chick embryo. The upper micrograph (*A*) shows a section taken through somites wherein the sclerotome has begun dispersion under a dermamyotomal cap. The lower micrograph (*B*) shows the effect of streptomyces hyaluronidase treatment in ovo. The somites have become individually condensed and closer together. In addition, the overlying ectoderm and underlying endothelium of the dorsal aorta are closely applied to the somites. Sclerotomal cells are no longer dispersed and appear clumped together. This experiment indicates that hyaluronate is essential for the separation of these cells. [Reproduced, with permission, from Solursh et al. (86).] (*C*) A model for expansion of intercellular spaces. Prior to expansion, hyaluronate is secreted between epithelial cells and held in a compact configuration due to confinement by cellular junctions or the basal lamina. On breakage or rearrangement of these restraints, the swelling pressure exerted by the hyaluronate meshwork (represented by the tightly coiled springs) causes expansion of the intercellular spaces and separation of the cells. [Reprinted, with permission, from Toole et al. (105).]

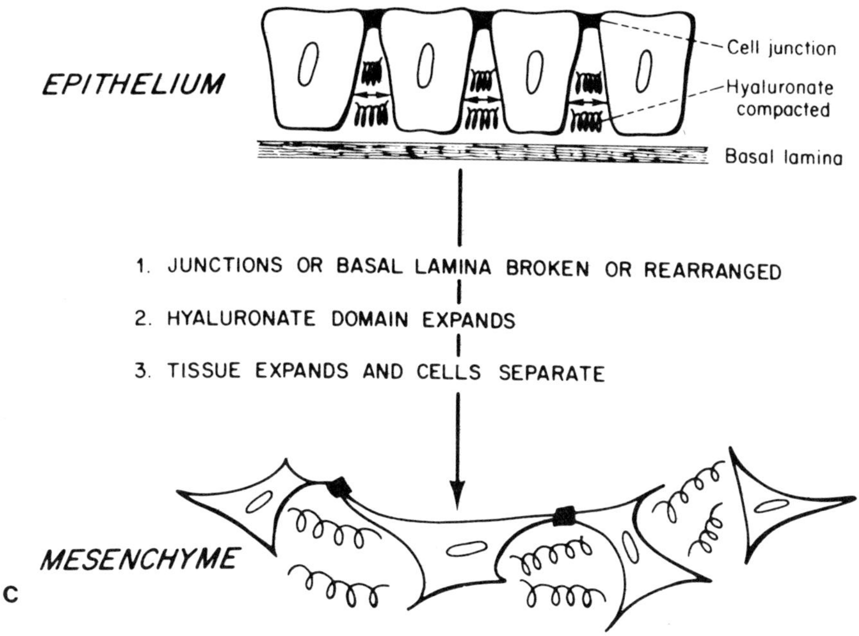

FIGURE 12. *(continued)*

tiation (82–85). Cell movement appears to take place by two different morphogenetic strategies, both of which may be regulated in part by hyaluronate-containing pericellular matrices. These are (1) displacement of cells by expansion of hyaluronate-containing inter- or pericellular spaces, e.g., during the epithelium-to-mesenchyme transition in the sclerotome (86) and the primitive streak (87); (2) migration of groups of cells into a hydrated matrix enriched in hyaluronate, e.g., neural crest cells (88–90), cardiac cushion cells (91), and corneal mesenchyme (83).

During the epithelium-to-mesenchyme transition in the somite, hyaluronate-rich matrix deposited between the epithelial cells expands at the same time as intercellular junctional complexes are reduced in number and rearranged, and the basal lamina becomes disorganized. This results in larger spaces between cells and their displacement from one site to another (86) (Fig. 12A). Alterations in the configuration of a hyaluronate meshwork similar in structure to that depicted in Figure 5 could give rise to this expansion of the pericellular matrix by the mechanism illustrated in Figure 12C. The hyaluronate between the epithelial cells would be physically restricted to that compartment due to the cell junctions and basal lamina or bound there by interactions with the epithelial surface or other macromolecules as dis-

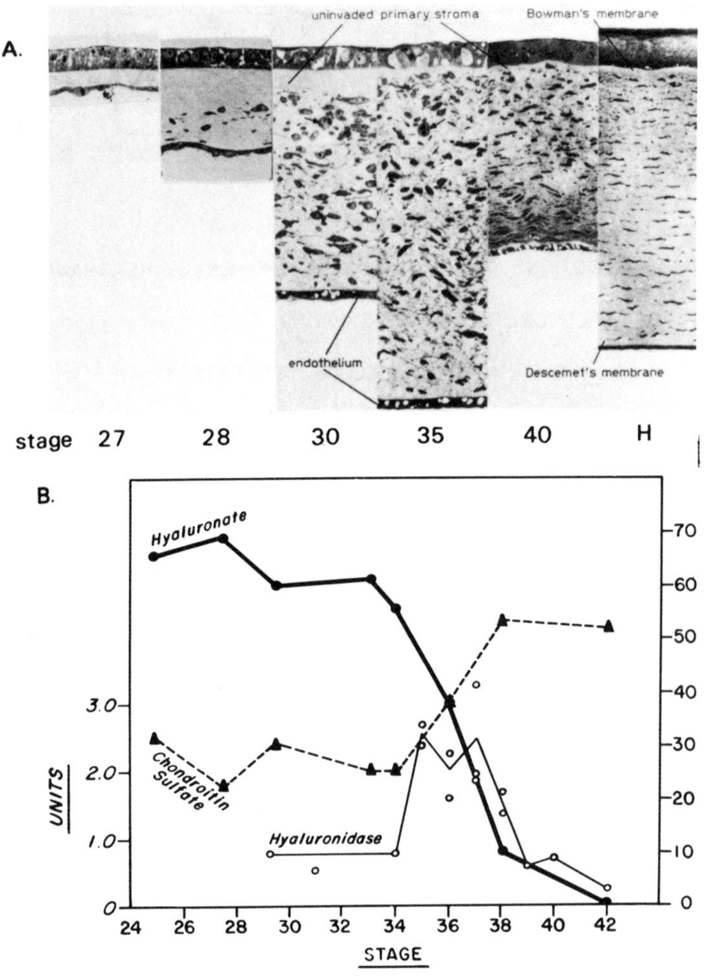

FIGURE 13. Comparison of morphology (*A*) and GAG metabolism (*B*) in the embryonic chick cornea. Hyaluronate synthesis is predominant before and during stromal swelling and the migration of mesenchymal cells into the stroma (stages 27–35). Hyaluronidase activity appears at the end of the migratory stage, concomitant with stromal deswelling (stages 35–40). [Reprinted, with permission, from Toole (85).]

cussed in the previous sections. When confined in this manner, the hyaluronate meshwork exerts a swelling pressure on adjacent boundaries due to its highly negatively charged nature and consequent tendency to occupy a very large volume in solution (3). On alteration of the boundaries, e.g., rearrangement of cell junctions or disruption of the basal lamina, the hyaluronate meshwork would expand resulting in large pericellular spaces, overall tissue expansion, and cell translocation (86).

During cell migration, an analogous expansion of extracellular spaces occurs in the regions into which the cells move. A dramatic example is found in the invasion of the chick embryo cornea by mesenchymal cells between days 5½ and 8 of development (Fig. 13) (92–93). The beginning of this migration coincides precisely with increased hydration and swelling of the hyaluronate-rich stroma lying between the corneal epithelium and endothelium, as well as with an increase in the space between the orthogonal layers of collagen fibers within the stroma. Subsequently, when the stroma loses water, condenses, and the migration ceases, hyaluronidase activity within the cornea increases and the hyaluronate content decreases (83). In a similar fashion, the movement of embryonic neural crest cells (88–90) and cardiac cushion cells (91) takes place in highly hydrated, hyaluronate-containing matrices. Also, cell movement during tissue remodeling and repair (82, 94, 95) and in certain cases, tumorigenesis (96) has been shown to be associated with high hyaluronate concentrations in the extracellular matrix. It seems that whenever mesenchymal cells move, they do so within a highly hydrated, hyaluronate-rich milieu.

Once again, we would speculate that a rapid change in configuration of the hyaluronate meshwork present around or between the cells in these tissues might cause the tissue to swell. In this case, the initial restraints could be physical crossbridges holding tissue compartments at a constant volume and shape (97, 98), binding of cationic materials (39), or chain interactions giving rise to ordered packing of molecules (99, 100). If these restraints were then broken or altered, such as to be less restrictive, then the hyaluronate meshwork would expand until a new state of equilibrium was reached (Fig. 13). This swelling would thus open up spaces between collagen fibers or cell layers that may in turn facilitate migration of cells into these areas. Subsequent degradation of the hyaluronate would cause a reduction in swelling pressure and result in recondensation (Fig. 13).

6. CONCLUSION

Based on the preceding discussion, we would propose that hyaluronate may contribute to morphogenesis in three ways. First, physical changes in the hyaluronate-rich peri- and extracellular matrices may generate forces that in some cases directly cause translocation of cells and in other cases create spaces between cells and fibrous layers that constitute an appropriate environment for promoting or even guiding cell movement. Second, hyaluronate-containing pericellular coats may inhibit adhesion of moving cells to other cellular or fibrous structures, thus preventing precocious immobilization or differentiation. Third, cell-substratum interactions may be regulated in part by hyaluronate.

In addition to hyaluronate, which has been the major focus of this chapter, the morphogenesis of various tissues and organs during development is accompanied by dramatic transitions in the molecular nature and proportions of each class of extracellular macromolecules, viz., the proteoglycans, collagens, and other glycoproteins. These materials interact in specific ways that contribute to the structural properties of the pericellular matrices characteristic of each morphogenetic phase, as well as to their influence on the cells within these matrices.

Three general types of investigation have formed the basis of evolving ideas concerning the role of extracellular macromolecules in morphogenesis: (1) descriptive studies correlating changes in these materials with morphogenetic events in vivo; (2) experimental studies of their relationship to specific types of cell behavior in vitro and biochemical analysis of mechanisms controlling this relationship; (3) experimental manipulations in vivo, based on the above, that probe directly their role in cellular events. Our knowledge arising from each of these types of investigations has been derived from diverse tissue and cell systems and is consequently somewhat fragmented. It has, however, served well to stimulate a growing awareness among biologists of the potential importance of the pericellular matrix in influencing cell behavior and the dynamic changes within these matrices that may be crucial to epigenetic phenomena.

Future experimentation will need to focus more incisively on questions such as the following: (1) the detailed molecular mechanisms of interaction between the pericellular matrix components; (2) the regulation of synthesis and degradation of individual components in relation to changes in the pericellular matrix; (3) the relationship of a given set of interactions to the superstructure of a specific pericellular matrix and the exact spatial localization of the components in relation to one another and to the cell surface; (4) convincing demonstrations of the role of a given pericellular matrix in specific aspects of cell behavior in vitro and in vivo, and analysis of the contribution of individual components.

ACKNOWLEDGMENTS

The recent work described in this article was supported by NIH grants DE05838 and AM03564. This is publication No. 867 of the Robert W. Lovett Group for the Study of Diseases Causing Deformities, Medical Services, Massachusetts General Hospital.

REFERENCES

1. V. C. Hascall and G. Hascall, in E. Hay, ed., *Cell Biology of Extracellular Matrix*, Plenum Press, New York, 1981, p. 39.
2. U. Lindahl and M. Hook, *Ann. Rev. Biochem.* **47**, 385 (1978).
3. W. D. Comper and T. C. Laurent, *Physiol. Rev.* **58**, 255 (1978).
4. S. Gay and E. J. Miller, *Collagen in the Physiology and Pathology of Connective Tissue*, Fischer, New York, 1978.
5. D. J. Prockop, K. I. Kivirikko, L. Tuderman, and N. A. Guzman, *New Engl. J. Med.* **301**, 13 (1979).
6. T. F. Linsenmayer, in E. Hay, ed., *Cell Biology of Extracellular Matrix*, Plenum Press, New York, 1981, p. 1.
7. R. O. Hynes, in E. Hay, ed., *Cell Biology of Extracellular Matrix*, Plenum Press, New York, 1981, p. 295.
8. H. K. Kleinman, R. J. Klebe, and G. R. Martin, *J. Cell Biol.* **88**, 473 (1981).
9. E. D. Hay, *Cell Biology of Extracellular Matrix*, Plenum Press, New York (1981).
10. Y. Mikuni-Takagaki and B. P. Toole, *J. Biol. Chem.* **256**, 8463 (1981).
11. J. R. Hassell, D. A. Newsome, and V. C. Hascall, *J. Biol. Chem.* **254**, 12346 (1979).
12. K. Yamada, D. W. Kennedy, K. Kimata, and R. M. Pratt, *J. Biol. Chem.* **255**, 6055 (1980).
13. S. Sakashita, E. Engvall, E. Ruoslahti, *FEBS Lett.* **116**, 243 (1980).
14. T. M. Chiang, A. E. Postlethwaite, E. H. Beachey, J. M. Seyer, and A. H. Kang, *J. Clin. Invest.* **62**, 916 (1978).
15. L. Kjellen, A. Oldberg, and M. Hook, *J. Biol. Chem.* **255**, 10407 (1980).
16. C. B. Underhill and B. P. Toole, *J. Cell Biol.* **82**, 475 (1979).
17. B. J. Clarris and J. R. Fraser, *Exp. Cell Res.* **49**, 181 (1968).
18. B. J. Clarris, J. R. Fraser, and S. J. Rodda, *Ann. Rheum. Dis* **33**, 240 (1974).
19. R. L. Patterson, D. A. Peterson, F. Dienhart, and F. Howard, *Proc. Soc. Exp. Biol. Med.* **149**, 594 (1975).
20. M. Lipmann, *Nature* **219**, 33 (1968).
21. Z. Darzynkiewicz and E. A. Balazs, *Exp. Cell Res.* **66**, 113 (1971).
22. A. R. Poole, I. Pidoux, A. Reiner, L. H. Tang, H. Choi, and L. Rosenberg, *J. Histochem. Cytochem.* **28**, 621 (1980).
23. W. H. McBride and J. B. Bard, *J. Exp. Med.* **149**, 507 (1979).
24. R. H. Cohn, J. J. Cassiman, and M. R. Bernfield, *J. Cell Biol.* **71**, 280 (1976).

25. P. M. Kraemer and B. J. Barnhart, *Exp. Cell Res.* **114**, 153 (1978).
26. Y. Mikuni-Takagaki and B. P. Toole, *J. Biol. Chem.* **254**, 8409 (1979).
27. C. B. Underhill and B. P. Toole, *J. Biol. Chem.* **255**, 4544 (1980).
28. C. B. Underhill and B. P. Toole, *Exp. Cell Res.* **131**, 419 (1981).
29. R. W. Orkin and B. P. Toole, *J. Biol. Chem.* **255**, 1036 (1980).
30. R. W. Orkin and B. P. Toole, *J. Cell Biol.* **85**, 248 (1980).
31. R. W. Orkin, C. B. Underhill, and B. P. Toole, *J. Biol. Chem.* **257**, 5821 (1982).
32. L. Hakansson, R. Hallgren, and P. Venge, *J. Clin. Invest.* **66**, 298 (1980).
33. B. P. Toole, *Am. Zool.* **13**, 1061 (1973).
34. O. W. Wiebkin and H. Muir, *FEBS Lett.* **37**, 42 (1973).
35. C. Handley and D. A. Lowther, *Biochim. Biophys. Acta* **444**, 69 (1976).
36. M. Solursh, T. E. Hardingham, V. C. Hascall, and J. H. Kimura, *Dev. Biol.* **75**, 121 (1980).
37. P. M. Kraemer, *Biochemistry* **10**, 1437 (1971).
38. P. M. Kraemer, *Biochemistry* **10**, 1445 (1971).
39. P. M. Kraemer, in R. O. Hynes, ed., *Surfaces of Normal and Malignant Cells*, Wiley, New York, 1979, p. 149.
40. L. Kjellen, A. Oldberg, K. Rubin, and M. Hook, *Biochem. Biophys. Res. Commun.* **74**, 126 (1977).
41. P. M. Kraemer, *Biochem. Biophys. Res. Commun.* **78**, 1334 (1977).
42. L. Kjellen, I. Pettersson, and M. Hook, *Proc. Natl. Acad. Sci. USA* **78**, 5371 (1981).
43. C. B. Underhill and J. M. Keller, *J. Cell. Physiol.* **89**, 53 (1976).
44. K. G. Vogel and V. F. Kendall, *J. Cell. Physiol.* **103**, 475 (1980).
45. L. A. Culp, B. A. Murray, and B. J. Rollins, *J. Supramol. Struct.* **11**, 401 (1979).
46. H. Muir, in J. W. Lash and M. M. Burger, eds., *Cell and Tissue Interactions*, Raven Press, New York, 1977, p. 87.
47. T. R. Oegema, V. C. Hascall, and R. Eisenstein, *J. Biol. Chem.* **254**, 1312 (1979).
48. J. McMurtrey, B. Radhakrishnamurthy, E. R. Dalferes, G. S. Berenson, and J. D. Gregory, *J. Biol. Chem.* **254**, 1621 (1979).
49. B. Radhakrishnamurthy, F. Smart, E. R. Dalferes, and G. S. Berenson, *J. Biol. Chem.* **255**, 7575 (1980).
50. B. Norling, B. Glimelius, B. Westermark, and A. Wasteson, *Biochem. Biophys. Res. Commun.* **84**, 914 (1978).
51. L. Coster, I. Carlstedt, and A. Malmstrom, *Biochem. J.* **183**, 669 (1979).
52. K. G. Vogel and R. O. Kelley, *J. Cell. Physiol.* **92**, 469 (1977).
53. M. B. Matthews, *Biochem. J.* **96**, 710 (1965).
54. B. P. Toole and D. A. Lowther, *Biochem. J.* **109**, 857 (1968).
55. B. Obrink, *Eur. J. Biochem.* **34**, 129 (1973).
56. T. R. Oegema, J. Laidlaw, V. C. Hascall, and D. D. Dziewiatkowski, *Arch. Biochem. Biophys.* **170**, 698 (1975).
57. K. Nakao and R. I. Bashey, *Exp. Mol. Pathol.* **17**, 6 (1972).
58. R. Lam, W. J. Claffey, and P. H. Geil, *Biophys. J.* **24**, 613 (1978).
59. R. L. Trelstad, K. Hayashi, and B. P. Toole, *J. Cell Biol.* **62**, 815 (1974).

60. E. D. Hay and S. Meier, *J. Cell Biol.* **62**, 889 (1974).
61. R. H. Cohn, S. D. Banerjee, and M. R. Bernfield, *J. Cell Biol.* **73**, 464 (1977).
62. B. P. Toole and D. A. Lowther, *Arch. Biochem. Biophys.* **128**, 567 (1968).
63. B. P. Toole, *J. Biol. Chem.* **251**, 895 (1976).
64. R. Greenwald, C. Schwartz, and J. Cantor, *Biochem. J.* **145**, 601 (1975).
65. S. H. Love, B. T. Shannon, Q. N. Myrvik, and W. S. Lynn, *J. Reticuloendothel. Soc.* **25**, 269 (1979).
66. B. Pessac and V. Defendi, *Science* **175**, 898 (1972).
67. A. Wasteson, B. Westermark, U. Lindahl, and J. Ponten, *Int. J. Cancer* **12**, 169 (1973).
68. C. B. Underhill and A. Dorfman, *Exp. Cell Res.* **117**, 155 (1978).
69. C. B. Underhill, *J. Cell Sci.* (1982), in press.
70. C. B. Underhill and B. P. Toole, *J. Cell. Physiol.* **110**, 123 (1982).
71. J. R. Fraser and B. J. Clarris, *Clin. Exp. Immunol.* **6**, 211 (1970).
72. M. M. Burger and G. S. Martin, *Nature New Biol.* **237**, 9 (1972).
73. B. J. Rollins and L. A. Culp, *Biochemistry* **18**, 141 (1979).
74. L. A. Culp, R. Ansbacher, and C. Domen, *Biochemistry* **19**, 5899 (1980).
75. E. A. Turley and S. Roth, *Cell* **17**, 109 (1979).
76. B. W. Mayer, E. D. Hay, and R. O. Hynes, *Dev. Biol.* **82**, 267 (1981).
77. Y. Mikuni-Takagaki and B. P. Toole, *J. Cell Biol.* **85**, 481 (1980).
78. A. G. Atherly, B. J. Barnhart, and P. M. Kraemer, *J. Cell Physiol.* **90**, 375 (1977).
79. B. J. Barnhart, S. H. Cox, and P. M. Kraemer, *Exp. Cell Res.* **119**, 327 (1979).
80. M. Fisher and M. Solursh, *Exp. Cell Res.* **123**, 1 (1979).
81. D. H. Bernanke and R. R. Markwald, *Texas Rep. Biol. Med.* **39**, 271 (1979).
82. B. P. Toole and J. Gross, *Dev. Biol.* **25**, 57 (1971).
83. B. P. Toole and R. L. Trelstad, *Dev. Biol.* **26**, 28 (1971).
84. B. P. Toole, *Dev. Biol.* **29**, 321 (1972).
85. B. P. Toole, in E. Hay, ed., *Cell Biology of Extracellular Matrix,* Plenum Press, New York, 1981, p. 259.
86. M. Solursh, M. Fisher, S. Meier, and C. T. Singley, *J. Embryol. Exp. Morphol.* **54**, 75 (1979).
87. M. Fisher and M. Solursh, *J. Embryol. Exp. Morphol.* **42**, 195 (1977).
88. R. M. Pratt, M. A. Larsen and M. C. Johnston, *Dev. Biol.* **44**, 298 (1975).
89. M. A. Derby, *Dev. Biol.* **66**, 321 (1978).
90. J. E. Pintar, *Dev. Biol.* **67**, 444 (1978).
91. R. R. Markwald, T. P. Fitzharris, H. Bank, and D. H. Bernanke, *Dev. Biol.* **62**, 292 (1978).
92. E. D. Hay and J. P. Revel, in A. Wolsky and P. S. Chen, eds., *Fine Structure of the Developing Avian Cornea*, Karger, Basel, 1969.
93. E. D. Hay, *Int. Rev. Cytol.* **63**, 263 (1980).
94. T. Reid and M. H. Flint, *J. Embyol. Exp. Morphol.* **31**, 489 (1974).
95. C. N. Bertolami and R. B. Donoff, *J. Surg. Res.* **25**, 256 (1978).
96. B. P. Toole, C. Biswas, and J. Gross, *Proc. Natl. Acad. Sci. USA* **76**, 6299 (1979).
97. T. Ebendal, *Cell Tiss. Res.* **175**, 439 (1977).

98. J. Lofberg, K. Ahlfors, and C. Fallstrom, *Dev. Biol.* **75**, 148 (1980).

99. W. T. Winter and S. Arnott, *J. Mol. Biol.* **117**, 761 (1977).

100. E. R. Morris, D. A. Rees, and E. J. Welsh, *J. Mol. Biol.* **138**, 383 (1980).

101. B. P. Toole, in S. H. Barondes, ed., *Neuronal Recognition*, Plenum Press, New York, 1976, pp. 275–329.

102. J. Gross, in *The Harvey Lectures*, Series 68, Academic Press, New York, 1974, pp. 351–432.

103. R. Goldberg, C. B. Underhill, and B. P. Toole, *Anal. Biochem.* (1982), in press.

104. D. Heinegard and V. C. Hascall, *J. Biol. Chem.* **249**, 4250 (1974).

105. B. P. Toole, C. B. Underhill, Y. Mikuni-Takagaki, and R. W. Orkin, in R. M. Pratt and R. L. Christiansen, eds., *Current Research Trends in Prenatal Craniofacial Development*, Elsevier/North-Holland, New York, 1980, pp. 263–275.

9

Fibronectin in Cell Interactions

KENNETH M. YAMADA
Membrane Biochemistry Section
Laboratory of Molecular Biology
National Cancer Institute, National Institutes of Health,
Bethesda, Maryland 20205

The cell surface and the extracellular matrix play important roles in many aspects of cell behavior including cell adhesion, cell recognition, embryonic differentiation, and growth regulation (see also Chapters 1–3 and 8) (1–4). A molecule frequently located in these regions in vivo and in vitro is the adhesive glycoprotein fibronectin. Studies of purified fibronectin are beginning to provide insight into some of these events (for review see references 5–15).

This glycoprotein has advantages as a model system for the analysis of cell interactions, because it is a major cellular and plasma protein that can be purified in quantities large enough for extensive biochemical analyses. Recent experiments now permit us to assign specific functions of fibronectin in various cellular interactive events to specific regions of the molecule. Different combinations of these regions appear to act to mediate different complex events.

This chapter will review the biological activities of fibronectin, and then will show examples of how its actions have been analyzed at the molecular level.

1. LOCATION OF FIBRONECTIN

There are at least two different types of fibronectin, known as cellular fibronectin and plasma fibronectin. The cellular form is present in substantial quantities on the cell surface as insoluble fibrils, whereas plasma fibronectin circulates as a soluble protein in blood at a concentration of approximately 0.3 mg/ml. Although both types of fibronectin are very similar in composition and structure, they are not identical (discussed in Section 3).

Cellular fibronectin exists on the cell surface and in extracellular spaces as fibrils and aggregates (Fig. 1). On cells in tissue culture, fibronectin can often be seen to bridge between cells and the substrate or between cells and neigh-

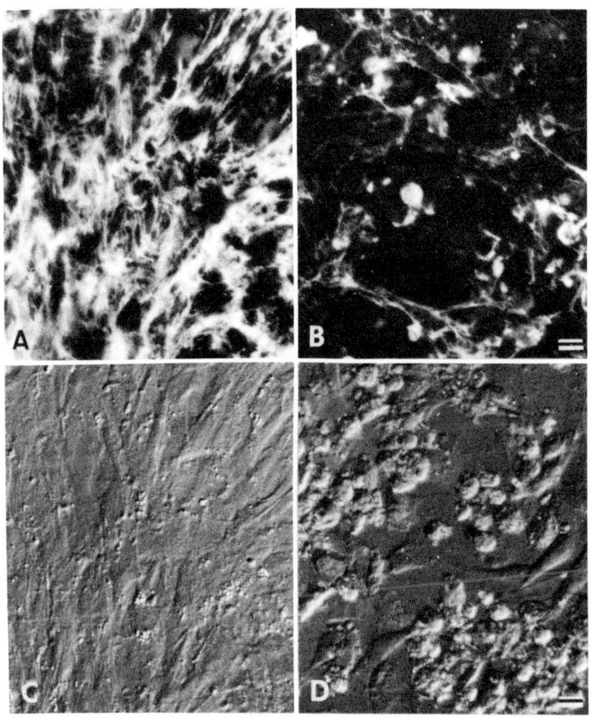

FIGURE 1. Fibronectin on the surfaces of normal and malignant embryonic fibroblasts. The location of fibronectin was determined by the binding of anti-fibronectin antibodies labeled with the fluorescent dye fluorescein and examined by immunofluorescence microscopy. Note the fibrillar organization of fibronectin on the normal chick embryo fibroblasts (*A*) and its decrease on the surfaces of malignant cells in a parallel culture of cells infected by Rous sarcoma virus (*B*). The corresponding views of these cells by interference contrast microscopy are shown in (*C* and *D*). Bars = 20 μm.

boring cells. Fibronectin is present in vivo in connective tissue, in basement membranes that underlie epithelia, and in a number of embryonic tissues, including the regions through which certain cell migrations occur. For example, fibronectin is found in the cell-free spaces through which neural crest cells migrate; it is also found in the outer layers of the kidney glomerular basement membrane in a location consistent with an adhesive function in the attachment of epithelium to the basement membrane (6, 16–18).

2. FUNCTION OF FIBRONECTIN

Fibronectin is thought to function in a number of important biological events (Table 1). As with other recently discovered cellular macromolecules, the function of fibronectin in vivo must be inferred from data obtained from in vitro biological assays and from in vivo immunological studies. It should be remembered that such studies can suggest, but cannot as yet prove, that fibronectin acts in vivo in each of the biological events described in the following Section.

2.1. Fibronectin in Adhesion

A general function of fibronectin appears to be to mediate several important adhesive events. There are, of course, many other types of adhesive mol-

TABLE 1. Biological Activities of Fibronectin

1. Cell–substrate adhesion
   a. Adhesion and spreading on plastic and glass
   b. Adhesion to collagen (types I-V)
   c. Adhesion to fibrin
2. Cell–cell adhesion
   a. Aggregation of living cells
   b. Agglutination of aldehyde-fixed erythrocytes
3. Maintenance of normal cell morphology
   a. Parallel alignment of cells
   b. Cell shape
   c. Cell surface microvilli and ruffles
   d. Organization of actin-containing microfilaments
4. Cell motile and chemotacticlike activity
5. Embryonic differentiation (inhibition or stimulation)
6. Nonimmune opsonization by macrophages

ecules besides fibronectin, as discussed in Chapter 3; many of these are much more tissue-specific than fibronectin. Fibronectin can promote both cell–cell and cell–substrate adhesiveness of fibroblastic and certain epithelial cells. The role of fibronectin in adhesion of cells to substrates such as collagen has been documented the most extensively.

2.1.1. Cell-Substrate Adhesion. Fibronectin stimulates or in some cases is actually required for cell attachment and spreading on substrates of collagen, plastic, or glass (7, 12). This requirement for fibronectin has been demonstrated for many cells in experiments in which they are placed into cell culture media lacking fibronectin and are then examined for the effects on adhesion of exogenously added purified fibronectin. There are two types of cells that do not require the addition of exogenous fibronectin for adhesion to substrates: (1) those that already synthesize large amounts of their own cell-surface fibronectin and (2) certain epithelial cells. Recent experiments suggest that many epithelial cells may instead utilize the glycoprotein laminin to attach to type-IV collagen, which is the type of collagen most likely to serve as an epithelial cell substrate in basement membranes (see Section 6).

The overall process of fibronectin-mediated cell attachment and spreading requires divalent cations, such as Ca^{2+}, and metabolic energy; this energy requirement suggests that the processes are more complex than simple sticking of an inert cell to a substrate (7, 19). These cell–substrate adhesive activities of fibronectin are displayed with equal activity by fibronectins isolated from either plasma or cultured cells (20, 21).

2.1.2. Cell–Cell Adhesion. Fibronectin can also stimulate the adherence of cells to other cells. It increases the aggregation or agglutination of both living and formaldehyde-fixed cells in test systems for cell adhesion (22, 23), although there is no evidence as yet that fibronectin is located in any particular specialized intercellular adhesive structure such as the desmosome.

Specifically, cellular fibronectin increases the adhesion of fibroblastic cells to one another, and it agglutinates fixed sheep and human red blood cells. This hemagglutinating activity suggests that the protein could be a lectin (see Chapter 7). The hemagglutinating activity can be inhibited by a specific type of complex sugar unit, the lipid-linked oligosaccharides on glycolipids such as G_{T1b}. This sugar specificity further suggests that fibronectin is a lectinlike molecule (24), although the relative contributions of sugar sequence and overall charge remain to be evaluated. Interestingly, cellular fibronectin is over a hundredfold more active than plasma fibronectin in mediating hemagglutination (21), indicating important biological differences between these two molecules.

2.2. Cell Morphology and Alignment

Another class of fibronectin biological activity involves maintaining normal cell morphology and cell "social" behavior, especially of fibroblastic cells. When these cells become cancerous, they often lose the fibronectin present on the cell surface (Fig. 1). This loss is the result of a combination of factors that includes decreased biosynthesis, increased proteolytic degradation, and decreased capacity of the cells to bind fibronectin (5, 6, 8). Many aspects of cell behavior are altered in these cells, and the experimental restoration of fibronectin to the cell surface can restore many but not all parameters of normal cell behavior (5, 8, 15).

2.2.1. Cell Social Interactions. If purified cellular fibronectin is reconstituted back onto fibronectin-deficient transformed cells, these cells can regain a commonly observed sign of normal cell–cell "social" interactions—the capacity to respond to neighboring cells by aligning in parallel arrays (Figs. 1, 2). The mechanisms by which fibronectin restores this behavior is not known, although it may restore normal cell–cell and cell–substrate adhesion, which permits the transfer of other information. However, fibronectin-treated cells do not regain normal growth control, suggesting that this aspect of cell–cell interaction is not regulated by fibronectin (22).

2.2.2. Cell Shape and Cytoskeleton. Fibronectin treatment can also at least partially restore other properties to normal, including cell morphology, morphology of the cell surface, and even the intracellular organization of microfilament bundles (Fig. 2) (25–28). Conversely, normal cells treated with

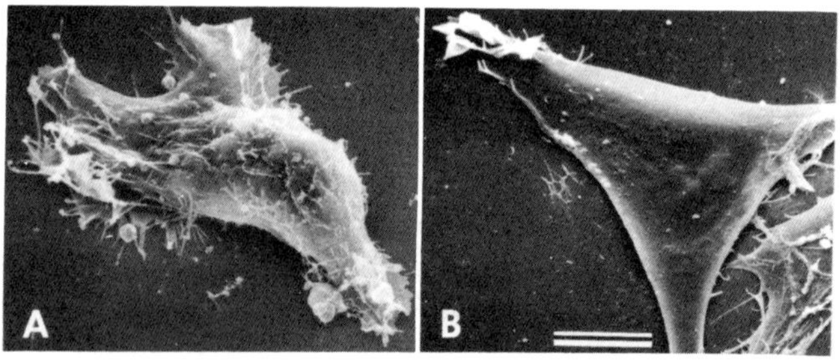

FIGURE 2. Morphology of malignant and fibronectin-treated fibroblasts. Chick embryo fibroblasts were transformed by Rous sarcoma virus, then examined by scanning electron microscopy before and after treatment with purified cellular fibronectin. Note the abnormally rounded shape and increased numbers of projecting structures on the surfaces of the malignant cells (*A*), and the restoration of a more normal flattened, fibroblastic shape and a smoother cell surface to the cells treated for 24 hr with 25 μg/ml purified fibronectin (*B*). Bar = 10 μm.

anti-fibronectin antibodies become similar in morphology to transformed and malignant cells (29). These studies implicate fibronectin in the maintenance of normal cell morphology and social behavior.

2.2.3. Cell Motility. Fibronectin treatment also stimulates the motility of certain cells and it even displays chemotactic activity toward fibroblasts and neural crest cells (22, 30–32). Since fibronectin can be demonstrated by immunological procedures in certain pathways of cell migration in embryos (see Section 2.3 and Chapter 6), it may function to promote or to guide certain cell migratory events in embryonic development.

2.3. Developmental Roles of Fibronectin

Immunological studies have documented a series of developmentally regulated changes in the quantities of fibronectin in different tissues (6). For example, embryonic chick blastoderm development is accompanied by changes in fibronectin localization, and neural crest cell migration is accompanied by the appearance of fibronectin in the cell-free spaces through which these cells migrate (17, 33, 34). Conversely, muscle and cartilage differentiation are accompanied by a dramatic loss of fibronectin (6, 16, 35–37). These findings suggest that fibronectin might play important roles in embryonic cell migrations or possibly even in certain aspects of cytodifferentiation.

2.3.1. Effects on Embryonic Differentiation. The latter hypothesis has been examined in vitro in three different experimental systems in which quantities of fibronectin are known to change (Table 2). The addition of purified fibronectin to chondrocytes causes a reversion to a more mesen-

TABLE 2. Effects of Fibronectin on Embryonic Differentiation

Undifferentiated Cell	Differentiated Function	Differentiated Cell	Effect of Fibronectin
Mesenchyme cell	Sulfated proteoglycan and type-II collagen synthesis	Chondrocyte	Decreased proteoglycan and increased type-I collagen synthesis
Myoblast	Fusion into syncytium	Myotube	Inhibition
Neural crest cell	Catecholamine synthesis	Adrenergic cell	Stimulation

chymal phenotype: the cells flatten and become similar in morphology to mesenchymal cells, their synthesis of sulfated proteoglycans as a differentiated product is inhibited, and they convert from the synthesis of type-II collagen to type-I collagen (38, 39). Although it is not known whether these results are due to changes in adhesion and cell shape or to more complex events, they do suggest that if abnormally high levels of cellular fibronectin were to be maintained during the development of this tissue, differentiation would be impaired.

Similarly, although fibronectin is involved in initial attachment and growth of myoblasts (40), the addition of exogenous cellular fibronectin to myoblasts ultimately inhibits myoblast fusion. These results also suggest that fibronectin may be playing a negative regulatory role; that is, that the removal or inactivation of fibronectin may be required in certain tissues for the orderly progression of certain events in differentiation (41).

In contrast, the neural crest cell may provide an example of a positive regulatory role for fibronectin in development. Isolated neural crest cells generally do not synthesize fibronectin (42, 43). Yet these cells come into contact with this glycoprotein during migration, providing an opportunity for it to affect not only migration but also cytodifferentiation (see Chapter 6 for a more extensive discussion).

The exogenous addition of cellular fibronectin to neural crest cell cultures causes an increase in the number of neuronal cells containing catecholamines (43). Complex mixtures of extracellular materials also have similar effects (44). Fibronectin therefore appears to promote differentiation to this phenotype. Although this hypothesis requires more rigorous testing in vivo, it may provide yet another example of the regulation of development by an extracellular molecule (see also Chapter 8). The possibility of other positive regulatory functions for fibronectin in development should be explored.

2.3.2. Function in Embryonic Migration. Recent immunological studies that are still preliminary and ongoing in a number of laboratories are beginning to provide evidence that fibronectin may be involved in important cell migratory events during embryonic development. These descriptive studies show that fibronectin appears in migratory pathways at the appropriate times during development to be able to play a role in stimulating or directing certain cell migrations.

For example, immunofluorescence studies with antifibronectin antibodies show that during early development of the chick blastoderm, zones of fibronectin appear around the perimeter of the blastoderm and in the germinal crescent in patterns suggesting that the migratory movements of adjacent masses of cells may be directed by these depositions of fibronectin (33).

Another example occurs later during development, when neural crest cells migrate along fibronectin-enriched regions of the developing embryo; these cells may even utilize fibronectin stored in granules called interstitial bodies for migration (17, 34). An increasing number of examples are being investigated in other developmental systems.

As these studies are published, it will be important to consider in each case whether the timing of fibronectin appearance and disappearance is consistent with a primary, causal role, or whether such changes simply correlate with a remodeling of the extracellular matrix as a secondary consequence of cell migration through these areas.

It will be particularly important to determine by experimental means whether fibronectin is necessary for any particular aspect of these developmentally regulated cell migrations. Possible approaches to this problem include in vivo microinjection studies using antibodies against fibronectin and against other extracellular molecules to disrupt their functions experimentally, studies of the effects of artificially establishing alternative pathways of in vivo migration by injection of purified fibronectin, and in vitro studies of the mechanisms by which fibronectin promotes cell migration.

A final possible developmental function of fibronectin may occur in the morphogenesis of certain tissues. Such morphogenetic events are notoriously difficult to analyze because of their complexity and their requirements for three-dimensional control (see Chapter 4). Nevertheless, possible roles for fibronectin have been suggested in certain model systems that have been established to dissect complex morphogenetic events experimentally. For example, fibronectin appears to promote the establishment of the basal lamina of epithelia in embryonic tooth development, which may influence the subsequent development of the tooth (45). In a model system for tissue-tissue interactions, the in vitro movements of groups of heart cells relative to other heart cells can be modified by treatment of cells with a partially purified preparation of fibronectin (46). It has also been suggested that the aggregation events in somite formation may be effected by fibronectin (23). Further exploration of these very preliminary findings will be of considerable interest because they may provide insights into morphogenesis, perhaps the most complex process in embryonic development (see also Chapter 4).

2.4. Nonimmune Opsonic Activity

A final class of cellular activity mediated by fibronectin involves the uptake of certain materials by the reticuloendothelial system. For example, macrophages utilize fibronectin to bind and to endocytose denatured collagen and actin. Fibronectin appears to act as an "opsonin," which binds to certain

TABLE 3. Binding Activities of Fibronectin

Cell surface
Collagen and gelatin
Fibrinogen and fibrin
Heparin and heparan sulfate
Hyaluronic acid
Transglutaminase
Complement component C1q
Actin
DNA
Staphylococcus aureus
Gangliosides

materials to mediate their uptake, although by mechanisms unrelated to classical antibody opsonization (9, 47, 48). It has been suggested that plasma fibronectin functions as a part of the body's defense against certain molecules; if tissues are damaged by trauma, fibronectin may bind and remove cellular debris from the circulation before it impairs blood flow (9, 49).

2.5. Binding Activities

Fibronectin also binds to a number of biologically important macromolecules (Table 3), although many other materials that have been tested do not bind. In several cases, the binding has been shown to be specific and of relatively high affinity. For example, fibronectin binds to collagen and to two glycosaminoglycans with repeating sugar units—heparin and hyaluronic acid. The binding has been shown to occur at separate sites on fibronectin, since maximal binding of one molecule will not block the binding of another (e.g., reference 50). However, in some cases, the binding of one molecule can modulate the strength of binding of another molecule; for example the binding of heparin or hyaluronic acid modifies the binding of plasma fibronectin to collagen (51–53).

The current view of fibronectin is that it is a multifunctional molecule that can mediate a number of cellular-adhesive and molecular-binding interactions; it may, of course, also have more complex effects in events such as embryonic development. Analyses of the molecular mechanisms by which fibronectin acts in these events have concentrated on comparisons of the different types of fibronectin and especially on a molecular dissection of the adhesive and binding activities.

3. CELLULAR VERSUS PLASMA FIBRONECTINS

Although fibronectins isolated from plasma and from the cell surface are very similar in structure and composition, they are not identical (21, 54–57).

For example, cellular fibronectin is an insoluble protein at neutral pH, whereas plasma fibronectin is readily soluble. In addition, cellular fibronectin is substantially more active than plasma fibronectin in restoring a normal morphological phenotype to malignant cells and especially in agglutinating erythrocytes. Yet several other biological activities are shared equally by the two forms of fibronectin; for example, the cell attachment and spreading activities are identical for equal amounts of each protein, and both proteins are equally active in promoting the uptake of gelatin by macrophages (20, 21, 57, 58).

There are subtle structural differences that accompany these differences in biological activity. For example, the polypeptide chains of cellular and plasma fibronectins have at least three regions that differ in size due to apparent deletions or insertions (56). These different regions can account for the significant difference in overall size between these molecules. These postulated small deletions may eventually explain how these two forms of fibronectin can be so similar in certain biological activities, yet can differ markedly in others. There may therefore be at least two genes or two differentially spliced messenger RNA molecules coding for these two fibronectins. A third type of fibronectin has been found in amniotic fluid that has unusual amounts of carbohydrate; it is not known whether the polypeptide chains are related to those of cellular or plasma fibronectins (59, 60).

Antibodies to either plasma or cellular fibronectin usually cross-react with reactions of complete immunological identity for both molecules. Attempts to inactivate or absorb such polyclonal antibodies with one form of fibronectin in order to leave behind only those antibodies specific for the other form have been uniformly unsuccessful. However, it is now possible to obtain mouse hybridomas that produce monoclonal antibodies reacting only with cellular fibronectin and not with plasma fibronectin (61, 62). These antibodies should prove useful in determining the locations of the cellular form of fibronectin in vivo.

4. STRUCTURE OF FIBRONECTIN

In order to determine how fibronectin acts, it is necessary to understand its structure. Both cellular and plasma fibronectins are composed of subunits of 220–250,000 daltons that are linked by disulfide bonds located only at the

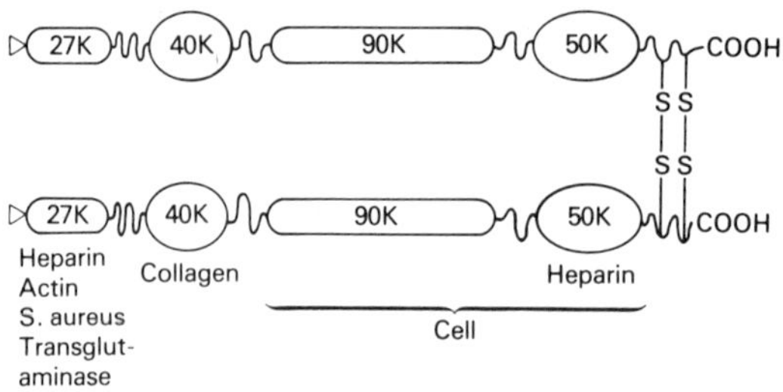

FIGURE 3. Current model of the structural and functional domains of fibronectin. Both cellular and plasma fibronectins are composed of a series of protease-resistant domains represented by the ovoid structures, each of which often contains a specific binding site for a ligand. These structured domains are interconnected by protease-sensitive regions that are thought to be composed of flexible, nonstructured polypeptide. The numbers indicate molecular weight (K = 1 kilodalton), and the macromolecules listed below each domain are those that bind to each isolated domain. The amino end is at the left, and is blocked (triangles) by a pyroglutamate group; the carboxy end is at the right.

carboxy end of the molecule (Fig. 3). The monomers are very similar, but they may not be entirely structurally identical (63, 64). The overall shape of the molecule is slightly to substantially elongated rather than globular.

The carbohydrates on fibronectin are of only a single class, the "complex" oligosaccharide containing mannose, sialic acid, and other sugars linked to asparagine residues in the polypeptide chain (65, 66). The newly discovered inhibitor tunicamycin specifically interferes with the glycosylation of proteins containing such asparagine-linked sugars; cells treated with this drug produce carbohydrate-free fibronectin. Surprisingly, the rates of fibronectin synthesis and secretion remain normal, and the carbohydrate-free molecule is as biologically active as normal fibronectin. However, such nonglycosylated fibronectin is substantially more susceptible to proteolytic degradation, suggesting that the carbohydrates on fibronectin function primarily to protect the molecule against proteases and abnormal rates of protein turnover (67, 68).

5. SPECIFIC BINDING SITES

By the use of various proteases, fibronectin can be cleaved into separate polypeptide regions that still contain the binding sites for different molecules. In several cases, it has been possible to isolate the binding site by means of its specific binding activity.

5.1. Affinity Chromatography

One approach used for affinity chromatographic isolation of specific binding regions of fibronectin is shown in Figure 4. A molecule to which fibronectin is known to bind, called the ligand, is attached covalently to agarose beads. The beads are placed into a column, and fibronectin is poured onto the column and is allowed to bind to the ligand.

A protease is then introduced into the column. Proteases have a strong propensity to attack regions of polypeptides that are flexible and exposed, and they tend to spare structured regions of the molecule in which most of the polypeptide is buried; fibronectin is known to contain such structural

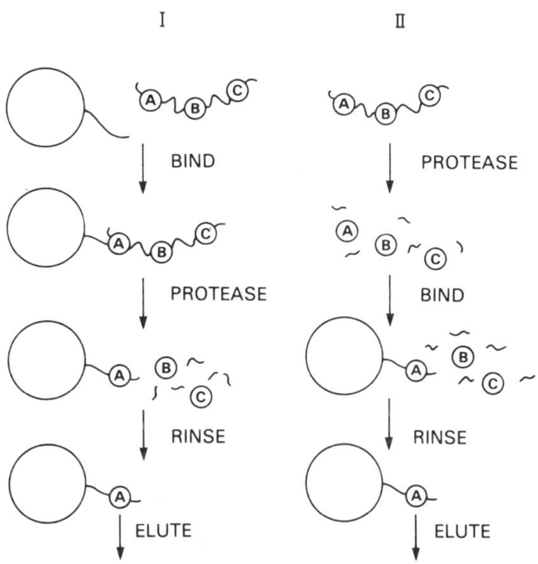

FIGURE 4. Methods for the purification of specific binding regions from fibronectin molecules. Both methods depend upon the fact that only specific regions of fibronectin bind to specific ligands such as collagen or heparin. The ligand is first covalently linked to agarose beads by cyanogen bromide, then the beads are poured into a column to make an affinity column. In method I, intact fibronectin is permitted to bind to the affinity column, then is digested by a protease. After extensive washing, only the protease-resistant domain that is specifically bound to a ligand such as collagen remains bound and can be recovered from the column. In method II, fibronectin is cleaved by the protease prior to chromatography. The resulting complex mixture of various fragments is then poured onto the column. Only the fragments that bind to the covalently linked ligand are retained on the column. The fragments bound on the columns are recovered after release by treatments that disrupt their interactions with the ligand, e.g., by high concentrations of salts or urea.

domains (69). If a binding site is located within such a structured region, it is spared from proteolytic cleavage, and it remains intact and bound to the ligand even after extensive washing of the column to remove other fragments of the protein. An alternate approach is to predigest the fibronectin with a protease, inhibit the enzyme with a specific inhibitor, then attempt to isolate protease-resistant structural and functional domains by affinity chromatography (Fig. 4). By these approaches, it has been possible to isolate regions of fibronectin that bind to collagen (70–74), heparin (75–77), actin (78), fibrin (79), and transglutaminase (80, 81). In addition, another fragment of fibronectin has been purified that contains a region of the molecule that binds to cells (82, 83) (Fig. 3).

5.2. Analysis of Purified Fragments

The biological activities of a number of such fragments have been analyzed to examine how the intact molecule acts in a variety of cellular events. Figure 5 summarizes the analysis of three cell adhesive events using such fragments (82–85).

The effects of the intact molecule, a fragment containing both cell- and collagen-binding sites, and the separated cell- and collagen-binding sites are shown on cell–cell and cell–substrate adhesion. Only the intact molecule is effective in mediating cell–cell alignment of transformed cells or hemagglutination; these events presumably require a multivalent molecule capable of binding to more than one cell to promote alignment of cells or to agglutinate them by linking cells together to form clusters (Fig. 5).

However, a large fragment still containing cell- and collagen-binding sites retains the ability to mediate cell attachment to collagen and plastic. A further proteolytic cleavage that separates these sites destroys the capacity to attach cells to collagen; yet one piece still retains the capacity to attach cells to simple plastic substrates and therefore still retains a cell binding site (Fig. 5).

In these experiments, the collagen-binding region remains intact and can always bind whatever fragments contain this site to collagen substrates; but when its connection to the cell-binding fragment is cleaved, it can no longer by itself mediate cell attachment to collagen. These and other biological experiments indicate that fibronectin is a multifunctional molecule with multiple, specific interaction sites for the cell surface, collagen, and other ligands (Figs. 3 and 5) (82–85).

It should now be possible to begin to analyze more complex cellular and developmental effects of fibronectin in terms of the activity of specific fragments of the molecule. For example, recent studies show that out of all of the fibronectin fragments tested for chemotactic activity toward fibroblasts or

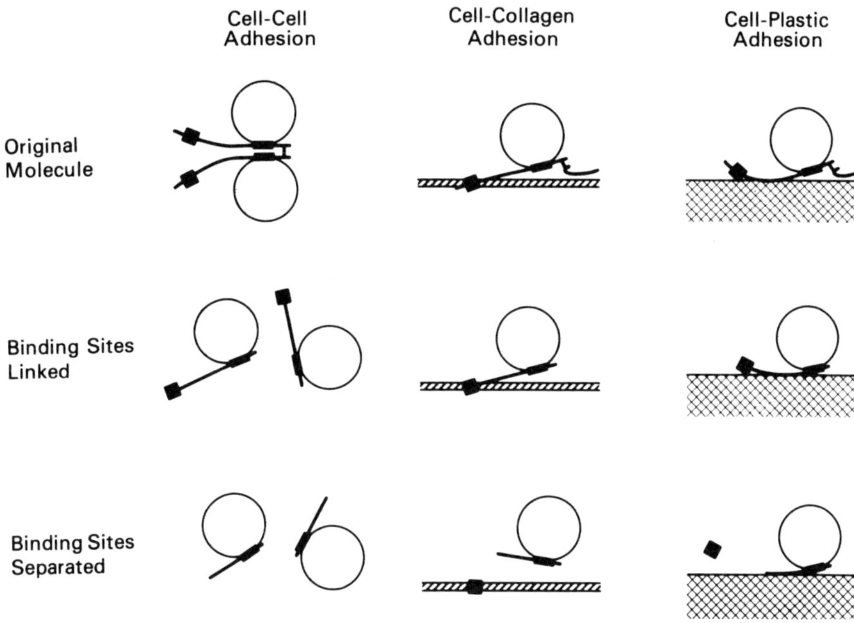

FIGURE 5. Analysis of the cell- and collagen-binding sites of fibronectin using specific, chymotrypsin-digested fragments of fibronectin. The open circles represent cells, the closed squares represent the collagen-binding site on the fibronectin molecule, and the closed rectangles are the cell-binding sites of fibronectin. Assays of fragments in various biological assays for adhesion are shown from left to right. From top to bottom are comparisons of the original intact fibronectin molecule, a large chymotryptic fragment of 205,000 daltons that still has both binding sites, and subsequent fragments of 40,000 and 160,000 daltons in which only one of the sites is present. This figure is discussed in detail in Section 5.2; data supporting each aspect of the model can be found in reference 83.

neural crest cells, only the cell-binding fragment of fibronectin is active (32, 86, 87). These chemotacticlike events can therefore be mediated by only this region of the molecule, and do not require either collagen-binding activity or multivalency. Possible mechanisms of such chemotactic events are discussed in Chapter 5.

6. LAMININ

Laminin is a newly discovered, extracellular molecule that has some similarities and several important differences when it is compared to fibronectin (88, 89). Both molecules are of high molecular weight and are linked by disulfide bonds to form multimeric molecules. Both can bind readily to collagen and heparin, can mediate cell attachment to collagen, and are hemagglutinins (90–93). However, they are biochemically and biologically distinct. They have different amino acid and carbohydrate compositions, and they are immunologically unrelated. In biological assays, laminin appears to be specific for the attachment of certain epithelial cells to type-IV collagen of basement membranes (91, 92). Fibronectin shows much less specificity; it is less effective than laminin in mediating epithelial cell attachment to type-IV collagen, but much more effective in mediating fibroblastic cell attachment to a variety of collagen types.

The existence of adhesive molecules such as fibronectin and laminin, as well as other less well-understood molecules that can help to attach cells to collagen (12) raises the interesting possibility that there exist multiple molecules with different binding specificities to different substrates and cells, and that many more may be discovered. In addition, detailed comparisons of the proteolytic fragments and binding sites of molecules such as laminin with those of fibronectin should be of interest.

7. PROSPECTS

The biochemical framework is now established for beginning to understand the binding and adhesive activities of the fibronectins and related extracellular adhesive molecules. These studies should continue to provide insights into the molecular basis of this class of cell adhesion, as well as other more complex events. For example, the nature of each active site region can now be analyzed regarding structure, substrate specificity, and mode of action in terms of hydrophobic or electrostatic mechanisms. The nature of the postulated receptor to which fibronectin binds in the plasma membrane requires rigorous examination beyond the preliminary studies that suggest it may contain glycoprotein and (or) glycolipid constituents (24, 94, 95, 96).

In addition, the role of fibronectin in development can now be examined more analytically. For example, in vitro studies may indicate which regions of the fibronectin molecule are required for its effects on development and may indicate how they act in modulating development. In vivo studies with antibodies to specific regions of the fibronectin molecule should also provide important information about the function of this molecule in the very complex events of differentiation in vivo.

Other adhesive molecules, including laminin, can probably be analyzed using approaches similar to those that have been successful for fibronectin. These types of studies should help to unravel the molecular mechanisms of the complex actions of cell-surface and extracellular matrix glycoproteins on a variety of cell interactions in development.

REFERENCES

1. W. A. Frazier and L. Glaser, *Annu. Rev. Biochem.* **48**, 491 (1979).
2. S. Subtelny and N. K. Wessells, eds., *The Cell Surface: Mediator of Developmental Processes*, Academic Press, New York, 1980.
3. E. D. Hay, ed., *Cell Biology of Extracellular Matrix,* Plenum Press, New York, 1981.
4. G. Poste and G. L. Nicholson, eds., *Cell Surface Reviews*, North-Holland Publishing, New York, 1976.
5. K. M. Yamada and K. Olden, *Nature* **275**, 179 (1978).
6. A. Vaheri and D. F. Mosher, *Biochim. Biophys. Acta* **516**, 1 (1978).
7. F. Grinnell, *Int. Rev. Cytol.* **53**, 65 (1978).
8. R. O. Hynes, A. T. Destree, M. E. Perkins, and D. D. Wagner, *J. Supramol. Struct.* **11**, 95 (1979).
9. T. M. Saba and E. Jaffe, *Am. J. Med.* **68**, 577 (1980).
10. M. W. Mosesson and D. L. Amrani, *Blood* **56**, 145 (1980).
11. E. Pearlstein, L. Gold, and A. Garcia-Pardo, *Mol. Cell Biochem.* **29**, 103 (1980).
12. H. K. Kleinman, R. J. Klebe, and G. R. Martin, *J. Cell Biol.* **88**, 473 (1981).
13. E. Ruoslahti, E. Engvall, and E. G. Hayman, *Coll. Res.* **1**, 95 (1981).
14. R. O. Hynes, in E. D. Hay, ed., *Cell Biology of the Extracellular Matrix*, Plenum Press, New York, 1981, p. 295.
15. K. M. Yamada, in M. Horowitz, ed., *The Glycoconjugates*, Vol. III, Academic Press, New York, 1981, p. 331.
16. S. Stenman and A. Vaheri, *J. Exp. Med.* **147**, 1054 (1978).
17. D. Newgreen and J. P. Thiery, *Cell Tissue Res.* **211**, 269 (1980).
18. P. J. Courtoy, Y. S. Kanwar, R. O. Hynes, and M. G. Farquhar, *J. Cell Biol.* **87**, 691 (1980).
19. R. J. Klebe, J. R. Hall, P. Rosenberger, and W. D. Dickey, *Exp. Cell Res.* **110**, 419 (1977).
20. S. D. J. Pena and R. C. Hughes, *Cell Biol. Int. Rep.* **2**, 339 (1978).
21. K. M. Yamada and D. W. Kennedy, *J. Cell Biol.* **80**, 492 (1979).
22. K. M. Yamada, K. Olden, and I. Pastan, *Ann. N.Y. Acad. Sci.* **312**, 256 (1978).
23. C. M. Cheney and J. W. Lash, *Eur. J. Cell Biol.* **22**, 422 (abstract) (1980).
24. K. M. Yamada, D. W. Kennedy, G. R. Grotendorst, and T. Momoi, *J. Cell. Physiol.* **109**, 343 (1981).
25. K. M. Yamada, S. S. Yamada, and I. Pastan, *Proc. Natl. Acad. Sci. USA* **73**, 1217 (1976).
26. K. M. Yamada, S. H. Ohanian, and I. Pastan, *Cell* **9**, 241 (1976).
27. M. C. Willingham, K. M. Yamada, S. S. Yamada, J. Pouyssegur, and I. Pastan, *Cell* **10**, 375 (1977).
28. I. U. Ali, V. Mautner, R. Lanza, and R. O. Hynes, *Cell* **11**, 115 (1977).
29. K. M. Yamada, *J. Cell Biol.* **78**, 520 (1978).
30. I. U. Ali and R. O. Hynes, *Cell* **14**, 439 (1978).
31. V. Gauss-Müller, H. K. Kleinman, G. R. Martin, and E. Schiffman, *J. Lab. Clin. Med.* **96**, 1071 (1980).
32. J. H. Greenberg, S. Seppa, H. Seppa and A. T. Hewitt, *Dev. Biol.* **87**, 259 (1981).
33. D. R. Critchley, M. A. England, J. Wakely and R. O. Hynes, *Nature* **280**, 498 (1979).

34. B. W. Mayer, E. D. Hay and R. O. Hynes, *Dev. Biol.* **82**, 267 (1981).
35. C. A. Lewis, R. M. Pratt, J. P. Pennypacker, and J. R. Hassell, *Dev. Biol.* **64**, 31 (1978).
36. W. Dessau, J. Sasse, R. Timpl, F. Jilek, and K. von der Mark, *J. Cell Biol.* **79**, 342 (1978).
37. L. B. Chen, *Cell* **10**, 393 (1977).
38. J. P. Pennypacker, J. R. Hassell, K. M. Yamada, and R. M. Pratt, *Exp. Cell Res.* **121**, 411 (1979).
39. C. M. West, R. Lanza, J. Rosenbloom, M. Lowe, H. Holtzer, and N. Avdalovic, *Cell* **17**, 491 (1979).
40. M. Chiquet, E. C. Puri, and D. C. Turner, *J. Biol. Chem.* **254**, 5475 (1979).
41. T. R. Podleski, I. Greenberg, J. Schlessinger, and K. M. Yamada, *Exp. Cell Res.* **122**, 317 (1979).
42. J. Loring, C. Erickson, and J. Weston, *J. Cell Biol.* **75**, 71a (abstract) (1977).
43. M. Sieber-Blum, F. Sieber, and K. M. Yamada, *Exp. Cell Res.* **133**, 285 (1981).
44. M. Sieber-Blum and A. M. Cohen, *Dev. Biol.* **80**, 96 (1980).
45. A. G. Brownell, C. C. Bessem, and H. C. Slavkin, *Proc. Natl. Acad. Sci. USA* **78**, 3711 (1981).
46. P. B. Armstrong, *J. Cell Sci.* **44**, 263 (1980).
47. F. A. Blumenstock, T. M. Saba, P. Weber, and R. Laffin, *J. Biol. Chem.*, **253**, 4287 (1978).
48. P. W. Gudewicz, J. Molnar, M. Z. Lai, D. W. Beezhold, G. E. Siefring, R. B. Credo, and L. Lorand, *J. Cell Biol.* **87**, 427 (1980).
49. T. M. Saba, F. A. Blumenstock, W. A. Scovill, and H. Bernard, *Science* **201**, 622 (1978).
50. K. M. Yamada, D. W. Kennedy, K. Kimata, and R. M. Pratt, *J. Biol. Chem.* **255**, 6055 (1980).
51. F. Jilek and H. Hörmann, *Hoppe-Seyler's Z. Physiol. Chem.* **360**, 597 (1979).
52. E. Ruoslahti and E. Engvall, *Biochim. Biophys. Acta* **631**, 350 (1980).
53. S. Johansson and M. Höök, *Biochem J.* **187**, 521 (1980).
54. K. M. Yamada, D. H. Schlesinger, D. W. Kennedy, and I. Pastan, *Biochemistry* **16**, 5552 (1977).
55. M. Vuento, M. Wrann, and E. Ruoslahti, *FEBS Lett.* **82**, 227 (1977).
56. M. Hayashi and K. M. Yamada, *J. Biol. Chem.* **256**, 11292 (1981).
57. R. O. Hynes, I. V. Ali, A. T. Destree, V. Mautner, M. E. Perkins, D. R. Senger, D. D. Wagner, and K. K. Smith, *Ann. N.Y. Acad. Sci.* **312**, 317 (1978).
58. D. Marquette, J. Molnar, K. M. Yamada, D. Schlesinger, S. Darby, and P. Van Alten, *Mol. Cell Biochem.* **36**, 147 (1981).
59. E. Crouch, G. Balian, K. Holbrook, D. Duksin, and P. Bornstein, *J. Cell Biol.* **78**, 701 (1978).
60. E. Ruoslahti, E. Engvall, E. G. Hayman, and R. G. Spiro, *Biochem. J.* **193**, 295 (1981).
61. K. D. Noonan, N. E. Noonan, and K. Yamada, *J. Supramol. Struct. Suppl.* **5**, 302 (abstract) (1981).
62. B. T. Atherton and R. O. Hynes, *Cell* **25**, 133 (1981).
63. M. Kurkinen, T. Vartio, and A. Vaheri, *Biochim. Biophys. Acta* **624**, 490 (1980).
64. C. R. Birdwell, A. R. Brasier, and L. A. Taylor, *Biochem. Biophys. Res. Commun.* **97**, 574 (1980).

65. M. Fukuda and S. Hakomori, *J. Biol. Chem.* **254**, 5451 (1979).
66. S. Takasaki, K. Yamashita, K. Suzuki, S. Iwanaga, and A. Kobata, *J. Biol. Chem.* **254**, 8548 (1979).
67. K. Olden, R. M. Pratt, and K. M. Yamada, *Cell* **13**, 461 (1978).
68. K. Olden, R. M. Pratt, and K. M. Yamada, *Proc. Natl. Acad. Sci. USA* **76**, 3343 (1979).
69. S. S. Alexander, G. Colonna, K. M. Yamada, I. Pastan, and H. Edelhoch, *J. Biol. Chem.* **253**, 5820 (1978).
70. G. Balian, E. M. Click, E. Crouch, J. Davidson, and P. Bornstein, *J. Biol. Chem.* **254**, 1429 (1979).
71. L.-H. E. Hahn and K. M. Yamada, *Proc. Natl. Acad. Sci. USA* **76**, 1160 (1979).
72. E. Ruoslahti, E. G. Hayman, P. Kuusela, J. E. Shively, and E. Engvall, *J. Biol. Chem.* **254**, 6054 (1979).
73. L. I. Gold, A. Garcia-Pardo, B. Frangione, E. C. Franklin, and E. Pearlstein, *Proc. Natl. Acad. Sci. USA* **76**, 4803 (1979).
74. D. D. Wagner and R. O. Hynes, *J. Biol. Chem.* **255**, 4304 (1980).
75. M. Hayashi, D. H. Schlesinger, D. W. Kennedy, and K. M. Yamada, *J. Biol. Chem.* **255**, 10017 (1980).
76. K. Sekiguchi and S. Hakomori, *Proc. Natl. Acad. Sci. USA* **77**, 2661 (1980).
77. E. Ruoslahti, E. G. Hayman, E. Engvall, W. C. Cothran, and W. T. Butler, *J. Biol. Chem.* **256**, 7277 (1981).
78. J. Keski-Oja and K. M. Yamada, *Biochem. J.* **193**, 615 (1981).
79. K. Sekiguchi and S. Hakomori, *Biochem. Biophys. Res. Commun.* **97**, 709 (1980).
80. F. Jilek and H. Hörmann, *Hoppe-Seyler's Z. Physiol. Chem.* **358**, 1165 (1977).
81. D. F. Mosher, P. E. Schad, and J. M. Vann, *J. Biol. Chem.* **255**, 1181 (1980).
82. M. D. Pierschbacher, E. G. Hayman and E. Ruoslahti, *Cell* **26**, 259 (1981).
83. L.-H. E. Hahn and K. M. Yamada, *Cell* **18**, 1043 (1979).
84. J. A. McDonald and D. G. Kelley, *J. Biol. Chem.* **255**, 8848 (1980).
85. R. Ehrismann, M. Chiquet, and D. C. Turner, *J. Biol. Chem.* **256**, 4056 (1981).
86. A. E. Postlethwaite, J. Keski-Oja, G. Balian, and A. H. Kang, *J. Exp. Med.* **153**, 494 (1981).
87. H. E. J. Seppä, K. M. Yamada, S. T. Seppä, M. H. Silver, H. K. Kleinman, and E. Schiffmann, *Cell Biol. Int. Rpts.* **5**, 813 (1981).
88. A. E. Chung, R. Jaffe, I. L. Freeman, J.-P. Vergnes, J. E. Braginski, and B. Carlin, *Cell* **16**, 277 (1979).
89. R. Timpl, H. Rhode, P. G. Robey, S. I. Rennard, J.-M. Foidart, and G. R. Martin, *J. Biol. Chem.* **245**, 9933 (1979).
90. S. Sakashita, E. Engvall, and E. Ruoslahti, *FEBS Lett.* **116**, 243 (1980).
91. V. P. Terranova, D. H. Rohrbach, and G. R. Martin, *Cell* **22**, 719 (1980).
92. I. Vlodavsky and D. Gospodarowicz, *Nature* **289**, 304 (1981).
93. D. W. Kennedy, D. H. Rohrbach, G. R. Martin, T. Momoi, and K. M. Yamada (submitted).
94. H. K. Kleinman, G. R. Martin, and P. H. Fishman, *Proc. Natl. Acad. Sci. USA* **76**, 3367 (1979).
95. F. Grinnell, *J. Cell Biol.* **86**, 104 (1980).
96. J. D. Aplin, R. C. Hughes, C. L. Jaffe, and N. Sharon, *Exp. Cell Res.* **134**, 488 (1981).

10

Plant Cell–Cell Interactions

FRANK B. DAZZO
Department of Microbiology and Public Health
and
Department of Crop and Soil Sciences
Michigan State University
East Lansing, Michigan 48824

Plants participate in a variety of cellular recognition processes through biochemical, physiological, and morphological responses. Cellular recognition processes will occur whenever a foreign cell becomes metabolically active on the surface of plant cells. If the foreign cell is a male pollen gamete, the recognition process directs the outcome of pollen germination, growth, and successful fertilization of the female ovary in the flower. If the foreign cell is from a different plant species that comes into contact at the interface of a graft, the cellular recognition processes determine whether the graft will be accepted or rejected. If the foreign cell is an invasive microbe—either a symbiont or a pathogen—the cellular recognition process will determine whether the infection is successful or abortive. Thus, in many ways, survival of a higher plant species depends on its ability to participate effectively in cellular recognition.

Most of what little is known about the underlying biochemical basis for cellular recognition in higher plants has been discovered since 1973. The fundamental working hypothesis upon which most subsequent investigations were based was that positive recognitions between cells arise from a specific union, reversible or irreversible, between chemical groupings on the surface of interacting cells (1). This hypothesis implies that important communication events occur when cells that recognize one another come into contact, and therefore, the complementary components of the cell surfaces have naturally been the center of focus for most biochemical studies. The recent widespread interest in this subject had led to the organization of numerous symposia and critical reviews (2–24).

1. PLANT CELL CONSTITUENTS RELATED TO RECOGNITION

1.1. The Plant Cell Wall

Unlike animal cells, the plasma membrane of plant cells is enclosed in a rigid cell wall. The structure of cell walls of higher plants varies considerably, not only among different species, but also between particular cell types and their stages of differentiation (12). The primary wall, characteristic of actively growing cells, consists of a negatively charged gel matrix of hemicelluloses, polyuronides (pectins), and glycoproteins, which are held together between a rigid and highly structured network of cellulose microfibrils.

The fine details of the biochemical structure are not fully understood. In primary walls of sycamore (25) the pectic matrix consists of polyuronides, arabinogalactans, xyloglucans, rhamnogalacturonans, and hydroxyproline-rich glycoproteins which form a matrix polymer through covalent glycosidic linkages. This matrix is a macromolecular glycoprotein that associates with the cellulose microfibrils by hydrogen bonding with the xyloglucan. When primary walls of lupine (26) and kidney bean (27) were examined, differences in saccharide bonding within the matrix were found. These data indicate that there is a structural variability of the cell walls of plants, as with cell walls of eubacteria, that may be superimposed upon a basic common structure.

An inherent problem in applying these structural data to explain recognition is that they only describe what is either covalently linked or irreversibly hydrogen-bonded together as the cell wall. Specific molecules available in situ for complementary binding and cellular recognition on the cell wall outer surface may be removed during the purification of the walls. It is also possible that plasma membrane or protoplasmic determinants of recognition, which are shed from the membrane and passed through the wall matrix to lie on the outer cell wall face during growth (12), could be lost during purification.

1.2. The Plasmodesmata

Communication between adjacent plant cells may also occur through plasmodesmata, cylinders of plasma membrane that extend through the cell wall and connect the cytoplasmic components of adjacent plant cells.

1.3. Carbohydrates and Lectins

There is a vast possibility for variation in fine structure of carbohydrate-containing components of plant cells. This variation arises from the multitude of different possibilities of sequences, branching, and structure of the monosaccharides (28). For example, compare the variations possible in structure of di- and tripeptides of identical amino acids, and contrast the possible permutations with two or three identical monosaccharides. A peptide of two or three identical amino acids can exist only in one form, whereas two identical monosaccharides may combine in glycosidic linkage to form eleven different disaccharides and three identical monosaccharides can form 176 different trisaccharides. Three different amino acids can be linked to form six different peptides, whereas three different monosaccharides may yield 1076 different trisaccharides! Thus, carbohydrate moieties in oligosaccharides, polysaccharides, glycoproteins, and glycolipids on plant surfaces and the cells that interact with them have great informational potential for recognition processes.

The ubiquitous occurrence in plant tissues of special proteins or glycoproteins, called lectins, which can discriminate between different saccharide constituents on cell surfaces, has prompted a great deal of research designed to determine if they are involved in cellular recognition (see also Chapter 7). The three-dimensional conformation of the lectin polypeptide creates pockets that are specifically complementary with sugar residues. Therefore, the lectin can interact specifically with saccharide residues on cell surfaces in a manner that is analogous to an antigen-antibody interaction. Accordingly, the glycoprotein-saccharide interaction can be inhibited specifically by "hapten" sugars, which at high concentrations can occupy the binding sites on the lectin normally occupied by the polysaccharide residues. In addition, some lectins undergo conformational changes upon saccharide binding.

The size and shape of the pocket may vary among different lectins, thus giving rise to classes of lectins that have different sugar specificities. Some lectins preferentially accommodate terminal sugar residues. Others may accommodate internal sugar residues in the polysaccharide at points of particular branched linkages, while other lectins may be specific for oligomers of the same repeating monosaccharide although the monosaccharide itself may not be an effective hapten inhibitor. An example of the latter case is the potato lectin, which recognizes internal linkages of *N*-acetylglucosamine, but requires chitobiose or chitotriose (oligomers of glucosamine) for effective hapten inhibition. It should be recognized that hapten inhibitors may be analogs and not necessarily identical to the saccharide receptor of the polysaccharide which binds the lectin. Thus, inhibition studies will yield clues to the nature of this determinant. The final definitive structure comes only after detailed analyses of the minimal saccharide sequence that binds the lectin.

2. RECOGNITION PROCESS IN PLANTS

2.1. Fertilization Process in Flowering Plants

Many different types of male pollen are deposited by wind or insects onto the female stigmatic surfaces of a flower, yet only that of the appropriate type is able to germinate, penetrate, grow down the style, and eventually reach the ovule. A variety of cellular recognition processes ensure that compatible male gametes will unite with female gametes to preserve the plant species as a breeding group, yet encouraging outcrossing within the species.

2.1.1. Compatible Fertilization Events. The compatible pollen-stigma interaction proceeds with the following generalized sequence of events (3, 12, 13, 28) (Fig. 1). After opening of the flower, a sticky exudate is secreted onto the papillar surface of the stigma. Immediately after deposition of the pollen, components of the outer pollen wall layer (exine) make contact with receptors on the sticky surface of stigma papillae (event 1). These stigmatic receptors, presumably glycoproteins, are glycosylated (28). The compatible pollen then becomes hydrated, immediately releasing outer wall (exine) proteins, and within minutes releases inner-wall (intine) proteins. These proteins may affect cell recognition events, as described later. There is little biochemical information on the basis for this hydration and leakage process. The release of these pollen wall proteins is followed by germination of the hydrated pollen and growth of the germ tube (event 2). This process is followed by the appearance of enzymes that can digest the cuticular layer of the stigma and provide a portal of entry for the penetrating germ tube (event 3). Recognition of the glycosylated receptors on the stigmatic surface by the pollen tube is apparently necessary for penetration, since prior binding of the lectin concanavalin A to these glycoslyated residues on the papillar cell surface will block tube penetration (28). (It would be interesting to determine if the surface of the germ tube has a lectin that can specifically recognize these glycosylated receptors on the stigma surface.) After the germ tube of the compatible pollen penetrates the papillar cell, it continues a remarkable, directed growth down the transmitting tissue of the style (event 4). It has been suggested that this guided growth is due to a chemotropic attraction towards the ovaries (3), although "contact guidance" may also be involved (17). Activation of protein metabolism commences in the ovaries before the

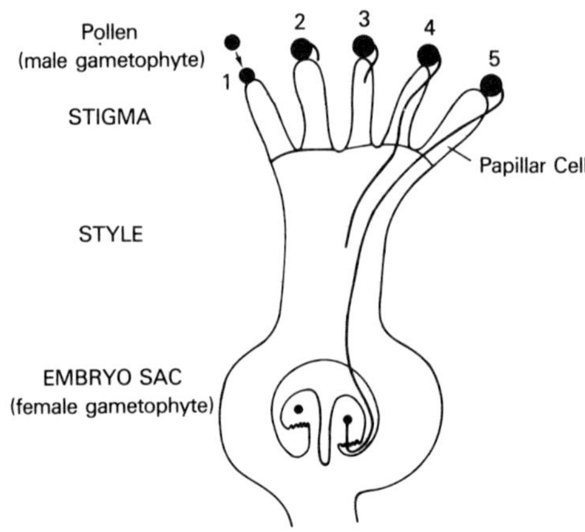

	Compatible Pollen Events	Fate of Foreign or Incompatible Pollen
1.	Pollen deposition on stigma papilla. Exine proteins contact stigma surface.	
2.	Hydration, germination, tube growth.	No hydration, germination, or recognition.
3.	Cutinase activation, tube penetration of cuticle	No cutinase activation or stigma penetration. Sporophytic incompatibility.
4.	Tube grows through style.	Tube growth stops in style. Gametophytic incompatibility.
5.	Tube penetrates embryo sac, tube discharges, sperm fertilizes egg.	Failure of sperm to react with the egg.

FIGURE 1. Pollination events in flowering plants. Redrawn from Clarke and Knox (12).

tips of the pollen tube are halfway to the ovule (15). This suggests that long-distance communication is occurring between the ovule and the pollen tubes. Finally, the pollen tube tip penetrates the embryo sac and ruptures at a critical oxygen partial pressure, discharging the male gametes into the egg apparatus (15). The male gamete contacts and fuses with the egg plasma membrane, followed by fusion of the male and female gamete nuclei (event 5).

2.1.2. Recognition Barriers to Fertilization. The first recognition step rejects some foreign pollen, since they remain dry after deposition on the stigma surface (12). In most flowering plants, one of two developmental systems, either the sporophytic or the gametophytic system prevents self-pollination or inbreeding. This self-incompatibility occurs in about 3000 species from more than 50 families of higher plants (15).

In the sporophytic self-incompatability system, rejection is accomplished by poor germ tube growth and little or no penetration of the stigmatic papillar cell wall. Incompatibility is accompanied by deposition of material tentatively identified by histochemistry as callose, a β-1,3 glucan, at the site where the germ tube contacts the stigmatic surface (3). This tentative identification of callose is obtained by fluorescence microscopy using fluorochromes that fluoresce when they bind to callose. However, the commonly used fluorochromes, aniline and primuline, are not specific for β-1,3 glucans (30), and further studies are needed to better identify this material. The deposition of this new material can be used as a bioassay for the presence of pollen-borne recognition factors from incompatible pollen. Using this method, evidence has been obtained suggesting that the recognition factors for the sporophytic self-incompatibility system are proteins carried in the exine layer of the pollen grain (3), which are released during the first few minutes when hydration occurs. It should be possible to isolate and characterize these recognition proteins. There is also evidence for preexisting pollen germination inhibitors that could contribute to the incompatibility reaction (15).

In the gametophytic incompatability system, the germ tube penetrates the stigma cell wall but its growth is arrested within the transmitting tissues of the style. Sometimes the wall of the tube bursts with the release of cytoplasmic materials, and in other cases, the tube tip becomes occluded with callose (?), as in the sporophytic system. It has been suggested that the recognition factors responsible for rejection in the gametophytic self-incompatibility system are in the female matrix materials through which the tube grows in the transmitting tract of the style (3). The pollen tube of petunia, a gametophytic self-incompatibility plant, is inhibited in vitro by extracts from self-incompatible styles, but not by extracts from styles from compatible petunias (3).

In some cases of incompatibility, the pollen tube reaches the embryo sac but is unable to fuse with the egg apparatus, and the ovule degenerates (15). In these cases, the growth of the germ tube is not arrested in the style because the latter is hollow.

2.1.3. Genes and Gene Products of Self-Incompatibility. Genetic studies indicate that molecular components for specificity of the self-incompatibility systems are encoded for by multiple alleles of a multigene complex called the S genes (3). Compatibility requires that the pollen and ovaries be of different S alleles, and identical alleles prevent self-pollination (31). Macromolecules detected serologically and electrophoretically from stigmas have been identified and specifically correlated with specific S-allele self-incompatibility genotypes (33, 34). Recently, S-allele-specific glycoproteins have been purified from stigmas of *Brassica* sp (35, 36) that specifically prevent male pollen of this species from germinating on the stigma surface, which would otherwise be compatible (35). Such experimental evidence suggests that S-allele-specific glycoproteins function in plant cell-cell recognition and control of sporophytic self-incompatibility (35).

2.2. Cell Recognition on Grafts

2.2.1. Terminology. Grafting is a joining of somatic tissue of plants to develop a tissue union that functions as a single plant. The transferred donor organ, usually a shoot, is called the *scion* and the lower component recipient is the *stock*. When the components belong to the same plant, the combination is an *autograft*; when they belong to a same species the combination is a *homograft*; and when they belong to different species it is a *heterograft*. A graft that "takes" is called compatible and one that is rejected is incompatible. Grafting can be artificial, as when the two tissues are intentionally cut and placed together, or it can occur naturally as exemplified by the banyan tree, which drops aerial roots that develop autografts with stems and roots at the base of the tree.

2.2.2. Development of a Graft Union. The progress of a graft union can be measured by an increase in its tensile strength (37). This can be accomplished by applying increasing amounts of force in the form of weights to the scion until the union is broken. This approach revealed that in a compatible graft, there was an initial phase (typically 4 days) of intensifying cohesion between large, callus parenchymatous cells of the union, followed by a second phase of rapidly increasing strength, which reaches a maximum 3 to 4 days later (3, 37).

In both compatible and incompatible grafts, the cohesion begins to increase during the first 12 hr after the two surfaces are brought together (37). During the first phase, the walls of the cells at the junction begin to thicken considerably. There is intense cell division on both sides of the junction, accompanied by an increase in prominence of dictyosomes, which probably participate in the wall thickening process (37). To date, the molecules involved in the cohesion of cells at the junction of stock and scion have not been isolated or identified. There are speculations that during the first phase (which is accomplished by both compatible and incompatible grafts) cohesion is initiated by excreted pectic substances and other wall polysaccharides (3). The second phase of cohesion is accomplished only by compatible grafts. It has been proposed that secreted glycoproteins and lectins interact with these polysaccharides to intensify the cohesion during this phase of development (37).

2.2.3. Recognition and Rejection of Grafts. In a compatible graft, the union of cells and tissues in the neighborhood of the graft must clear the wounded zone of debris and restore the continuity of the vascular conducting systems and tissues (3). In an incompatible graft, there is mutual rejection by stock and scion; the cells that form vascular elements never fuse across the junction, and continuity is lost.

The ability of the plant to discriminate between compatible and incompatible grafts provides evidence for the existence of a recognition mechanism by the somatic tissues. In general, the more genetically related the scion and stock are, the greater probability that the graft union will be accepted (12). Does this mean that compatibility is a consequence of recognition and acceptance of self, and incompatibility of nonself?

Data that are compatible with this theory come from studies which examine the structural (immunological) relatedness of macromolecular constituents from cells of different plant varieties. In beans (38), the strongest immunological cross-reactivity of seed antigens occurs with closely related varieties in which stem-stem grafts are most readily accepted. Similar results have been reported for graft-compatible varieties of apples (39, cited in 12).

The need to examine if lectins participate in this cellular recognition process (3) is an important direction research may take in this new field.

2.3. Cellular Recognition in Plant–Microbe Interactions

Plants are constantly coming into contact with microorganisms, and the events that ensue have profound influences on plant morphogenesis, nutrition, and pathogenesis (18). Specificity is the hallmark of most plant-microbe interactions, which implies that recognition systems function during certain stages of the development of the relationship. Recognition is most obvious when there is a restricted host range, for example, the nitrogen-fixing *Rhizobium*-legume symbiosis, but also occurs in those interactions where the host-range is broad, for example, *Agrobacterium tumefaciens* crown-gall disease. In the latter case, recognition may involve molecules that are commonly found in the cell walls of many different plants.

When a plant encounters an imcompatible pathogen, the host cell may (1) form structures that restrict further invasion of the parasite; (2) accumulate inhibitory substances; or (3) collapse, thus effectively preventing the spread of the parasite (10). The latter phenomenon is called the hypersensitive response and makes an important contribution to the plant's resistance to disease.

There are many recognition phenomena that occur when a microbe infects a plant host, and the attachment that follows the contact of the microbe with the host cell is one good example of a recognition event studied at the biochemical level. This is not to imply that accomplishment of firm attachment need be the key recognition step, but rather that it is only one of the keys needed to unlock the many doors blocking the way to successful infection.

2.3.1. Surface Chemistry of Bacteria That Infect Plants. Most bacteria that infect plants are Gram-negative. The unique feature of the Gram-negative cell is its outer envelope (also called the outer membrane) on which a mosaic of macromolecules is inserted, some of which may extend outward from the surface. The outer envelope contains a small variety of hydrophobic proteins, phospholipid, and a lipopolysaccharide (LPS). The latter is a specific component localized exclusively in the outer leaflet of the outer membrane (Fig. 2). Its hydrophobic end, called lipid A, is a disaccharide of glucosamine acylated with long-chain fatty acids that insert into the outer membrane and anchor the macromolecule to the cell. The lipid A is covalently attached to an eight-carbon acid called 2-keto-3-deoxyoctanoic acid (KDO), which is only found in LPS. KDO is commonly linked to a seven-carbon sugar (heptose), hexosamines, and a few hexoses. This region of the LPS is called the R core. Attached to the R core and extending out from the surface of the membrane is a polymerized repeating oligosaccharide called the somatic O antigen, which varies tremendously in composition and structure among bacterial strains. The LPS is of importance here because of its involvement in cellular recognition with higher plants.

Also of importance to cellular recognition is the bacterial capsule, which may extend several micrometers out from the surface of the cell and is a fibrillar network of acidic heteropolysaccharides (Fig. 3). The anionic nature of the capsular polysaccharides can arise from the carboxyl groups of uronic acids and/or from pyruvate and succinate group substitutions on some of the

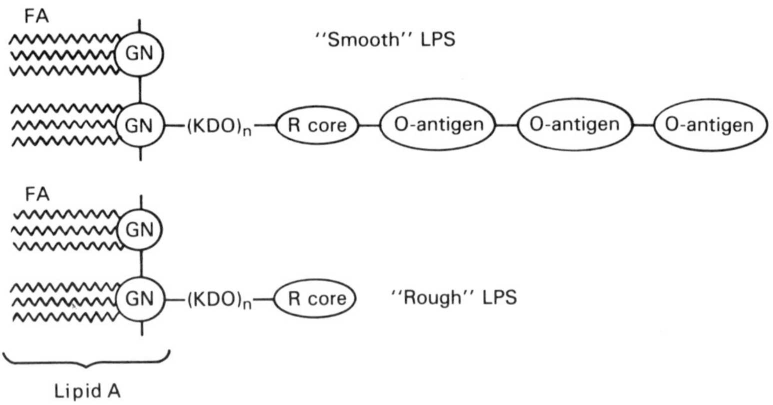

FIGURE 2. Schematic structure of bacterial lipopolysaccharide. In "smooth" LPS, long-chain fatty acids (FA) are acylated to glucosamine (GN) disaccharide to form the lipid A, which is covalently bound to 2-keto-3-deoxyoctanoate (KDO), the R core, and a repeating oligosaccharide side chain called the O antigen. LPS is considered "rough" if the repeating O antigen and sometimes portions of the R core are not added on during synthesis by genetically altered strains.

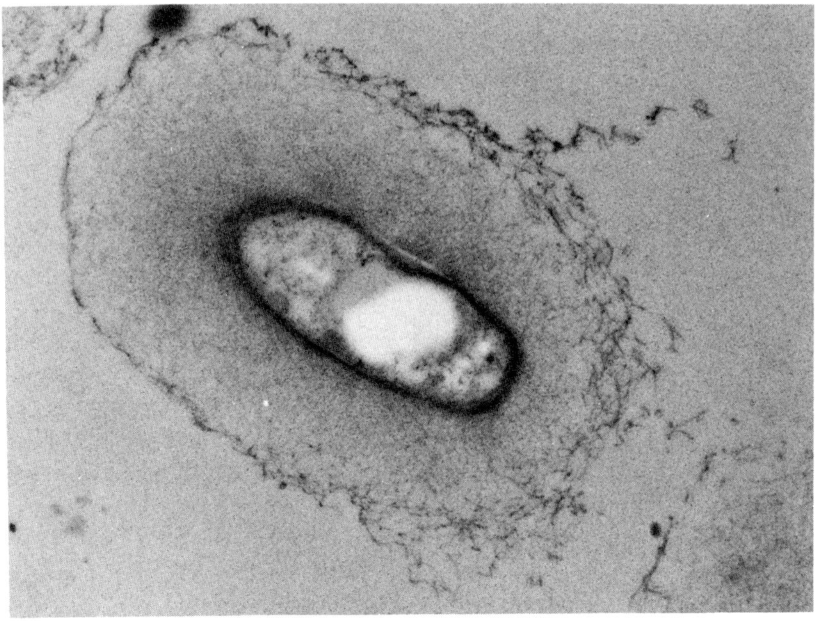

FIGURE 3. Staining of *Rhizobium trifolii* 0403 with ruthenium red reveals the dense fibrillar capsule surrounding cells which can bind the clover lectin, trifoliin A [Reprinted, with permission, from Dazzo and Brill (58). Courtesy of the American Society for Microbiology.]

sugars. The exact nature by which the capsule associates with the outer membrane components is not known. Capsular polysaccharides, LPS, and, in bacteria that infect plants, neutral homopolysaccharides of glucose (β-1,2 and β-1,4 glycosidically linked; references 40, 41), may be excreted into the environment. The complexity of some polysaccharides elaborated by these bacteria is illustrated by elucidation of the repeating unit structure of an exopolysaccharide from *Rhizobium meliloti* strain U27, which infects alfalfa roots (42). The octasaccharide was the largest repeating unit ever demonstrated for a bacterial polysaccharide.

2.3.2. Adherence of *Agrobacterium* to Its Host. *Agrobacterium tumefaciens* is a pathogen that initiates crown-gall tumors in a variety of dicots when introduced into wounded areas. For example, various nursery plants often acquire such tumors, which may eventually kill the host. The adherence of virulent *Agrobacterium* to a specific site on the host cell wall exposed at wounding is an essential early step of the infection process (43). The number of tumors formed on wounded leaf tissue is proportional to the inoculum size and the adherence step is completed within 6 hr. Specificity in adherence was shown by the inability of a variety of other bacteria to inhibit the formation of tumors when they were applied before or with virulent tumor-forming bacteria. The O antigen of lipopolysaccharide isolated from the outer envelope of these Gram-negative bacteria was an effective inhibitor of tumorigenesis (44) (Whatley and Lippincott, personal communication), indicating that the externally projecting LPS polysaccharide chains function in the adherence of these bacteria to the host. The carbohydrate structure on the O antigen that is compatible with adherence to the host is not yet known. The genetic information that dictates the polysaccharide structure capable of specifying adherence is both chromosomal and plasmid-borne (45).

What is the nature of the host site for attachment? This problem was approached by examining cell wall components that bind to the bacteria and block their attachment to wounded host sites prior to tumor formation (46). Purified cell wall fragments from dicotyledonous plants were as inhibitory as the bacterial LPS. Neither cell walls of monocotyledonous plants nor embryos of dicots were effective inhibitors. The monocots are not as susceptible to crown-gall disease as are the dicots. This correlation suggests that the presence or absences of adherence sites on the host may be an important factor that determines host specificity in this plant–microbe interaction.

Clues to understanding the biochemical nature of the adherent site on the cell walls were obtained with the use of plant cell wall fractions (46). Polygalacturonic acid, pectin, and arabinogalactan were inhibitory cell wall fractions. The importance of a free carboxyl group at carbon-6 in the galacturonic acid residues of pectin was suggested by the lack of inhibition of binding of preparations of cell walls of embryos and monocots. These are high in pectin methyl groups, which mask the C-6 carboxyl groups. This hypothesis was confirmed by converting cell walls of corn (a monocot) and bean embryo cell walls into inhibitory materials by treating them with pectin methyl esterase. This change was reversible by subsequent methylation with pectin methyl transferase and *S*-adenosyl-methionine. These studies provide the evidence that inhibition of tumor initiation is decreased with an increase in methylation in the cell wall binding component. Lectins (carbohydrate-binding proteins) that combine with α-glucosyl, mannosyl, fucosyl, and internal *N*-acetylglucosamine residues were not inhibitory to the tumor formation, perhaps because they did not combine with the sites on the polysaccharides necessary for site-specific attachment of the bacteria. Lippincott and Lippincott (46) proposed that the adherence step essential for *Agrobacterium* infection is mediated by a direct polysaccharide-polysaccharide binding of the O antigen of LPS to polygalacturonic acid residues in plant cell walls exposed by wounding.

2.3.3. Interaction of Systemic Bacterial Pathogens and Host Cell Walls.

The adsorption and envelopment of bacterial pathogens by plant cell walls have received considerable attention, particularly in view of the importance of this process in the mechanism of host resistance to disease (10, 47). After infiltration into leaves, only avirulent or nonpathogenic bacterial cells adsorb to the plant cell walls and are enveloped by electron-dense vesicles and fibrillar materials. In contrast, virulent cells remain unadsorbed and free to multiply in the intercellular fluid, where they eventually become systemic. Membrane damage and electrolyte leakage of the host cells follow soon after the envelopment and immobilization of the avirulent cells, and these cytological changes are viewed as part of a hypersensitive response by the plant that is important to its resistance to infectious disease (48, 49).

Tobacco leaves have more bacterial agglutinating activity if they are previously infiltrated with the avirulent incompatible organism *Pseudomonas pisi* (50). Little or no agglutinating activity was recovered from leaves previously infiltrated with the virulent (compatible) organism *Pseudomonas tabaci*. Similar in vivo agglutination of avirulent, unencapsulated *Erwinia amylovora* in the xylem vessels of apple leaves has been found. A recent study has shown the positive correlation between extracellular polysaccharide production and virulence of *Erwinia amylovora* (51). The agglutinating substance obtained from either apple or tobacco leaf tissue was suggested to be a glycoprotein lectin with agglutinating activity that combined with glycosylated receptors on the bacterial cell surface (50).

Sequeira and Graham (48) studied the interactions of lectins isolated from potato and tobacco with virulent and avirulent strains of the pathogen *Pseudomonas solanacearum*. Their hypothesis is that adsorption and host-pathogen recognition may involve a complementary interaction of lectins on the host cell wall with carbohydrate receptors on the bacterial cell surface. Their studies showed that lectins purified from these hosts agglutinate only avirulent cells. Failure of bacterial cells to agglutinate with the purified lectins was correlated with the presence of extracellular polysaccharide (EPS) produced in copious amounts by the virulent strains of *Pseudomonas solanacearum*. Virulent cells were agglutinated by potato lectin if their EPS was removed by washing, and avirulent cells would not agglutinate if the EPS from the virulent strains was added to the lectin preparation. The potato lectin can bind with the EPS of the virulent strains and with the LPS of both the virulent and the avirulent strains. The glycosylated receptor on the LPS is apparently internal *N*-acetylglucosamine residues in the R-core lipid A of the LPS, and inhibition studies indicate that oligomers of *N*-acetylglucosamine (e.g., chitotriose) are potent inhibitors of this lectin-polysaccharide interaction. The O antigen of the LPS is apparently unnecessary for this lectin interaction, as LPS from rough strains also are reactive. In experiments with potato lectin and ^{14}C-labeled polysaccharides from *Pseudomonas solanacearum*, both EPS and LPS were shown to bind (10). EPS interfered with binding of LPS to the lectin. Since EPS is readily soluble and diffuses into the medium surrounding the invading bacterium, Sequeira hypothesizes (10) that the soluble EPS covers all of the available lectin sites on the host wall, thus allowing the bacterium to multiply freely in the intercellular spaces. The potato and tobacco lectins are found to be localized on the host cell walls.

The lipid A:R core region of several Gram-negative bacteria will induce a protective, nonspecific resistance to bacterial pathogens in potatoes and tobacco if applied 24 hr prior to inoculation with virulent *Pseudomonas solanacearum* (53). Consistent with this is the finding that intact LPS from *Pseudomonas solanacearum* binds strongly to tobacco mesophyll cell walls (53), and this induction of disease resistance may involve the interaction between the host cell wall lectin and the same components of the bacterial cell wall involved in recognition and attachment. A survey of many strains of *Pseudomonas solanacearum* has shown that, indeed, rough LPS is necessary in order to produce this hypersensitive response (54). It is possible that the O antigen in smooth strains masks the portion of the LPS that is involved in the recognition necessary for induction of the hypersensitive response. Studies on the biochemical basis of induced disease resistance carry a great deal of significance, since they may lead to important applications that reduce the burden of plant pathogenesis on crops.

2.3.4. Attachment of *Rhizobium* to Legume Root Hairs. *Rhizobium* is a genus of Gram-negative bacteria that selectively infects legume roots and then forms root nodules that fix atmospheric nitrogen into ammonia (18). The infection process is very selective for certain combinations of rhizobia and legume, and this high degree of host-range specificity currently defines the taxonomy of the bacterium. For example, *Rhizobium trifolii* infects clover and *Rhizobium meliloti* infects alfalfa. Successful infection of legumes by these nitrogen-fixing bacteria is of immense importance in the nitrogen cycle of terrestrial life.

2.3.4.1. Phase-I Attachment. Adhesion of the symbionts to the root surface constitutes one of the many critical "recognition" events that occur before successful infection and nodular development of the symbiotic state. Quantitative microscopic assays (18, 55) and numerous electron microscopic studies (17–22, 56–59) have revealed multiple mechanisms of rhizobial attachment to clover root hairs. A nonspecific mechanism allows all rhizobia to attach in low numbers (2–4 cells per 200 μm root hair length per 12 hr). In addition, a specific mechanism that has been studied in the *Rhizobium trifolii*-clover symbiosis allows selective attachment in significantly ($P = 0.005$) larger numbers (22–27 cells per 200 μm root hair length per 12 hr) (55). Electron microscopy (57) disclosed that the initial bacterial attachment step consisted of contact between the fibrillar capsule of *Rhizobium trifolii* and electron-dense globular aggregates lying on the outer periphery of the fibrillar clover root hair cell wall (Fig. 4). This "docking" stage is the first step of phase-I attachment and occurs within minutes after inoculation of encapsulated cells of *Rhizobium trifolii* on the host clover.

What cell surface molecules are involved in phase-I attachment? We have approached this problem by examining the surface components of the bacterium and the host that interact with the same order of specificity as is observed with the adhesion of the bacterial cells. Immunochemical and genetic studies have demonstrated that the surfaces of *Rhizobium trifolii* and of clover epidermal cells contain a unique antigen that is immunochemically cross-reactive (57, 58), suggesting its structural relatedness on both symbionts. This antigen contains receptors that bind noncovalently to a multivalent lectin called trifoliin A (originally trifoliin), which has recently been purified from clover seeds and seedling roots and has been partially characterized (60). A specific hapten inhibitor of trifoliin A is 2-deoxyglucose (57, 61). The first clue that trifoliin A on the root may be involved in rhizobial attachment came from the observation that 2-deoxyglucose specifically inhibited the attachment of *Rhizobium trifolii* to clover root hairs (55), reducing the high level of bacterial adhesion to that characteristic of background

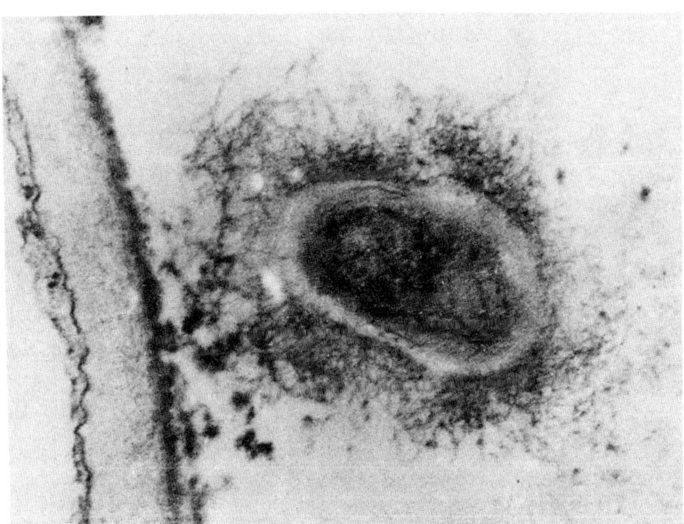

FIGURE 4. Transmission electron micrograph of *Rhizobium trifolii* NA 30 in the docking stage of phase-I attachment to a clover root hair. The fibrillar capsule of the bacterium is in contact with electron-dense globular aggregates on the outer periphery of the root hair cell wall [Reprinted, with permission, from Dazzo and Hubbell (57). Courtesy of the American Society for Microbiology.]

levels. Subsequent studies showed that 2-deoxyglucose specifically facilitated the elution of trifoliin A from the intact clover root (60, 61) and inhibited the binding of *Rhizobium trifolii* capsular polysaccharide to clover root hairs (61). As a negative control, 2-deoxyglucose did not inhibit adsorption of *Rhizobium meliloti* or its capsular polysaccharide to alfalfa root hairs (55, 61). Consistent with the above results in the *Rhizobium*-clover system, hapten-facilitated elution of *Rhizobium*-binding lectins from intact pea, soybean, and alfalfa roots has also been reported (62, 63) (W. Kamberger, personal communication).

We proposed a model to explain this early recognition event of phase-I attachment on the clover root hair surface prior to infection (57). According to this hypothesis, trifoliin A recognizes similar saccharide residues on *Rhizobium trifolii* and clover and cross-bridges them in a complementary fashion to form the correct molecular interfacial structure that initiates the preferential and specific adsorption of the bacteria to the root hair surface (Fig. 5).

This model proposes that there are host-specific receptor sites on the legume root where recognition of the rhizobial symbiont occurs. A key

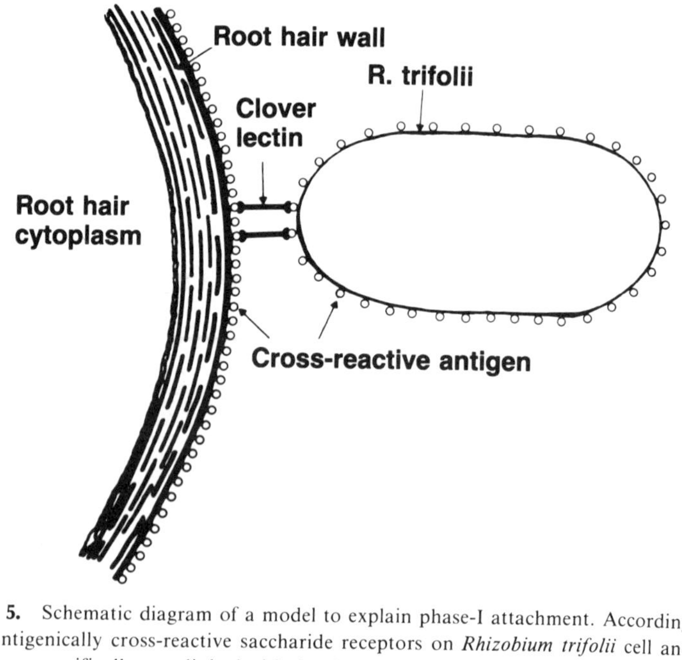

FIGURE 5. Schematic diagram of a model to explain phase-I attachment. According to this model, antigenically cross-reactive saccharide receptors on *Rhizobium trifolii* cell and clover root hair are specifically crosslinked with the clover lectin, trifoliin A. [Reprinted, with permission, from Dazzo and Hubbell (57). Courtesy of the American Society for Microbiology.]

experiment that demonstrated these receptor sites and localized them on the root surface was performed by first labeling the trifoliin A-binding capsular polysaccharide of *Rhizobium trifolii* with the fluorescent dye fluorescein isothiocyanate, incubating the conjugate with sterile seedling roots for a brief period, and then examining the root by fluorescence microscopy (61). The results are shown in Figure 6. The receptor sites on clover roots which immediately bound the *Rhizobium trifolii* capsular polysaccharide were located at discrete root hair sites that have differentiated on the epidermal root surface. They accumulate at root hair tips (single arrow) and diminish toward the base of the root hair (58, 61). This result highlights the importance of epidermal cell differentiation in the development of receptor sites that recognize rhizobia. Undifferentiated epidermal cells did not bind the bacterial

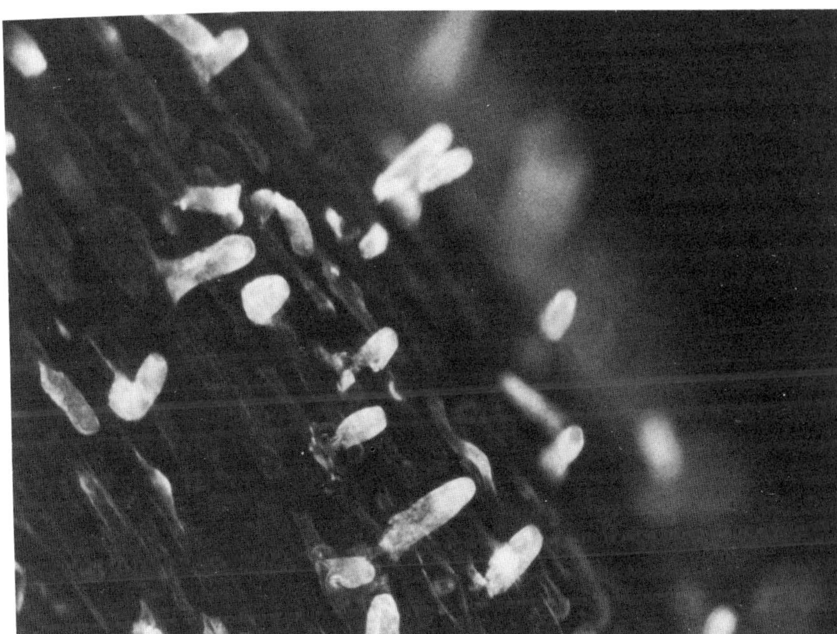

FIGURE 6. One of the exclusive relationships between a legume and a nitrogen-fixing bacterium is demonstrated by the binding of *Rhizobium trifolii* to root hairs of its plant host clover. A fluorescent dye was first conjugated to the bacterial capsular polysaccharide. This labeled polysaccharide was then incubated with clover roots. Fluorescence of the roots showed that binding of the bacterial polysaccharide was restricted to the root hairs or root hair primordia, the differentiated cells that serve as target cells for infection by *Rhizobium trifolii*. Similar results have been obtained with other rhizobia and their corresponding specific legume hosts. [Reprinted, with permission, from Dazzo and Brill (61). Courtesy of the American Society for Microbiology.]

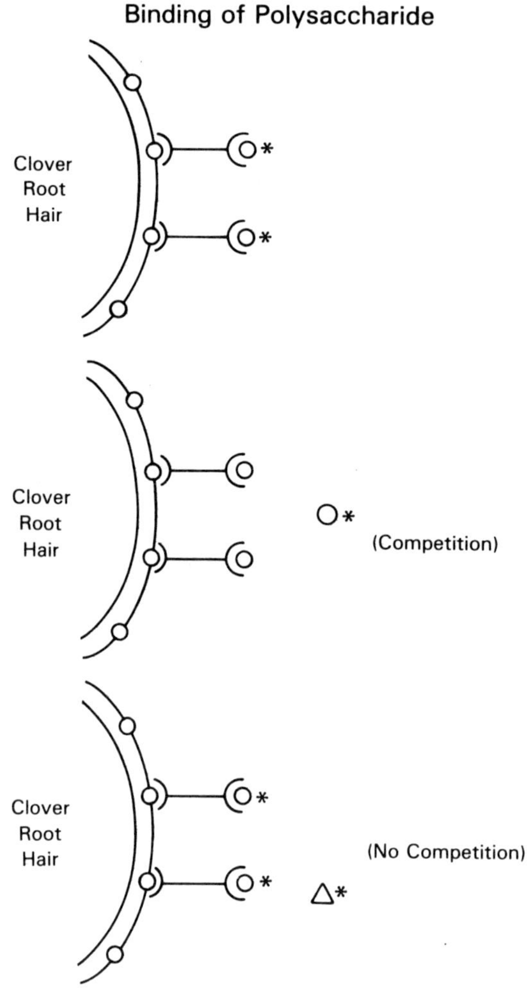

FIGURE 7. Schematic diagram illustrating specificity of receptor sites in the *Rhizobium*-legume symbiosis. Labeled polysaccharide from *Rhizobium trifolii* (O*) binds to receptor sites on clover root hair tips. Unlabeled polysaccharide from the same bacterial species (O) has a conserved structural component allowing it to compete with the labeled polysaccharide for the same receptor sites. Polysaccharide from another species, *Rhizobium meliloti* (Δ*) has a different structure, which is recognized by receptor sites on its host, alfalfa, but not on clover.

polysaccharide, whereas epidermal root hair primordia gained this cell surface property (double arrow, Fig. 6).

Specificity of these receptor sites was demonstrated by the ability of unlabeled capsular polysaccharide from *Rhizobium trifolii*, but not from *Rhizobium meliloti*, to block the binding of the labeled polysaccharide (Fig. 7). Similar experiments have shown that the specific binding of homologous rhizobial polysaccharides to alfalfa, pea, and soybean hosts was selective for the root hairs, where this recognition occurs (61, 64–66). In the case of clover, soybean and peas, the sugar haptens for the corresponding host lectin specifically inhibited the binding of the labeled rhizobial polysaccharide as evidence that the root lectins were involved in this recognition process (61, 65, 66). In addition, these receptor sites match the distribution of the root-hair lectin and the primary sites of rhizobial attachment (60) (Fig. 8).

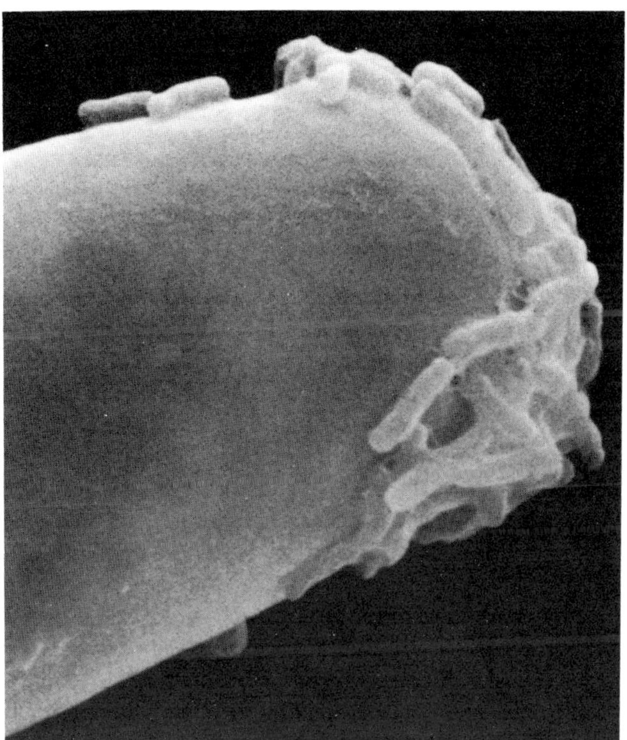

FIGURE 8. Phase-I attachment of *Rhizobium trifolii* 0403 to the tip of a clover root hair after 15 min of incubation as examined by scanning electron microscopy. [Reprinted, with permission, from Dazzo and Brill (58). Courtesy of the American Society for Microbiology.]

Rhizobium cells precoated with their corresponding plant lectin will adhere in higher numbers than noncoated cells to their host roots (7, 55, 67), indicating the accessibility of these lectin receptors on the root surface. Quantitative adsorption studies with seedling roots of several legumes indicate that the cross-reactive antigen on *Rhizobium trifolii* is host-specific, since antibody that cross-reacts with *Rhizobium trifolii* can be adsorbed by clover roots but not by roots of alfalfa or joint vetch, which are examples of other cross-inoculation groups (58).

Several experiments suggest that trifoliin A and the cross-reactive anti-clover root antibody bind to the same or similar overlapping determinants on *Rhizobium trifolii* (58). First, the antibody and the lectin bind specifically to the same isolated polysaccharides from *Rhizobium trifolii*. Second, this interaction is specifically inhibited by 2-deoxyglucose (Fig. 9). Third, the

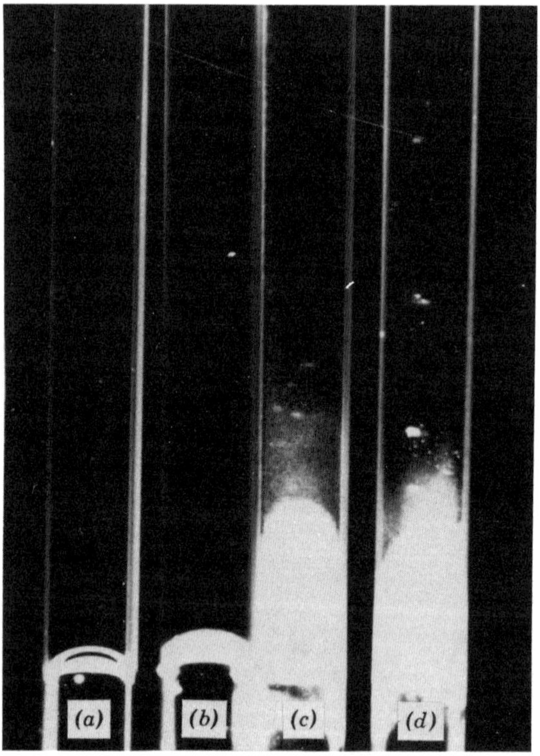

FIGURE 9. Capillary tube precipitation of capsular polysaccharides from *Rhizobium trifolii* 0403. Each tube contains capsular polysaccharide in buffer solution. Tubes A and C have rabbit clover root antiserum; tubes B and D have trifoliin A. Tubes A and B are hapten-inhibited with 2-deoxyglucose (30 m*M*).

genetic markers of *Rhizobium trifolii* that bind trifoliin A and the antibody cotransform into *Azotobacter vinelandii* with 100% frequency (68). And fourth, monovalent Fab fragments of IgG from anti-clover root antiserum strongly block the binding of trifoliin A to *Rhizobium trifolii* (58). Considered collectively, these studies suggest that *Rhizobium trifolii* and clover roots have similar saccharide receptors for trifoliin A. However, the definitive test of their identity as antigenically related structures will require knowledge of the minimal saccharide sequence that binds trifoliin A.

The model would predict that all infective strains of *Rhizobium trifolii*, regardless of their antigenic disparity, should possess the same saccharide sequence cross-reactive with clover. It was very important to test this prediction, since it is well known that the serology of surface polysaccharides on *Rhizobium* is very strain-specific because of compositional differences in the immunodominant determinants (69, 70). Immunological tests involving passive hemagglutination and its inhibition, nevertheless, showed that the saccharide sequence responsible for the antigenic cross-reactivity with clover roots was conserved in otherwise immunochemically distinct surface polysaccharides of *Rhizobium trifolii* strains (58), and therefore it is not the immunodominant structure of the polysaccharides.

The results of three experiments indicate that trifoliin A and antibody to the clover root cross-reactive antigen bind to the same *Rhizobium trifolii* saccharide determinants that bind these bacteria to clover root hairs (58). First, Fab fragments of anti-clover root antibody adsorbed on cells of *Rhizobium trifolii* blocked their attachment to clover root hairs. Second, only the *Azotobacter vinelandii* hybrid transformants that carried the trifoliin A receptor bound to clover root hairs. Third, competition assays using fluorescence microscopy indicated that the *Rhizobium trifolii* polysaccharides carrying the trifoliin A receptor had the highest affinity for clover root hairs.

Soybean root cells in culture selectively adhere to homologous *Rhizobium japonicum* cells, and bacterial adherence is inhibited by pectinase treatment of the root cells (71). Encapsulated cells of *Rhizobium japonicum* or cells associated with slime specifically bind to soybean lectin (72–75), and to soybean root hairs (73). The specific attachment of *Rhizobium japonicum* to root hairs of the small soybean (*Glycine soya*) is specifically inhibited by *N*-acetylgalactosamine, the specific hapten for the soybean root lectin (75). It has been reported that the lectin from bean (*Phaseolus vulgaris*) binds specifically to surface polysaccharides of the bean symbiont, *Rhizobium phaseoli* (76), and there is indirect evidence that this lectin may be on the bean root epidermis at sites where binding by rhizobia and symbiotic infection occur (77). As with *Rhizobium trifolii* (55), nonnodulating mutant strains of *Rhizobium leguminosarum* (78) adhere in smaller numbers to pea root hairs as compared with the wild-type nodulating strains (C. A. Napoli, personal communication). Clearly, selectivity in rhizobial adherence to legume roots may operate in several species.

At low inoculum densities, *Rhizobium trifolii* may be exposed to clover root exudate for a few hours before the bacterial cells contact the root hair surface. During this time, trifoliin A in root exudate (79) may bind to the cells (80). Recently, we have found that *Rhizobium trifolii* growing near clover roots has trifoliin A selectively deposited in situ at one cell pole (Fig. 10) (80). This alteration in distribution of trifoliin A receptors may possibly explain the distinctive polar attachment of the bacterial cells to the root hair surface towards the end of phase-I attachment (Dazzo et al., in press). Under certain growth conditions, the receptor sites on *Rhizobium japonicum* that bind soybean lectin also accumulate at one cell pole (75, 81, 82).

2.3.4.2. Phase-II Adherence. Phase-I attachment is followed by phase-II adherence, which firmly anchors the bacteria to the root hair surface during later preinfective stages (18–22). During phase-II adherence, fibrillar materials visible under the scanning electron microscope are associated with the adherent bacteria after prolonged incubation with the clover root (18–22, 56) (Fig. 11).

The composition of aggregated microfibrils characteristic of phase-II adherence is unknown. One possibility is that they are the cellulose microfibrils produced by many rhizobial species (40, 83). These neutral exopolysaccharides tend to flocculate cells, often when they are in stationary phase of growth in broth culture. Phase-II adherence may be important in maintaining the firm contact between the bacterium and the host root hair necessary for triggering the tight root hair curling (shepherd's crook formation) and successful penetration of the root hair cell wall during infection (40).

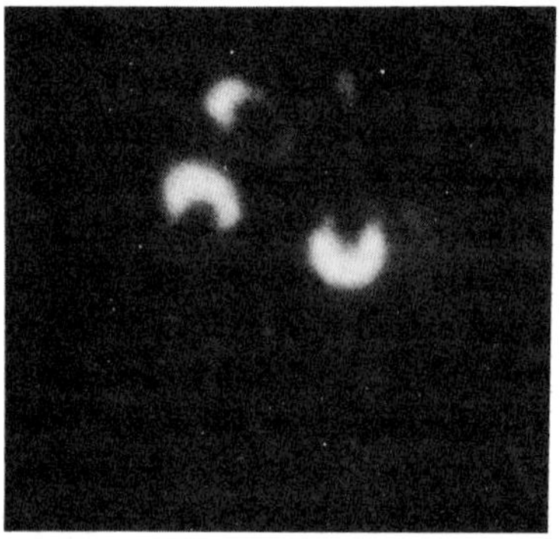

FIGURE 10. In situ binding of root-eluted trifoliin A to one cell pole of *Rhizobium trifolii* 0403 in the clover rhizosphere as detected by immunofluorescence.

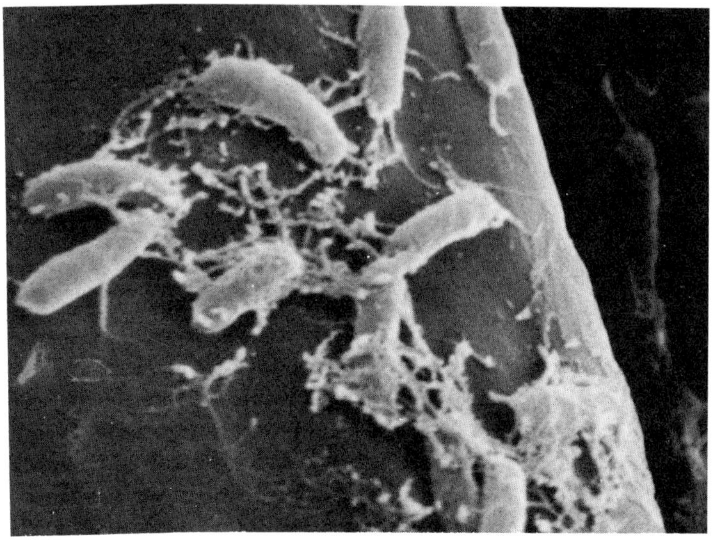

FIGURE 11. Phase-II adherence of *Rhizobium trifolii* 0403 after prolonged incubation on a clover root hair as examined by scanning electron microscopy. Note the fibrillar appendages associated with the adherent bacteria.

Although adherence of infective rhizobia to target root hairs is a prerequisite for infection, several observations indicate that other undefined events must occur to initiate root hair infection. Heterologous rhizobia can adhere in small numbers to root hairs but do not infect them (18, 55), and very few of the root hairs to which infective rhizobia adhere eventually become infected. Genetic hybrids of *Azotobacter vinelandii* (which carry the trifoliin A-binding saccharide receptor on their surface as a result of intergeneric transformation with DNA from *Rhizobium trifolii* [68]) have acquired the ability to adhere specifically to clover root hairs (58) but do not infect them. The latter experiment is particularly informative, since we put into *Azotobacter* what it takes *Rhizobium trifolii* specifically to attach and adhere, yet the hybrid cells lacked the ability to carry the infection process to the stage of root hair penetration. Possibilities for other genes or gene products that may not have been transferred or expressed include those controlling cell wall hydrolytic enzymes (84), inducers of host polygalacturonase (85, 86), root hair curling factors (87), and periplasmic extrinsic substance, ES-6000, which promotes root hair infection (88).

2.3.4.3. Regulation of Attachment in the Rhizobium-Legume Symbiosis. How is the attachment process regulated? Do both symbionts participate in regulating the components of the adhesive interface? Answers to these important questions are beginning to emerge. It has been known for many years that fixed nitrogen (e.g., NH_4^+, NO_3^-) is one of the many environmental factors that limit the development and the success of the *Rhizobium*-legume symbiosis in nature. For instance, root hairs are resistant to infection by rhizobia when the roots are supplied with critical concentrations of NO_3^-. In order to study this phenomenon in more detail, we developed an immunocytofluorimetric assay to measure levels of trifoliin A on the surface of clover roots grown in different levels of NO_3^- (89). We found that the immunologically detectable levels of trifoliin A and the specific binding of *Rhizobium trifolii* to root hairs decreased in a parallel fashion as the concentration of NO_3^- increased in the rooting medium to 15 m*M* (89). This experiment supported the hypothesis that trifoliin A is involved in binding *Rhizobium trifolii* to clover root hairs. The levels of NO_3^- that inhibited attachment were well below the levels that stunted seedling growth. Interestingly, the levels of trifoliin A and rhizobial attachment were increased at 1 m*M* NO_3^-. Thus, low levels of NO_3^- enhanced the recognition process and high levels shut it off.

We have begun to search for an explanation of this nitrate effect on recognition. Ligand-binding studies using radiolabeled $^{13}NO_3^-$ (generated in a cyclotron) have detected no direct interaction of trifoliin A and NO_3^-, and other studies show that NO_3^- does not interfere with the binding of trifoliin A to its specific antibody or *Rhizobium trifolii* cells to clover roots after brief periods of incubation (90). These studies indicate that the NO_3^- modulation of trifoliin A and of rhizobial accumulation on clover roots is in some way regulated by an intervening process rather than a direct interaction of receptors with this anion. Consistent with this hypothesis are two exciting new developments. The immunologically detectable levels of trifoliin A in root exudate and the accessibility of trifoliin A receptors on purified root cell walls are significantly reduced when developing clover seedlings are grown in 15 m*M* NO_3^- (90).

The selective ability of *Rhizobium trifolii* to adhere to clover root hairs should be influenced by conditions that affect the accumulation of trifoliin A on the host root surface and the saccharide receptor on the bacterium. Evidence supporting this hypothesis comes from data described above on the NO_3^- regulation of levels of trifoliin A on clover roots. Other studies also support this hypothesis by showing that, under certain growth conditions, the transient appearance of trifoliin A receptors on *Rhizobium trifolii* may influence the ability of these bacteria to attach to clover root hairs (91). Cells grown on agar plates of a defined medium were most susceptible to agglutination by trifoliin A when they were harvested at 5 days of growth. In broth culture, the antigenic determinants on the bacteria that are cross-reactive with clover roots were "exposed" for only short periods as cultures left their lag phase of growth and again as they entered stationary phase. Clover roots adsorbed in greatest quantity when the cells were harvested from plate culture incubated for 5 days and from broth cultures in early stationary phase. These studies draw attention to the basis of many anomalous lectin-binding results, and also indicate that the regulation of these lectin receptors may be very complex.

We recently found that growth-phase dependence for trifoliin A binding to *R. trifolii* is related to the appearance of a unique determinant in the lipopolysaccharide (LPS) of the bacteria (92). As the culture advances from exponential to early stationary phase, changes in the immunochemistry of the LPS were detected with antisera made specific for LPS of cells in early stationary phase by exhaustive adsorption with exponentially growing cells. Culture-phase-dependent differences were found in the quantities of several glycosyl components (e.g., quinovosamine, or 2-amino-2,6-dideoxyglucose) in the LPS that bound trifoliin A; the latter amino-dideoxyhexose was found to be an effective hapten inhibitor of trifoliin A when in the β-anomeric configuration. In addition, the LPS increase in apparent size. The new immunochemical determinants occurring in the LPS as cells enter stationary phase were apparently sites for trifoliin-A binding, since immune monovalent Fab fragments specific for these unique determinants blocked the agglutination of cells with trifoliin A.

There is a transient appearance and disappearance on *Rhizobium japonicum* of the receptor that specifically binds soybean lectin (9, 93). Most strains had their highest percentage of soybean lectin-positive cells and the greatest number of soybean lectin-binding sites per cell in the early and midlog phases of growth. The proportion of galactose residues in the capsular polysaccharide is high at a culture age when the cells bind the galactose-reversible soybean lectin (94). A decline in lectin-binding activity accompanying culture aging is concurrent with a decline in galactose content and a rise in 4-O-methyl galactose residues (94). The latter methylated sugar has low affinity for the galactose-binding soybean lectin. These results suggest that the galactose residues in the capsular polysaccharide become methylated and as a consequence the cell loses its ability to combine specifically with the soybean lectin.

With regard to genetic regulation, exciting evidence is beginning to emerge that genetic elements important to surface polysaccharides and symbiotic recognition are encoded on very large, transmissable plasmids of *Rhizobium* (95-99). For instance, the genes responsible for the 2-deoxyglucose-inhibitable attachment of *Rhizobium trifolii* to clover hairs are encoded on the large nodulation plasmid designated pWZ2 (99), and incorporation of the trifoliin A binding sugar, quinovosamine, into the LPS of *R. trifolii* is controlled by the clover nodulation plasmid (100).

In the past, a major controversy existed concerning the nature of the carbohydrate receptor for the lectin on the *Rhizobium* cell. Lectin receptor molecules have been identified as capsular polysaccharide (57), LPS (76), and a glycan (101). The picture now emerging is that there are multiple types of lectin receptor molecules on the rhizobial cell. Both host-specific lectin–LPS interactions (*Rhizobium meliloti*-alfalfa) and exopolysaccharide–lectin interactions (*Rhizobium japonicum*-soybean) have been found (102). Nonnodulating mutant strains of *Rhizobium japonicum* have a defective carbohydrate antigen associated with their LPS and acidic heteropolysaccharide (103). In the case of the fast-growing rhizobia that infect peas and clover, the host lectin will bind to both the LPS and capsular polysaccharides of the symbiont rhizobia at certain culture ages (58, 92, 102). Based on genetic and immunochemical studies (104), Kamberger has proposed the hypothesis that root hair attachment is mediated by lectins cross-bridging the capsular polysaccharides as a primary recognition event, followed by a more critical recognition event involving the host-specific binding of the lectin to the underlying bacterial LPS, which then triggers successful infection. One challenge of the next few years is to test the validity of this hypothesis.

The *Rhizobium*-legume symbiosis is a marvelous balance of many cell–cell communications that, in coordination, culminate in the formation of root nodules that fix nitrogen into ammonia fertilizer for the plant symbiont (for reviews, see references 7, 14, 18, 21). In addition to attachment, there is curling and branching of the root hair, penetration of the root hair cell wall by the rhizobia without host cell lysis, dome formation of the new infection thread, nuclear-directed growth and extension of the infection thread down the root hair shaft, infection-thread penetration of the cell wall at the base of the root hair, host cell proliferation in the inner cortex in front of the advancing infection thread, branching of the infection thread in the nodular cells, release and envelopment of the bacteria from the infection thread into peribacteroid membranes, transformation of the vegetative bacteria into bacteriods, differentiation of nodular tissue, leghemoglobin synthesis and assembly, nitrogenase expression, mechanisms for transport of photosynthates, oxygen, and nitrogen to respiring bacteriods, and excretion of newly fixed NH_3 from bacteriods and its incorporation into amino acids. Each of these events provides an excellent model to study the underlying biochemical mechanisms of plant-microbe cell interactions.

ACKNOWLEDGMENTS

Portions of the research described here were supported by NSF Grant PCM80-21906, the Science and Education Administration of the United

States Department of Agriculture Competitive Grant No. 78-59-2261-0-1-050-2, Grant SO7-RR07049-14 from the Biomedical Research Support Grant Program, Division of Research Resources, National Institutes of Health, the Toxicology Research Program of the Michigan Agricultural Experiment Station, and a Grant-in-Aid from Sigma Xi, the Scientific Research Society of North America. I thank Estelle Hrabak, John Sherwood, Georges Truchet, and Harold Miller for helpful suggestions on the manuscript. This is article No. 9725 of the Michigan Agricultural Experiment Station.

REFERENCES

1. F. M. Burnet, *Nature (London)* **232**, 230 (1971).
2. J. Hesolp-Harrison, *Annu. Rev. Plant Physiol.* **26**, 403 (1975).
3. J. Hesolp-Harrison, *Cellular Recognition Systems in Plants,* No. 100, University Park Press, Baltimore, Maryland, 1978, p. 60.
4. H. Kauss, *Prog. Bot.* **38**, 58 (1976).
5. H. Kauss, Lectins and their physiological role in slime molds and in higher plants, in *Encyclopedia of Plant Physiology, New Series, Plant Carbohydrates*, Vol. II, Springer-Verlag, Berlin, 1980.
6. P. Albersheim and A. J. Anderson-Prouty, *Annu. Rev. Plant Physiol.* **26**, 31 (1975).
7. B. Solheim and J. Praxton, Recognition in *Rhizobium*-legume systems, in *Alternative Plant Disease Management Concepts,* Rockefeller Foundation Conference, Como, Italy, 1980.
8. B. Solheim and J. Raa, eds., Cell wall biochemistry related to specificity in *Host-Plant Pathogen Interactions*, Universitetsforlanget, Norway, 1977.
9. W. D. Bauer, *Basic Life Sci.* **9**, 283 (1977).
10. L. Sequeira, *Annu. Rev. Phytopathol.* **16**, 453 (1978).
11. E. L. Schmidt, *Annu. Rev. Microbiol.* **33**, 355 (1980).
12. A. E. Clarke and R. B. Knox, *Quart. Rev. Biol.* **53**, 3 (1978).
13. J. A. Callow, *Curr. Adv. Plant Sci.* **18**, 181 (1975).
14. W. J. Broughton, *J. Appl. Bacteriol.* **45**, 165 (1978).
15. H. F. Linskens, Physiology of fertilization and fertilization barriers in higher plants, in S. Subtelny and N. Wessells, eds., *The Cell Surface: Mediator of Developmental Processes*, Academic Press, New York, 1980, p. 113.
16. T. E. Ferrari and D. H. Wallace, *Theor. Appl. Genet.* **50**, 211 (1977).
17. T. E. Ferrari, S. S. Lee, and D. H. Wallace, *Phytopathol.* **71**, 752 (1981).
18. F. B. Dazzo, Adsorption of microorganisms to roots and other plant surfaces, in G. Bitton and K. Marshall, eds., *Adsorption of Microorganisms to Surfaces*, John Wiley & Sons, New York, 1980, p. 253.
19. F. B. Dazzo, Lectins and the saccharide receptors as determinants of specificity in the *Rhizobium*-legume symbiosis, in S. Subtelny and N. Wessells, eds., *The Cell Surface: Mediator of Developmental Processes*, Academic Press, New York, 1980, p. 277.

20. F. B. Dazzo, Determinants of Host Specificity in the *Rhizobium*-Clover Symbiosis, in W. Orme-Johnston and W. E. Newton, eds., *Nitrogen Fixation*, Vol. 2, University Park Press, Baltimore, Maryland, 1980, p. 165.

21. F. B. Dazzo and D. H. Hubbell, Control of root hair infection, in W. J. Broughton, ed., *Ecology of Nitrogen Fixation*, Vol. 2, Oxford University Press, London, in press.

22. F. B. Dazzo, Adsorption of microorganisms to plant surfaces, in P. R. Rutter and D. Ellwood, eds., *Adsorption of Microorganisms to Surfaces*, Ellis Horwood, Ltd., United Kingdom, in press.

23. F. B. Dazzo, *Am. Soc. Microbiol. News* **45**, 238 (1979).

24. J. E. DeVay and H. E. Adler, *Annu. Rev. Microbiol.* **30**, 147 (1976).

25. K. Keegstra, K. W. Talmadge, W. D. Bauer, and P. Albersheim, *Plant Physiol.* **51**, 188 (1973).

26. J. A. Monro, D. Penny, and R. W. Bailey, *Phytochemistry* **15**, 1193 (1976).

27. B. M. Wilder and P. Albersheim, *Plant Physiol.* **51**, 889 (1973).

28. J. R. Champ, *Biochem. Soc. Symp.* **40**, 3 (1974).

29. R. B. Knox, A. Clarke, S. Harrison, P. Smith, and J. J. Marchalonis, *Proc. Natl. Acad. Sci. USA* **73**, 2788 (1976).

30. G. Faulkner, W. C. Kimmins, and R. G. Brown, *Can. J. Bot.* **51**, 1503 (1973).

31. D. Lewis, *Adv. Genet.* **6**, 235 (1954).

32. D. H. Wallace, *Theor. Appl. Genet.* **54**, 249 (1979).

33. M. E. Nasrallah, J. T. Barker, and D. H. Wallace, *Heredity* **25**, 23 (1970).

34. M. E. Nasrallah, D. H. Wallace, and R. M. Savo, *Genet. Res.* **20**, 151 (1972).

35. T. E. Ferrari, D. Burns, and D. H. Wallace, *Plant Physiol.* **67**, 270 (1981).

36. T. Nishio and K. Hinata, *Jpn. J. Genet.* **54**, 307 (1979).

37. M. M. Yoeman and R. Brown, *Ann. Bot.* **40**, 1265 (1976).

38. J. Kloz, and V. Turkova, *Serological Mas. Bull.* **29**, 6.8 (1963).

39. M. S. Dunin and L. I. Chefranova, *Rev. Roum. Biol. Ser. Bot.* **11**, 61 (1966) (in Russian).

40. C. A. Napoli, F. B. Dazzo, and D. H. Hubbell, *Appl. Microbiol.* **30**, 123 (1975).

41. L. P. T. Zevenhuizen and H. J. Scholten-Koerselman, *J. Microbiol. Serol.* **45**, 165 (1979).

42. P. E. Jansson, L. Kenne, B. Lindberg, H. Ljunggren, J. Lonngren, U. Ruden, and S. Svensson. *J. Am. Chem. Soc.* **99**, 3812 (1977).

43. B. B. Lippincott and J. A. Lippincott, *J. Bacteriol.* **97**, 620 (1969).

44. M. H. Whatley, J. S. Bodwin, B. B. Lippincott, and J. A. Lippincott, *Infect. Immun.* **13**, 1080 (1976).

45. M. H. Whatley, J. Margot, J. Schell, B. B. Lippincott, and J. A. Lippincott, *J. Gen. Microbiol.* **107**, 395 (1978).

46. J. A. Lippincott and B. B. Lippincott, *Science* **199**, 1078 (1978).

47. R. N. Goodman, *Mycopathologia* **65**, 107 (1978).

48. R. N. Goodman, P. Y. Huang, and J. A. White, *Phytopathology* **66**, 754 (1976).

49. L. Sequeira, G. Gaard, and G. A. DeZoeten, *Physiol. Plant Pathol.* **10**, 43 (1977).

50. R. N. Goodman, P. Y. Huang, J. S. Huang, and V. Thipanich, in K. Tomyama, ed., *Biochemistry and Cytology of Plant-Parasite Interactions*, Kodensha Ltd., Tokyo-Elsevier Scientific Publication, Amsterdam, 1976, p. 35.

51. A. R. Ayers, S. B. Ayers, and R. N. Goodman, *Appl. Environ. Microbiol.* **38**, 659 (1979).

52. L. Sequeira and T. L. Graham, *Physiol. Plant. Pathol.* **11**, 43 (1977).
53. T. L. Graham, L. Sequeira, and T. S. Huang, *Appl. Environ. Microbiol.* **34**, 424 (1977).
54. M. H. Whatley, N. Hunter, M. Cantrell, C. Hendrick, K. Keegstra, and L. Sequeira, *Plant Physiol.* **65**, 557 (1980).
55. F. B. Dazzo, C. A. Napoli, and D. H. Hubbell, *Appl. Environ. Microbiol.* **32**, 166 (1976).
56. P. J. Dart, *J. Exp. Bot.* **22**, 163 (1971).
57. F. B. Dazzo, and D. H. Hubbell, *Appl. Microbiol.* **30**, 1017 (1975).
58. F. B. Dazzo and W. J. Brill, *J. Bacteriol.* **137**, 1362 (1979).
59. C. A. Napoli and D. H. Hubbell, *Appl. Microbiol.* **30**, 1003 (1975).
60. F. B. Dazzo, W. E. Yanke, and W. J. Brill, *Biochim. Biophys. Acta* **539**, 276 (1978).
61. F. B. Dazzo and W. J. Brill, *Appl. Environ. Microbiol.* **33**, 132 (1977).
62. J. W. Kijne, I. A. M. Van der Schaal, and G. E. DeVries, *Plant Sci. Lett.* **18**, 65 (1980).
63. W. Gade, M. A. Jack, J. B. Dahl, E. L. Schmidt, and F. Wold, *J. Biol. Chem.* **256**, 12905 (1981).
64. T. A. Hughes, J. C. Leece, and G. H. Elkan, *Appl. Environ. Microbiol.* **37**, 1243 (1979).
65. T. A. Hughes and G. H. Elkan, *Plant Soil* **61**, 87 (1981).
66. G. Kato, Y. Maruyama, and M. Nakamura, *Plant Cell Physiol.* **22**, 759 (1981).
67. B. Solheim, A model of the recognition reaction between *Rhizobium trifolii* and *Trifolium repens*, in NATO Conferences on *Specificity in Plant Diseases*, Advanced Study Institute, Sardinia, 1975.
68. P. E. Bishop, F. B. Dazzo, E. R. Applebaum, R. J. Maier, and W. J. Brill, *Science* **198**, 938 (1977).
69. R. W. Carlson, R. E. Saunders, C. A. Napoli, and P. Albersheim, *Plant Physiol.* **62**, 912 (1978).
70. W. D. Dudman, *Rhizobium serology*, in R. Hardy and W. Silver, eds., *A Treatise on Dinitrogen Fixation*, Vol. IV, John Wiley & Sons, New York, 1977, p. 487.
71. M. Reporter, D. Raveed, and G. Norris, *Plant Sci. Lett.* **5**, 73 (1975).
72. B. B. Bohlool and E. L. Schmidt, *Science* **185**, 269 (1974).
73. A. K. Bal, S. Shantharam, and S. Ratnam, *J. Bacteriol.* **133**, 1393 (1978).
74. H. E. Calvert, M. Lalonde, T. V. Bhuvaneswari, and W. D. Bauer, *Can. J. Microbiol.* **24**,

75. G. Stacey, A. Paau, and W. J. Brill, *Plant Physiol.* **66**, 609 (1980).
76. J. S. Wolpert and P. Albersheim, *Biochem. Biophys. Res. Commun.* **70**, 729 (1976).
77. J. Hamblin and S. P. Kent, *Nature New Biol.* **245**, 28 (1973).
78. R. E. Saunders, R. W. Carlson, and P. Albersheim, *Nature* **271**, 240 (1978).
79. F. B. Dazzo and E. M. Hrabak, *J. Supramol. Struct. Cell Biochem.* **16**, 133 (1981).
80. F. B. Dazzo, E. Hrabak, J. Sherwood, and G. Truchet, Trifoliin in clover root exudate and its relationship to cell surface polarity of *Rhizobium trifolii*, in *4th International Symposium on Nitrogen Fixation*, Canberra City, Australia, 1980.
81. B. B. Bohlool and E. L. Schmidt, *J. Bacteriol.* **125**, 1188 (1976).
82. H. C. Tsien and E. L. Schmidt, *Can. J. Microbiol.* **23**, 1274 (1977).
83. M. H. Deinema and L. P. T. Zevenhuizen, *Arch. Mikrobiol.* **78**, 42 (1971).
84. E. Martinez-Molina, V. M. Morales, and D. H. Hubbell, *Appl. Environ. Microbiol.* **38**, 1186 (1979).

85. H. Ljunggren, and G. Fahraeus, *J. Gen. Microbiol.* **26**, 521 (1961).

86. A. Palomares, E. Montoya, and J. Olivares, *Microbios* **22**, 7 (1979).

87. P. Y. Yao, and J. M. Vincent, *Plant Soil* **45**, 1 (1976).

88. S. Higashi and M. Abe, *Appl. Environ. Microbiol.* **39**, 297 (1980).

89. F. B. Dazzo and W. J. Brill, *Plant Physiol.* **62**, 18-21 (1978).

90. F. B. Dazzo, E. Hrabak, M. R. Urbano, J. Sherwood, and G. Truchet. Regulation of recognition in the *Rhizobium*-clover symbiosis, in A. H. Gibson and W. E. Newton, eds., *Current Perspectives in Nitrogen Fixation*, Australian Academy of Science, Canberra, 1981, p. 292.

91. F. B. Dazzo, M. R. Urbano, and W. J. Brill, *Curr. Microbiol.* **2**, 15 (1979).

92. E. M. Hrabak, M. R. Urbano, and F. B. Dazzo, *J. Bacteriol.* **148**, 697 (1981).

93. T. V. Bhuvaneswari, S. G. Pueppke, and W. D. Bauer, *Plant Physiol.* **60**, 486 (1977).

94. A. J. Mort and W. D. Bauer, *Plant Physiol.* **66**, 158 (1980).

95. A. W. Johnston, J. Beyson, A. Buchanan-Woolsston, S. Setchell, P. Hirsch, and J. Beringer, *Nature* **276**, 634 (1978).

96. W. Zurkowski, and Z. Lorkeiwicz, *Genet. Res.* **32**, 311 (1978).

97. M. P. Nuti, F. Cannon, R. Prakash, A. Lepidi, and R. Schiperoort, The large plasmids of rhizobia: current status and their involvement in the expression of symbiotic functions, in A. H. Gibson and W. E. Newton, eds., *Current Perspectives in Nitrogen Fixation*, Australian Academy of Science, Canberra, Australia, 1981, p. 167.

98. I. Van Der Schaal and J. W. Kijne, Surface polysaccharides of a *Rhizobium leguminosarum* inf⁻ strain, in *Proceedings of the 7th North American Rhizobium Conference*, College Station, Texas, 1979, p. 17.

99. W. Zurkowski, *Microbios* **27**, 27 (1980).

100. R. Russa, T. Urbanik, E. Kowalczuk, and Z. Lorkiewicz, *FEMS Microbiol. Lett.* **13**, 161 (1982).

101. H. Planque′ and J. W. Kijne, *FEBS Lett.* **73**, 64 (1977).

102. W. Kamberger, *Arch. Microbiol.* **121**, 83 (1979).

103. R. J. Maier and W. J. Brill, *J. Bacteriol.* **133**, 1295 (1978).

104. W. Kamberger, *FEMS Microbiol. Lett.* **6**, 361 (1979).

Index